OPTICAL SPECTROSCOPIES of ELECTRONIC ABSORPTION

World Scientific Series in Contemporary Chemical Physics – Vol. 17

OPTICAL SPECTROSCOPIES OF ELECTRONIC ABSORPTION

J-R Lalanne

F Carmona and L Servant

Bordeaux I University and CNRS, France

Published by

World Scientific Publishing Co. Pte. Ltd.
P O Box 128, Farrer Road, Singapore 912805
USA office: Suite 1B, 1060 Main Street, River Edge, NJ 07661
UK office: 57 Shelton Street, Covent Garden, London WC2H 9HE

British Library Cataloguing-in-Publication Data
A catalogue record for this book is available from the British Library.

This work has been published with the help of the French Ministry for Culture.

OPTICAL SPECTROSCOPIES OF ELECTRONIC ABSORPTION

ISBN 981-02-3861-4

Printed in Singapore.

Preface

UV-visible absorption was initially employed by physicists from the end of the nineteenth century onwards (Balmer, 1885) to discover what would become the quantization of atomic electronic energy. Fifty years ago, however, the technique of UV-visible absorption was adopted as an ideal physico-chemical tool with which to study molecular structures in organic and inorganic chemistry. Complementary to infrared absorption spectroscopy, this technique was, for example, largely used in the fifties to determine the structure of steroids, a matter of vital — according to the original meaning of the word — importance. At that time, empirical rules, such as Woodward's rules, were employed for the interpretation of the results, and increments of wavelength absorption, characteristic of the groups substituted in the basic chromophores (multiple bonds, aromatic rings, etc.) were devised. From this a qualitative "mecano" was developed, offering the researcher a tailor-made tool with which to determine structures. *"Theory and Applications of Ultraviolet Spectroscopy"* by H.H. Jaffé and M. Orchin (J. Wiley, 1962) is, although basic, still an important textbook in this field.

With the more recent emergence of NMR spectroscopy, the interest in the UV and infrared spectroscopies then went into decline ... at least as far as the determination of molecular structures was concerned.

Then two successive and major events renewed their interest:

— The important development, twenty five years ago, of *photoelectron spectroscopy*, which measures the kinetic energy of the electrons that are emitted by molecules when they are irradiated by high energy photons, and which would eventually lead to Electronic Spectroscopy for Chemical Analysis (ESCA).
— The discovery of the *laser*, which allowed the setting up of so called Doppler-free spectroscopy, and more recently, spectroscopies of dense states involving the single molecule.

Then, very progressively, the interest in the electronic absorption spectroscopy moved from applied to fundamental research, in complete opposition to the more usual trend in science in which fundamentals induce applications.

Besides this double interest, however, the electronic absorption spectroscopies have constituted, since their origin, wonderful tools for illustrating the electronic structure of molecules and the application of quantum theories to their determination. Quite naturally then, these spectroscopies are largely introduced and studied at all levels of college education, essentially in chemistry, but also in physics. The authors have personally had the opportunity to teach them for the last ten years in undergraduate studies, in theoretical as well as in experimental lessons, both in Physical Chemistry and Physical Sciences courses. In the meantime, we have developed and, then, progressively organized a training course, which we were managing at that time, involving lessons, exercises and experimental projects for the self study of future teachers in the physical sciences in high schools.

In this book, we present the results of this collective thought, validated year after year in real teaching conditions, and largely improved thanks to numerous discussions with colleagues (often with passion!). It is complementary to the collective opus whose writing was supervised by one of us (JRL) and published by Masson (Structure électronique et liaison chimique, 1992) and World Scientific Publishers (Electronic Structure and Chemical Bonding, 1996).

The aim of this book is didactic, and involves, in one single volume, all the basic knowledge necessary to understand the subjects developed. It is illustrated by numerous appendices, largely tables, suggested bibliographies (voluntarily limited to basic works and review articles, the latter containing references to more specialized papers to which the interested reader may refer to), and about thirty problems taken from the large library of tests A, B and C of the French "Aggregation" examinations in chemistry and physics. We produce detailed solutions, sometimes proposed by students, whose presentation takes advantage of the abridged answers published each year by the "Union des Physiciens".

Chapter I is an *introduction*. The optical spectroscopy of electronic absorption is introduced and a short list of recent books in physical chemistry involving short presentations of various spectroscopies is given. The latter contains a few chapters devoted to the subjects developed in this book and could perhaps be read first.

Chapter II gives the classical electromagnetic description of light and its main features.

Chapter III gives in brief the knowledge that is required to understand the optical spectroscopies. It contains short and didactical successive presentations of *quantum mechanics, group theory and electronic structure of molecules.*

The last four chapters constitute the heart of the book.

Chapter IV is devoted to *UV-visible absorption spectroscopy*, and gives theoretical as well as experimental aspects. Numerous examples are given, in both organic and inorganic chemistry. They are illustrated by about twenty problems.

Chapter V is an introduction to *rotatory polarization* and to *circular dichroism*. Both effects belong to so called "optical activity spectroscopies". They were largely used in the past, since they were crucial to research in the fields of stereochemistry in inorganic chemistry, during the structural studies of transition metal complexes, for example. Their interest has increased again recently. The chirality of liquid crystals is a subject of fundamental research and there are many experiments based on optical activity and its variations with wavelength. Moreover, as in many other fields, the laser and the personal computer have completely changed our experimental set up. A few examples of recent realizations will be given.

Chapter VI discusses *photoelectron spectroscopy* which constitutes an experimental method well suited to the verification of theoretical results concerning electronic molecular structure.

Finally, chapter VII reports the essential features of the *laser* which are unavoidable in the set up of experiments in absorption spectroscopy (tunability, longitudinal single mode operation, frequency lock in), and next describes two Doppler-free spectroscopies (Saturated Absorption Spectroscopy and two-photon Absorption Spectroscopy). It concludes with a presentation of very recent so called "Selective Spectroscopies" which in the near future will be likely to provide original information on a *single* molecule in dense media.

Bordeaux, May 1997

Acknowledgements

We are most grateful to

— Mrs. L. Orrit, née Volker, who helped translate the book.

— Mrs. N. Robineau for preparing the final version of this book.

Orthevielle, January 1999.

Contents

PART I
Theoretical Frame

PART II
Optical Spectroscopies of Electronic Absorption

Symbols

Greek alphabet used in the book
(left: capital letter ; right: small letter)

Alpha		α	Theta	Θ	θ	Sigma	Σ	σ
Beta		β	Lambda	Λ	λ	Tau		τ
Gamma	Γ	γ	Mu		μ	Phi	ϕ	φ
Delta	Δ	δ	Nu		ν	Chi		χ
Epsilon		ε	Xi		ξ	Psi	Ψ	ψ
Zeta		ζ	Pi		π	Omega		ω
Eta		η	Rho		ρ			

Fundamental constants
(and their values in SI system)

Electron charge	q	– 1.602177	$\times 10^{-19}$	C
Avogadro constant	N	6.02214	$\times 10^{23}$	mol^{-1}
Planck constant	h	6.62608	$\times 10^{-34}$	Js
Rydberg constant	R	1.09737	$\times 10^{7}$	m^{-1}
Atomic mass unit	a.u	1.66054	$\times 10^{-27}$	kg
Electron rest mass	m_e	9.10939	$\times 10^{-31}$	kg
Proton rest mass	m_p	1.672623	$\times 10^{-27}$	kg
Permittivity of vacuum	ε_0	8.8541878	$\times 10^{-12}$	$C^2\ N^{-1}\ m^{-2}$
Speed of light in vacuum	c	2.99792458	$\times 10^{8}$	ms^{-1}

Conversion factors
(in SI system)

1 Angström	(Å)	=	10^{-10} m
1 Bohr	(B)	=	5.291772×10^{-11} m
1 Debye	(D)	=	3.3333×10^{-30} Cm
1 electronvolt	(eV)	=	1.602177×10^{-19} J
1 Hartree	(H)	=	4.35975×10^{-18} J

Typographical convention

italics	operator	H
bold type	vector	$\mathbf{r}$
brakets	matrix	(H)
square brackets	colum vector	$\begin{bmatrix} a \\ b \end{bmatrix}$
vertical dashes	determinant	$\begin{vmatrix} a & b \\ c & d \end{vmatrix}$
underlined letter	complex quantity	$\underline{P}$
Re	real part	$\mathrm{Re}\,(\underline{P})$
one point	first total derivative with respect to time	$\dot{r}$
two points	second total derivative with respect to time	$\ddot{r}$

PART ONE

THEORETICAL FRAME

Chapter I

Introduction

I.1. Generalities

As the title of this work suggests and as was pointed out in our preface, the subject of this book is the study of the *Optical Spectroscopies of Electronic Absorption*. It covers fundamental concepts in the following fields:

— *Spectroscopy:* This field is concerned with the study of spectra. A spectrum is a continuous distribution of radiant energy with a frequency dispersion, which can be obtained either from a white light source by using dispersing devices, or more frequently these last years, from a continuously tunable source of monochromatic radiation: The so-called tunable laser. The interaction between the radiative energy with matter is classically described by considering three fundamental processes — absorption, emission and scattering — all three of which act upon the microscopic system through its polarizability. These processes can be described within several different frameworks. In the course of this book, we will very often use the following adjectives: Microscopic or macroscopic (we also use the word phenomenological); classical, semi-classical or quantum.

— *Optics:* In this book, only the so-called "optical" range of the spectrum extending from the near ultraviolet to the near infrared will be studied. Optics, now a rapidly growing field, has an experimental origin. Optical spectroscopy is therefore associated to the now usual elements found on modern devices for signal detection and signal treatment such as phase detection systems, photon counting devices, box-cards, etc. Nor can these devices now be dissociated from the microcomputers, since they are the ones who monitor experiments and control the acquisition of the signal as well as its memorization and further treatment.

— The study of the *absorption* of radiation by matter. Absorption is an energy transfer from radiation to matter whereby the latter acquires additional energy to the detriment of the radiation whose intensity decreases. It is the first of the three previously noted processes. However, it is difficult to separate it from the emission process, since matter, though it will usually try to dissipate the extra energy increment by way of nonradiative processes, sometimes uses emission for this purpose.

— The study of the *electronic* properties of materials, since the energy of the radiation considered here is of the order the characteristic energies of the so-called "external" electrons in atoms and molecules (i.e. in the domain of the electron-volt). The knowledge of the molecular electronic structure is therefore indispensable to the understanding of the optical spectroscopies of absorption. Historically, our knowledge of molecular structure developed in three successive stages:

— The first stage is experimental: During the nineteenth century, physicists studied spectra of atoms and, particularly, the spectrum of the hydrogen atom, while Bohr was developing a planetary model leading to the notion of energy levels.

— The second stage is theoretical: In the thirties, the advent of quantum mechanics provided a fertile theoretical description of the structure of atoms and small molecules. Although many refinements were added to the description of that time, up to this end of century, no fundamental discovery arose to refute the results and the predictions of quantum theory.

— Last, the third stage is experimental again. In 1960, the discovery of lasers led to the development of many new outstanding and often original methods of experimental investigation, especially between 1970 and 1980. Presently, they stimulate new efforts in theoretical analysis.

We will therefore try to define the four above mentioned subjects and develop them in the order set forth in the preface.

Let us note that it would perhaps be profitable to read all or part of the three general books mentioned below.

I.2. Bibliography

1.2.1. General physical chemistry can be studied in:

Atkins P.W. — *Physical Chemistry*, 5th edition, Oxford University Press (1994).

Levine I.N. — *Physical Chemistry*, 3rd edition, Mc Graw-Hill Book Company (1988).

1.2.2. Atomic and molecular electronic structures are treated in:

Lalanne J.R. and *al.* — *Electronic Structure and Chemical Bonding,* World Scientific Publishers (1996).

Chapter II

Classical theory of light

II.1. Introduction: The different representations of light

There are three main ways which are frequently used to describe the interaction between light and matter.

— *Phenomenological approach (Einstein, 1905)*

Einstein's phenomenological approach is neither a quantum nor a classical approach. Light consists of "grains" of energy called *photons* which can be *absorbed* or *emitted* by atoms and by molecules.

— *Classical electromagnetic approach (19th century)*

Light is traditionally presented as a partial constituent of the spectrum of electromagnetic radiation (*cf.* Appendix II.4.1). Formalized during the 19th century, the classical electromagnetic approach consists in assimilating light to a *wave* which emanates from a source and propagates in a straight line, except when it is either *refracted* or *reflected*. With this approach, the phenomena of diffraction, of refraction at an interface, of interference, and of energy propagation are well accounted for. The wave carries energy, so-called light energy, which is distributed continuously over its surface and whose intensity is proportional to the square of the amplitude of the wave. *In a wave, a discontinuous distribution of energy (in grains or photons) is impossible.* The energy has electric and magnetic characteristics and therefore the wave is assumed to carry both *electric* and *magnetic* fields. In vacuum, light thus appears as an *electromagnetic wave* arising from a temporary modification of the structure of space brought about by the electric and the magnetic fields. This means that light has a *vectorial* character because the propagating perturbation of space itself has a vectorial character.

Maxwell's equations, which we are going to study more in detail in Section II.2, predict the properties of electromagnetic waves in vacuum and in matter (the "ether" imagined by physicists during the 19th century). These same equations prove that an electromagnetic wave consists of a very special superposition of an electric wave and a magnetic wave, as we shall see later. A monochromatic wave is also called a harmonic wave, or again a *radiation mode*. It is characterized by a *wavelength* and by entities which derive directly from the wavelength (period, frequency, wave number, wave vector).

— *Quantum approach (second half of the 20th century)*

In the quantification (*cf.* Section III.1) of the electromagnetic field a quantum harmonic oscillator is associated to each mode (of wave vector **k**) of the radiation field (*cf.* Bibliographic reference 4.6.3). This quantum oscillator describes the photon and is characterized by its creation (a^+) and annihilation (a) operators. The Hamiltonian associated to this mode is written as

$$H_k = \mathrm{h}\nu_\mathrm{k}(a^+a + 1/2) \tag{2.1.1}$$

The operator a^+a defines the operator "number of photons". If $|\mathrm{n_k}\rangle$ is an eigenvector of the Hamiltonian H_k, the eigenvalue equation is written as $H_k|\mathrm{n_k}\rangle = \mathrm{E_k}|\mathrm{n_k}\rangle$, and its solution yields the associated eigenvalue $(\mathrm{n_k}+1/2)\mathrm{h}\nu_\mathrm{k}$. $\mathrm{n_k}$ represents the number of photons in the state described by eigenvector $|\mathrm{n_k}\rangle$. The creation and annihilation operators satisfy the following equations:

$$\begin{aligned} a_k{}^+|\mathrm{n_k}\rangle &= \sqrt{\mathrm{n_k}+1}\,|\mathrm{n_k}+1\rangle \\ a_k|\mathrm{n_k}\rangle &= \sqrt{\mathrm{n_k}}\,|\mathrm{n_k}-1\rangle \end{aligned} \tag{2.1.2}$$

In fact, as we shall often see in Chapters III and IV, those states $|\mathrm{R_k}\rangle$ which are observed in reality consist of linear combinations of the different eigenmodes $|\mathrm{n_k}\rangle$ which form a basis for this decomposition. Using the density operator $\rho_k = |\mathrm{R_k}\rangle\langle\mathrm{R_k}|$ (*cf.* Section III.1.1), we can then define $\langle\mathrm{n_k}|\rho_k|\mathrm{n_k}\rangle$, the population of the state, i.e. the probability for state $|\mathrm{R_k}\rangle$ to be identical to the eigenstate $|\mathrm{n_k}\rangle$ with $\mathrm{n_k}$ photons.

This probability plays a very important role because it can be used to distinguish between the two main sources of radiation described later in this book; *spontaneous emission* (also called *chaotic* source) for which this probability reaches a maximum at $\mathrm{n_k} = 0$, and *stimulated emission* (i.e. the *laser*) for which this probability is greatest for a well determined value of $\mathrm{n_k}$ different from zero.

When the radiation consists of several modes — as is often the case — the Hamiltonian H_r for the total radiation is the sum of the above individual Hamiltonian's H_k.

We must warn however that in atomic spectroscopy the atom and the radiation must often be treated globally. In the simple case where the interaction between the atom and the radiation is neglected, the global Hamiltonian H_0 (Hamiltonian "without interaction") is just equal to the sum of the Hamiltonian of the atom (H_a) and the Hamiltonian of the radiation (H_r). If we simplify the problem by choosing only one radiation mode of wave vector **k** and an atom which has only two eigenlevels (named + and –) of H_0 (*cf.* TLSA approximation in Section IV.3.2), the global atom-radiation system is described by the eigenvectors $|\pm, n_k\rangle$ which are solutions of the eigenvalue equation

$$H_0 |\pm, n_k\rangle = E_{(\pm, n_k)} |\pm, n_k\rangle \tag{2.1.3}$$

The eigenvector $|\pm, n_k\rangle$ describes the *atom dressed by the photon*, (or "light dressed atom") though in this case the atom does not interact with the photon, and the associated eigenvalue (i.e. the energy $E_{(\pm, n_k)}$) is the sum of the energies of the atom and of the radiation. This fruitful formalism, due to Cohen-Tannoudji, is described in Bibliographic reference 2.5.3.

When completed by a Hamiltonian describing the atom-radiation interaction, this formalism leads to a very complete description of phenomena studied in laser absorption spectroscopy and especially of the radiation induced displacement of the energy of the atomic states (light shift or dynamic Stark effect) to which we shall come back in Chapter VII.

II.2. Classical theory of light

II.2.1. Electromagnetism and Maxwell's equations

II.2.1.1. Maxwell's equations

Maxwell's equations, which were established by Maxwell during the second half of the 19th century, constitute the foundation of *classical electromagnetism*. Their non-invariance by Galilean transformation was at the origin of the formulation of the Lorentz transformation, which itself lies at the basis of the restricted relativity theory imagined by Einstein. The remarkably compact formulation of Maxwell's equations in the time-space of restricted relativity can be found, among others, in Appendix II.4.2.

The equations link together the time-dependent (variable regime) electric field vector **E** and the magnetic field vector **B**. These relations are inferred partly from electrostatics and magnetostatics, and partly from the laws of electromagnetic induction.

Let us first limit ourselves to the case of a wave propagating in vacuum.

— *First law:* Gauss' theorem, which remains valid for time-dependent phenomena, implies that the flux of the electric field **E** through a closed surface is proportional to the electric charge enclosed within this surface. Let us take the divergence of **E** and integrate it over all space by using successively the divergence theorem and Gauss' law (*cf.* Bibliographic reference 2.5.2). We find

$$\iiint_V \nabla \mathbf{E} dv = \iint_S \mathbf{E} \mathbf{ds} = (1/\varepsilon_0)\iiint_V \rho dv \tag{2.2.1}$$

ε_0 is the electric permittivity of vacuum and ρ the charge density. From this we can deduce the so-called first law, or *Maxwell-Gauss* law:

$$\nabla \mathbf{E} = \rho / \varepsilon_0 \tag{2.2.2}$$

— *Second law:* By analogy with Gauss' theorem, and in the absence of magnetic load (Maxwell's treatment uses Ampere's theory of magnetism, not Coulomb's theory), the conservation of the flux of magnetic induction through a closed surface imposes the so-called second law of Maxwell-Gauss for stationary or time-dependent phenomena. This law is also called the equation of the *conservation of magnetic flux* and is written as

$$\nabla \mathbf{B} = 0 \tag{2.2.3}$$

— *Third law:* With Faraday's law we can calculate the electric field created in a region of space where a magnetic field **B** exists whose flux varies with time (the circulation of the electric field along a loop is equal to the opposite of the time derivative of the magnetic flux within this loop). By using successively Stockes' theorem and Faraday's law, we obtain

$$\iint_S (\nabla \times \mathbf{E})\mathbf{ds} = \oint_C \mathbf{E}\mathbf{dr} = -(\partial / \partial t)\iint_S \mathbf{B}\mathbf{ds} \tag{2.2.4}$$

from which we deduce the so-called third law, the *Maxwell-Faraday* law:

$$\nabla \times \mathbf{E} = -\dot{\mathbf{B}} \tag{2.2.5}$$

— *Fourth law:* It applies to the magnetic field created by electric currents. Ampere's law states that if an electric current of volume current density **J** flows through a loop, it induces a magnetic field **B** whose circulation along the loop is proportional to the flux of **J** through the loop. By using Stokes' theorem and Ampere's law, we obtain

$$\iint_S (\nabla \times \mathbf{B})\mathbf{ds} = \oint_C \mathbf{B}\mathbf{dr} = (1/\varepsilon_0 c^2)\iint_S \mathbf{J}\mathbf{ds} \tag{2.2.6}$$

c is the speed of light in vacuum. This equation shows that

$$\nabla \times \mathbf{B} = (1/\varepsilon_0 c^2)\mathbf{J} \tag{2.2.7}$$

J is proportional to the rotational of **B.** Its divergence should therefore vanish. However, in a variable regime, the equation of charge conservation is written as

$$\nabla \mathbf{J} = -\dot{\rho} \tag{2.2.8}$$

so that the divergence of **J** must clearly be different from zero. Therefore Ampere's theorem does not hold in a variable regime.

To elucidate this paradox, Maxwell had the brilliant idea to add a second source term to the second member of Ampere's equation (Equation 2.2.7). This term, written as $(1/c^2)\dot{\mathbf{E}}$, is called the *displacement current. Maxwell-Ampere's* equation thus becomes

$$\nabla \times \mathbf{B} = (1/\varepsilon_0 c^2)\mathbf{J} + (1/c^2)\dot{\mathbf{E}} \tag{2.2.9}$$

By taking the divergence of both members of the equation and by using the first Maxwell-Gauss equation, we can check that the electric charge conservation equation (Equation 2.2.8) is now satisfied. Please notice that introducing the displacement current has in fact "symmetrized" classical electromagnetism by creating an electric image to the law of Faraday. A variable electric field induces a magnetic field in the same way that a variable magnetic field induces an electric field. The electromagnetic entity (**E**,**B**) is thus reinforced and the introduction of the displacement current introduced by Maxwell — in spite of the absence of direct experimental evidence— undoubtedly constitutes his most important contribution to classical electromagnetism.

The second and third equations only imply the electromagnetic field (**E**, **B**). They can therefore be generalized without modification to a material medium. However, the first and the fourth laws associate sources (ρ, **J**), i.e. possible stationary (ρ) or mobile (**J**) electric charges, to the electromagnetic field (**E**, **B**). This means that the dielectric and magnetic properties of matter will have to be taken into account when writing these equations in a material medium. When submitted to an electric field, matter can acquire an electric polarization **P** proportional to the electric field (linear medium). A so-called displacement vector **D** is therefore introduced, which is defined as follows

$$\mathbf{D} = \varepsilon\mathbf{E} = \varepsilon_0\varepsilon_r\mathbf{E} = \varepsilon_0\mathbf{E} + \mathbf{P} \tag{2.2.10}$$

ε, ε_0, and ε_r are, respectively, the electric permittivity of the material medium, the electric permittivity of vacuum, and the relative electric permittivity of the material medium.

In the same way, when submitted to a magnetic field, matter can acquire a magnetization **M** proportional to the field (linear material). The magnetic field vector is traditionally written as

$$\mathbf{B} = \mu\mathbf{H} = \mu_0\mu_r\mathbf{H} = \mu_0\mathbf{H} + \mathbf{M} \tag{2.2.11}$$

μ, μ_0, et μ_r are, respectively, the magnetic permittivity of the material, the magnetic permittivity of vacuum, and the relative magnetic permittivity of the material (sometimes, the term "permeability" is used instead of "permittivity").

If we disregard the coefficient $(1/\mu_0)$, **H** represents the magnetic field of vacuum and μ_0 is the magnetic permeability of vacuum. The first and fourth of Maxwell's equations should therefore be modified in order to describe electromagnetic propagation in matter correctly. Their new formulations are given in Appendix II.4.2.

Please note that the third and fourth equations are called the *main equations*, while the two first ones are called the *constitutive* equations.

Comment: The electromagnetic field (**E**,**B**) is assumed to derive from an electromagnetic potential (V,**A**) by the relations $\mathbf{E} = -\nabla V - \dot{\mathbf{A}}$ and $\mathbf{B} = \nabla \times \mathbf{A}$. These relations do not define the electromagnetic potential in a unique way, and gauge changes are frequently used (*cf.* Bibliographic reference 2.5.2).

II.2.1.2. The wave equation

The wave equation, which is also called the propagation equation, can be derived from the two main equations. In vacuum and in the absence of sources ($\mathbf{J} = 0$, $\rho = 0$), we can take the rotational of the third equation and use the fourth equation. We obtain

$$\nabla \times (\nabla \times \mathbf{E}) = -\partial(\nabla \times \mathbf{B})/\partial t = -(1/c^2)\ddot{\mathbf{E}}$$

Remembering that the following equation $\nabla \times (\nabla \times \mathbf{E}) = \nabla(\nabla\mathbf{E}) - \Delta\mathbf{E}$ (where $\Delta\mathbf{E} = \Delta E_x \mathbf{u_x} + \Delta E_y \mathbf{u_y} + \Delta E_z \mathbf{u_z}$; the Cartesian coordinates are defined by the three unit vectors $\mathbf{u_x}$, $\mathbf{u_y}$, $\mathbf{u_z}$ and Δ is the delta- or Laplace operator) is fulfilled and combining this both with the relation just obtained and with the first of Maxwell's equations, we find (for $\mathbf{J} = 0$ and $\rho = 0$)

$$\Delta \mathbf{E} - (1/c^2)\ddot{\mathbf{E}} = 0 \qquad (2.2.12)$$

Equation 2.2.12 describes an electric wave propagating at speed $c = 1/\sqrt{\varepsilon_0\mu_0}$ (*cf.* Problem II.3.1). An identical procedure results in the same propagation equation for the magnetic field **B**. Last, the study of propagation in a material medium without source can be done by applying the same procedure to the equations given in Appendix II.3.2. Again, we find an equation similar to equation 2.2.12, except that the factor $1/c^2$ is replaced by $\varepsilon_r\mu_r/c^2$, where ε_r and μ_r are, respectively, the electrical permittivity and the relative magnetic permeability of the medium. The wave propagates with a phase velocity of $c/\sqrt{\varepsilon_r\mu_r}$. This phase velocity can be greater than the speed of light.

Comment: The very common case of the harmonic wave is interesting to study. The modulus of the fields of this wave can be written as

$$E(\mathbf{r}, t) = E_0(\mathbf{r})\cos[\omega t - \varphi(\mathbf{r})] \qquad (2.2.13)$$

r defines the position of the point in the space under consideration. ν is the frequency, $\omega = 2\pi\nu$ is the angular frequency (hereafter, we shall simply refer to ω as the "frequency") and $\varphi(\mathbf{r})$ the phase. Using complex notation, this relation writes

$$\underline{E}(\mathbf{r}, t) = \underline{E}(\mathbf{r})\exp(i\omega t) \qquad (2.2.14)$$

with

$$\underline{E}(\mathbf{r}) = E_0(\mathbf{r})\exp[-i\varphi(\mathbf{r})] \qquad (2.2.15)$$

Surfaces defined by functions like $\varphi(\mathbf{r})$ = constant define the wave surface of the harmonic wave. If a wave of this type is substituted in the wave equation, we obtain

$$\Delta\underline{E}(r) + k^2\underline{E}(r) = 0 \qquad (2.2.16)$$

This equation is called the Helmholtz propagation equation in which time does not appear explicitly, and which describes the variation of the amplitude of the field in the propagation space. k is the wave number and can also be written $k = 2\pi/\lambda$, where λ is the wavelength.

II.2.2. Three solutions for Maxwell's equations

We shall limit ourselves to the study of three particular solutions of the propagation equation in vacuum. The first one is isotropic in all space and corresponds to the *spherical* wave. The two others single out a specific direction of space; they are the *plane* wave and the *Gaussian* wave.

II.2.2.1. The spherical wave

Let r be the distance between a point of space and the origin of the coordinates of a Cartesian system x, y, z. We choose spherical coordinates and solve the wave equation 2.2.12 by writing the Laplace operator in spherical coordinates, where f designates one of the two fields **E** or **B**. An elementary calculation shows that the wave equation can be written as

$$\partial^2(\mathrm{rf})/\partial \mathrm{r}^2-(1/\mathrm{c}^2)\partial^2(\mathrm{rf})/\partial \mathrm{t}^2=0 \qquad (2.2.17)$$

It is the product rf which propagates. This equation can also take the form

$$[\partial/\partial \mathrm{r}-(1/\mathrm{c})(\partial/\partial \mathrm{t})][\partial/\partial \mathrm{r}+(1/\mathrm{c})(\partial/\partial \mathrm{t})]\mathrm{rf}=0$$

This relation shows that the general solution of the second-order differential equation 2.2.17 is the sum of two particular solutions which are revealed thanks to Equation 2.2.18:

$$\mathrm{rf}=\mathrm{a}_+(\mathrm{t}-\mathrm{r}/\mathrm{c})+\mathrm{a}_-(\mathrm{t}+\mathrm{r}/\mathrm{c})$$

that is

$$\mathrm{f}=(\mathrm{a}_+/\mathrm{r})(\mathrm{t}-\mathrm{r}/\mathrm{c})+(\mathrm{a}_-/\mathrm{r})(\mathrm{t}+\mathrm{r}/\mathrm{c}) \qquad (2.2.18)$$

We identify the sum of two progressive spherical waves propagating at speed c. The first wave is divergent and propagates toward large r (with an amplitude decrease of a_+/r). The second wave is convergent and propagates toward small r (with an amplitude increase of a_-/r).

In the case of the harmonic wave, the electric field of the divergent wave can be written as

$$\mathrm{E_x}=(\mathrm{A_0}/\mathrm{r})\cos[\omega(\mathrm{t}-\mathrm{r}/\mathrm{c})+\varphi] \qquad (2.2.19)$$

or, by using complex notations

$$\underline{\underline{\mathrm{E}}}_{\mathrm{x}}=(\mathrm{A_0}/\mathrm{r})\exp\{\mathrm{i}[\omega(\mathrm{t}-\mathrm{r}/\mathrm{c})+\varphi]\} \qquad (2.2.20)$$

where φ denotes the phase of the wave at the origin of space and time. This wave is not periodic in space because its amplitude decreases like 1/r. The fields are tangent to the spherical wave surface.

II.2.2.2. The plane wave

Let us now write the Laplace operator using Cartesian coordinates and let us consider a direction of unit vector **u**, defined in space by the directional cosines u_x, u_y and u_z. We can write

$$\begin{aligned}\partial^2 f/\partial x^2 &= (\partial/\partial x)(\partial f/\partial x) = (\partial/\partial x)[(\partial f/\partial u)(\partial u/\partial x)]\\ &= (\partial/\partial x)[(\partial f/\partial u)u_x] = u_x[(\partial/\partial u)(\partial f/\partial u)](\partial u/\partial x)\\ &= u_x^2(\partial^2 f/\partial u^2)\end{aligned}$$

The Laplace operator simply writes $\partial^2 f/\partial u^2$, and the wave equation takes the form

$$\partial^2 f/\partial u^2 - (1/c^2)\partial^2 f/\partial t^2 = 0 \qquad (2.2.21)$$

Or

$$[\partial/\partial u - (1/c)(\partial/\partial t)][\partial/\partial u + (1/c)(\partial/\partial t)]f = 0$$

Using this equation we can write the general solution in the following form:

$$f = a_+(t - u/c) + a_-(t + u/c) \qquad (2.2.22)$$

Different from the spherical wave, we see that in this case a direction is selected, direction **u**. Moreover, the attenuation in 1/r no longer exists. We are in presence of two progressive plane waves, propagating in opposite directions along **u**.

The plane wave is the most frequently used approximation to the light wave. Now, we shall describe some of its main properties which will be used all through this book, and particularly in Problem II.3.1.

— *Transverse structure of the electric and the magnetic fields*

To simplify the writing, let us assume direction **u** lies along the Oz axis. Then we can write, using the first of Maxwell's laws:

$$\nabla\mathbf{E} = \mathbf{u}(\partial\mathbf{E}/\partial u) = (\partial/\partial u)(\mathbf{uE}) = (\partial E_u/\partial u) = 0$$

E_u is either constant or zero. If we consider only variable fields (and this will always be the case in this section), we must choose the second condition. In this case, the electric field has no component along the propagation axis. We shall say the electric field is *transverse*.

Using the second of Maxwell's laws and replacing **E** by **B**, we arrive at a completely identical conclusion concerning **B**.

— *Orthogonal structure of the electric and magnetic fields*

The third of Maxwell's laws can be expressed as

$$\mathbf{u}\times(\partial\mathbf{E}/\partial u)+\dot{\mathbf{B}}=0$$

If we assume the presence of a progressive wave propagating in the positive **u** direction, we have

$$\partial\mathbf{E}/\partial u=-\dot{\mathbf{E}}/c$$

and, consequently

$$(\partial/\partial t)\left[\mathbf{B}-(1/c)(\mathbf{u}\times\mathbf{E})\right]=0 \qquad (2.2.23)$$

always excluding stationary fields, this assumes that

$$\mathbf{B}=(1/c)(\mathbf{u}\times\mathbf{E}) \qquad (2.2.24)$$

It results that (i) **E** and **B** are orthogonal, (ii) **u**, **E** and **B** form a direct reference system, and, (iii) B = E/c , i.e. $E/H=\sqrt{\mu_0/\varepsilon_0}$. This parameter is called the *impedance of vacuum* and its value is $376.6\ \Omega$. It should be remembered that we must write $B=E\sqrt{\varepsilon_r\mu_r}$ and $E/H=c\mu_0\sqrt{\mu_r/\varepsilon_r}$ in a material medium.

These three results are fundamental. The third result shows that the effects of the magnetic field will always be small compared to those induced by the electric field (the ratio between the amplitudes is c). We shall see some consequences of this important point in Chapter V devoted to optical activity.

— *Phase velocity*

Equation 2.2.22 indicates that the phase of the perturbation is the same for all the points in a plane normal to the propagation direction. This plane is called the *wave plane*. For a harmonic wave propagating in vacuum, the phase $\omega[t-(u/c)]+\varphi$ remains constant provided dt – (1/c)du = 0, i.e. if du/dt = c. This is the velocity of wave surfaces of equal phase and it is called the *phase velocity*. As we pointed out earlier, c must be replaced by v = c/n ($n=\sqrt{\varepsilon_r\mu_r}$ is the refractive index and is equal to 1 in vacuum) in a material medium. The phase velocity is equal to ω/k, where ω is the frequency and $k = 2\pi n/\lambda$ the modulus of the wave vector of the radiation (λ is the wavelength in vacuum).

— *Energy carried by the wave*

By calling ρ_e and ρ_m the densities of electric and of the magnetic energy carried by the wave, the total differentials of these energies can be written as

$$d\rho_e = -\mathbf{D}' d\mathbf{E}' \quad d\rho_m = -\mathbf{B}' d\mathbf{H}' \tag{2.2.25}$$

By integrating **E'** and **H'** from 0 to **E** and **H** respectively, and by using the relations

$$\mathbf{D}' = \varepsilon \mathbf{E}' \quad \mathbf{B}' = \mu \mathbf{H}'$$

we find

$$\rho_e = -(1/2)\varepsilon E^2 \quad \rho_m = -(1/2)\mu H^2 \tag{2.2.26}$$

These two energy densities, which are measured in joule per m^3, are equal (to demonstrate this, the relation $E/H = c\mu_0\sqrt{\mu_r/\varepsilon_r}$ can be used). Therefore the total energy density carried by the wave is

$$eE^2 = B^2/m = (\sqrt{\varepsilon_r\mu_r}/c)EH = (1/v)EH$$

where v is the phase velocity of the wave. To calculate the power per unit surface carried by the wave, we must multiply the above density by the volume defined as the volume swept by a unit surface of the wave during a unit of time, that is, by v. Thus, because of the transverse structure of the fields, we can define a surface-density vector which is normal to the wave surface in an isotropic medium and whose instantaneous power is given by

$$\varpi = \mathbf{E} \times \mathbf{H} \tag{2.2.27}$$

And the time average of this power can be written as

$$\overline{\varpi} = (1/2)E_0 \times H_0 = (1/2)\sqrt{\varepsilon'/\mu}E_0^2 \tag{2.2.28}$$

E_0 and H_0 are the field amplitudes. The modulus of the time average of the power is usually called the *intensity* I of the light source.

For a non magnetic dielectric medium — which we shall often consider in the rest of this book — the above relations lead to

$$\bar{\rho}/n^2 = \bar{I}/nc = (1/2)\varepsilon_0 E_0^2 \tag{2.2.29}$$

Let us indicate an order of magnitude of the entities we just defined: To an energy density of the order of the $mJ.cm^{-3}$ correspond a surface-power density of the order of the $MW.cm^{-2}$ and an electric field with an amplitude of the order of the $MV.m^{-1}$. These high densities, associated to very intense electric fields, are now frequently obtained thanks to lasers (*cf.* Chapter VII).

— *Wave polarization*

The electric field and the magnetic field are vectors. By convention, the polarization direction of a wave is the direction of its electric field. If the wave propagates along the positive Oz axis of an isotropic medium, its field **E** has two components, one along Ox and the other along Oy which we shall represent conventionally as

$$\underline{\mathbf{E}} = \begin{pmatrix} 1 \\ \underline{r} \end{pmatrix} \mathbf{E}_0 \exp(i\Phi) \tag{2.2.30}$$

The complex number $\underline{r}$ can be written as

$$\underline{r} = r\exp(i\Delta\Phi) \tag{2.2.31}$$

where the real number r is the ratio of the amplitudes y/x, and $\Delta\Phi$ is the phase difference $\Phi_y - \Phi_x$. This notation, which will be used in Chapter V, provide a very simple way to describe the three polarization states shown on Figure 2.2.1.

Please note that a linear vibration can be described as the sum of a right circularly- and a left circularly-polarized vibration with the same amplitude.

II.2.2.3. Gaussian wave

The Gaussian wave directly concerns the laser. A parallel beam which comes out of a laser and enters in an infinite isotropic material does not have the structure of spherical wave nor that of a plane wave. Indeed, the electromagnetic wave is produced in an amplifying medium which usually has a cylindrical symmetry around the Oz axis, thereby imposing a specific average propagation direction (namely, the Oz direction). In this case, we can look for a solution of the wave equation whose amplitude decreases when the distance to the axis increases. We write this solution in the form given by Equation 2.2.14 where we replace $\underline{E}(r)$ by $\underline{E}(x,y,z)$. It must satisfy Equation 2.2.16, that is

$$\Delta\underline{E}(x,y,z) + k^2\underline{E}(x,y,z) = 0 \tag{2.2.32}$$

More precisely, we must have

$$\underline{E}(x,y,z) = E_0\psi(x,y,z)\exp(-ikz) \tag{2.2.33}$$

i.e. it takes the appearance of a plane wave [exp(–ikz)] perturbed by the term ψ (x, y, z). This term must satisfy two important conditions:

— It must vary slowly along Oz, i.e. it must satisfy

$$\left|\partial^2\psi / \partial z^2\right| << k\left|\partial\psi / \partial z\right| \tag{2.2.34}$$

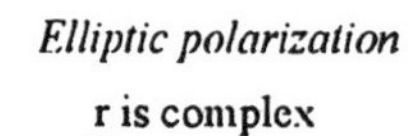

Elliptic polarization

$\underline{r}$ is complex

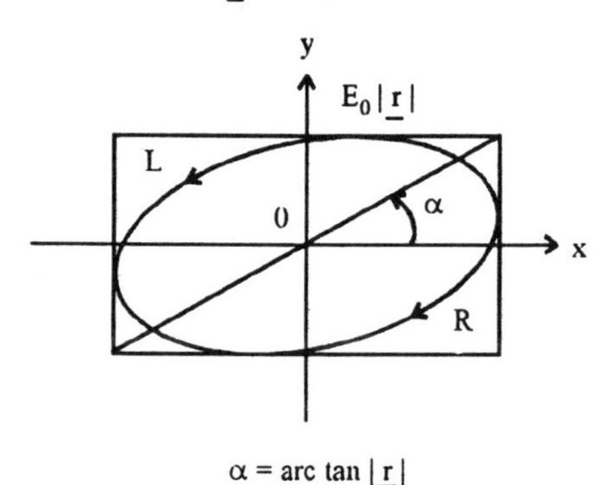

$\alpha = \text{arc tan}\, |\underline{r}|$

left polarized (L) if $\Delta\phi \in \,]\,0,\pi\,[$
right polarized (R) if $\Delta\phi \in \,]\pi,2\pi\,[$

Linear polarization

$\underline{r} = r$ is real

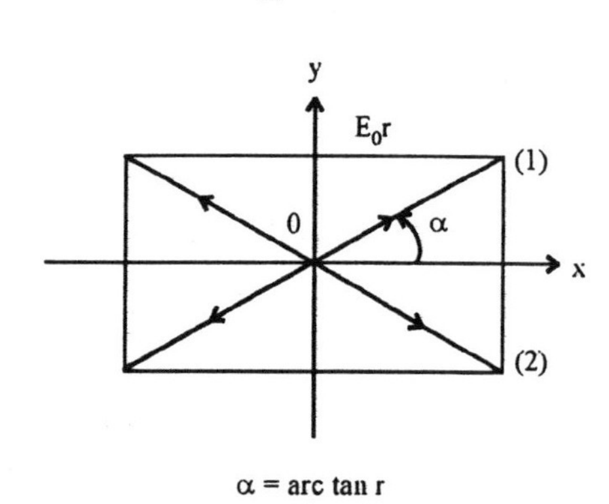

$\alpha = \text{arc tan}\, r$

polarized along (1) if $\Delta\phi = 2k\pi$
polarized along (2) if $\Delta\phi = (2k + 1)\,\pi$

Circular polarization

$\underline{r} = \pm\, i$

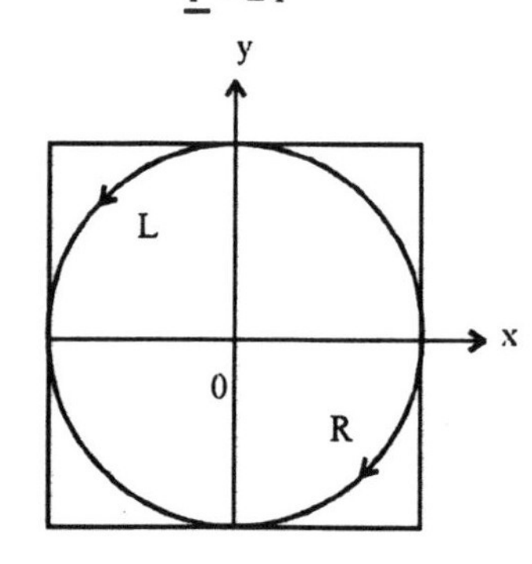

left polarized (L) if $\Delta\phi = \pi/2$
right polarized (D) if $\Delta\phi = 3\pi/2$

Figure 2.2.1: The three polarizations of an optical wave. The wave travels along the Oz axis(not shown) which points out of the page. The polarization is said to be levorotary or left polarized (respectively dextrorotary or right polarized) if the rotation of the point representing the rotation occurs in the trigonometric (respectively inverse trigonometric) direction.

— It must have a Gaussian radial variation, written as

$$\psi(r_\perp, z) = \exp[-iP(z)]\exp[-ikr_\perp^2 / 2q(z)] \tag{2.2.35}$$

in which we defined $r_\perp^2 = x^2 + y^2$.

This Gaussian shape may seem to be chosen completely arbitrarily. In fact, this is not so; it is imposed by the physical constraints which are due to the conception of the laser. In Section VII.2.1, the reader will find the physical explanations leading to such a choice.

Putting Equation 2.2.33 into Equation 2.2.32 and taking into account the approximation included in Equation 2.2.34, we obtain

$$\Delta_\perp \psi(x, y, z) - i2k\partial\psi / \partial z = 0 \tag{2.2.36}$$

with

$$\Delta_\perp = \partial^2 / \partial x^2 + \partial^2 / \partial y^2$$

By putting Equation 2.2.35 into Equation 2.2.36, we see that the plane wave with the Gaussian perturbation given by Equation 2.2.33 is indeed a possible solution of the wave equation. Moreover, we can specify the analytic forms of the factors P(z) and q(z) introduced in the test solution of Equation 2.2.35. Before giving the results of this calculation, whose details can be found in Bibliographic reference 7.6.3, we want to indicate the physical meaning of these parameters. P(z), called the *longitudinal wave parameter*, describes the longitudinal variations along Oz. Its real and imaginary parts represent, respectively, the variations of the phase and of the amplitude of the electric field along Oz. q(z), called the *radial wave parameter*, describes the radial variations of the field in the plane $z = \text{constant}$. The real and imaginary parts of $q(z)^{-1}$ represent, respectively, the variations of the wave surface along z and the distribution of the amplitude of the field around the Oz axis.

We arrive at the following expression for the electric field:

$$\begin{aligned} \underline{E}(x, y, z) = E_0[\omega(0)/\omega(z)]\exp - i\{k[z + r_\perp^2 / 2R(z)] - \phi(z)\} \\ \times \exp[-r_\perp^2 / \omega^2(z)] \end{aligned} \tag{2.2.37}$$

with

$$\begin{aligned} \omega(z) &= \omega(0)\sqrt{1 + [\lambda z / \pi\omega^2(0)]^2} \\ R(z) &= z\{1 + [\pi\omega^2(0) / \lambda z]^2\} \\ \phi(z) &= \text{Arc tg}[\lambda z / \pi\omega^2(0)] \\ \omega(0) &= \omega(z = 0) \end{aligned} \tag{2.2.38}$$

These two equations represent the structure of the Gaussian wave, illustrated on Figure 2.2.2, and whose properties are investigated in Problems II.3.2 and II.3.3.

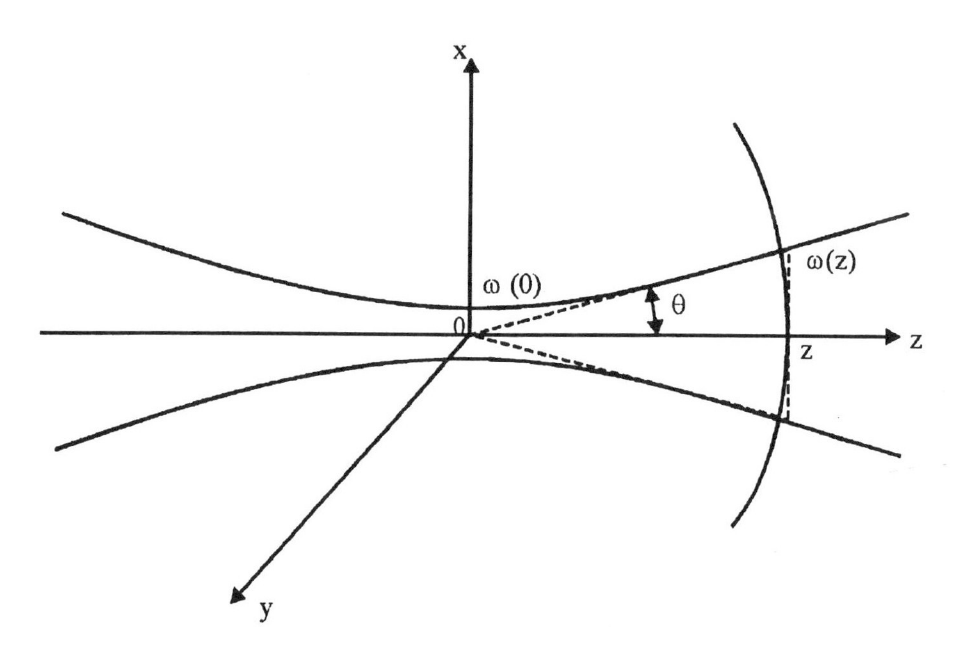

Figure 2.2.2: Geometry of the Gaussian wave.

The term $\exp\left[-r_{\perp}^{2}/\omega^{2}(z)\right]$ indeed confers a Gaussian radial attenuation to the wave, with a characteristic diameter of $\omega(z)$ which increases with z and which has a minimal extension $\omega(0)$ located at the coordinate origin.

This smallest diameter is called the *beam-waist* of the wave. The wave is not plane because the electric field decreases as $1/\omega(z)$, that is, as $1/z$ far away from the beam-waist. This behavior is characteristic of a spherical wave and R(z), which far from the beam-waist becomes equal to z, represents its radius of curvature. But because of the Gaussian attenuation it is not a pure spherical wave. Such a wave is called a *quasi-spherical wave*.

Table 2.2.1 summarizes the three above defined waves and compares their properties. Appendix II.4.3 gives their respective mathematical descriptions.

Table 2.2.1 — **Compared properties of the spherical, the plane and the Gaussian waves**

	Spherical wave	Plane wave	Gaussian wave (or quasi-spherical) far from the beam-wait
Wave surface	Sphere	Plan	Sphere
Energy distribution on the wave surface	Uniform	Uniform	Radial Gaussian attenuation
Direction of the electric and magnetic fields	Tangent to the wave surface	In the wave surface	Tangent to the wave surface
Amplitude of the fields	Attenuation inversely proportional to the radius of curvature radius	Constant	Attenuation inversely proportional to the radius of curvature and with a Gaussian radial behavior

II.3. Exercises and problems

II.3.1. Electromagnetic wave propagation in vacuum (from Agreg. P; A; 1995)

1. Wave equation

Recall Maxwell's equations in vacuum and in the absence of any charge density ($\rho = 0$) or currents ($j = 0$).

Give the definition of the d'Alembertian operator $\Box$ and state the propagation equations of the fields **E** and **B**.

Consider any one of the six components $E_x(x, y, z, t)$, $E_y(x, y, z, t)$, $E_z(x, y, z, t)$, $B_x(x, y, z, t)$, $B_y(x, y, z, t)$ and $B_z(x, y, z, t)$ and call it $f(x, y, z, t)$. Of what equation is $f(x, y, z, t)$ a solution?

For $f(x, y, z, t)$, define the concept of a plane wave propagating along the $\mathbf{u}_x$ direction. Express its velocity c as a function of ε_0 and μ_0. What is the present status of c in metrology? What unit in the SI system is determined by the choice of μ_0?

2. Structure of plane linearly-polarized monochromatic progressive waves.

In this part, we want to study a particular case of a plane electromagnetic progressive wave of the form

$$\mathbf{E} = \mathbf{E}_0 \cos(\omega t - kx) \quad \mathbf{B} = \mathbf{B}_0 \cos(\omega t - kx - \phi)$$

where $\mathbf{E}_0$ and $\mathbf{B}_0$ are two constant vectors, ϕ is an arbitrary constant and $k = \omega/c$. Define the propagation direction $\mathbf{u}$ of this wave. Define the polarization of this wave.

By using Maxwell's equations in complex notation, determine the structure of this wave. Do the waves have a transversal structure? Give the relation between $\mathbf{E}$, $\mathbf{B}$ and $\mathbf{u}$. Draw a figure. Define the wave impedance $\underline{Z}$ for this wave. Calculate the numerical value of $|\underline{Z}|$. Give a common example of longitudinal waves in another domain of physics.

1. Maxwell's equations are

$$\nabla \times \mathbf{E} = -\dot{\mathbf{B}} \qquad \nabla\, \mathbf{E} = 0 \quad ; \qquad \nabla \times \mathbf{B} = \left(1/c^2\right)\dot{\mathbf{E}} \qquad \nabla\, \mathbf{B} = 0$$

The d'Alembertian operator is defined by $\Box = \Delta - (1/c^2)\,\partial^2/\partial t^2$. Δ is the Laplace or delta operator.

The propagation equations are $\Box\mathbf{E} = 0 \quad \Box\mathbf{B} = 0$

$f(x, y, z, t)$ is the solution of one of the three projections of one of the two propagation equations on the Ox, the Oy, or the Oz axis.

A plane wave with frequency ω which propagates along $\mathbf{u}_x$ is written as

$$f(x - ct) = f_0 \cos\left\{\omega\left[t - (x/c)\right] + \varphi\right\}$$

φ is the phase and c the speed of light in vacuum whose value is equal to $1/\sqrt{\varepsilon_0 \mu_0}$.

Since 1973, c is determined from the relation $c = \lambda\nu$. The wavelength λ (3.39 μm) is that of a He-Ne laser. It is measured by interferometry (Michelson interferometer). The corresponding frequency ν is measured by comparison with a standard maser frequency after successive divisions.

The choice of μ_0 determines the value of the Tesla.

2. The wave propagates along the $\mathbf{u}_x$ axis. The phase planes of the wave are normal to the Ox axis (the phase only depends on the x coordinate). But the relation div $\mathbf{E} = 0$ imposes $\partial E_x/\partial x = 0$ (E_y and E_z only depend on x). However, E_x depends on x through its phase. This component is therefore necessarily equal to zero. The wave is transversal and lies in the Oyz plane.

Let us assume for simplicity that the electric field carried by the optical wave is linearly-polarized along Oy ($E_z = 0$). The equation $\nabla \times \mathbf{B} = (1/c^2)\,\dot{\mathbf{E}}$ shows that $\mathbf{B}$ is carried by Oz with the component $-\nabla_x B_z = (1/c^2)\dot{E}_y$. By appropriately choosing the phase origin so as to annul the integration constant, we obtain

$$B_z = \left(\omega / c^2 k\right) E_y \quad \text{with} \quad k = (2\pi/\lambda)|\mathbf{u}_x|$$

and therefore: $|\mathbf{B}| = (1/c)|\mathbf{E}|$

$$|\mathbf{H}| = (1/\mu_0 c)|\mathbf{E}|$$

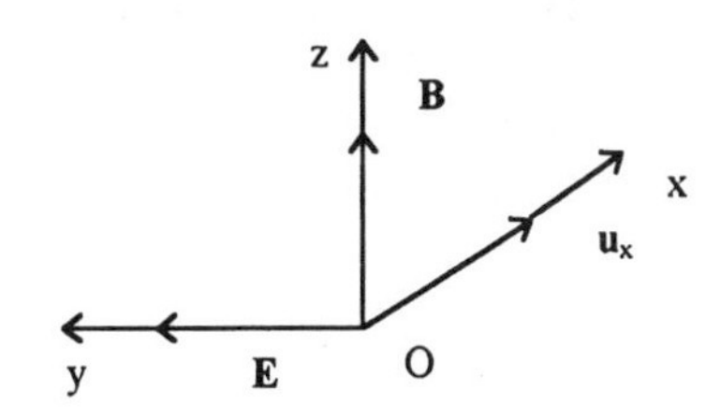

The parameter $|\underline{Z}|$, called wave impedance, can be written

$|\underline{Z}| = |\underline{E}|/|\underline{H}| = \sqrt{\mu_0}/\sqrt{\varepsilon_0}$

Numerically: $|\underline{Z}| \approx 376.6\ \Omega$

II.3.2. Search for a particular Gaussian solution of the wave equation

1. Let us choose the classical theory of radiation. Starting from Maxwell's equations, derive the wave equation which must be satisfied by all light waves.

2. Find a particular solution of this equation, starting from the perturbed plane wave $E(x,y,z) = \psi(x,y,z) \exp(-ikz)$, where ψ is a slowly-varying complex function of z which expresses how the wave differs from a plane wave.

3. Discuss and interpret the final result.

1. Applying the ∇ operator to the third of Maxwell's equations (*cf.* Equation 2.2.5) leads to $\nabla \times \nabla \times \mathbf{E} = -\nabla \times \dot{\mathbf{B}}$. And by using the fourth of Maxwell's equation, we find $\nabla \times \nabla \times \mathbf{E} = -\left(1/\varepsilon_0 c^2\right)\dot{\mathbf{J}} + \left(1/c^2\right)\ddot{\mathbf{E}}$. Remembering that $\mathbf{J} = 0$ in vacuum and that the first constitutive Maxwell's relation $\nabla \mathbf{E} = -\rho/\varepsilon_0$ leads to $\nabla \mathbf{E} = 0$ because $\rho = 0$, and by using the well-known relation $\nabla \times \nabla = \Delta + \nabla\nabla$, we obtain

$\Delta \mathbf{E} - \left(1/c^2\right)\ddot{\mathbf{E}} = 0$. This is the wave equation.

2. The wave equation can also be written as $\Delta \mathbf{E} + k^2\mathbf{E} = 0$ with $k = 2\pi/\lambda = \omega/c$, where λ and ω are, respectively, the wavelength and the frequency of the optical wave. Now we can write the suggested particular solution into the wave equation, and, assuming that ψ slowly varies along Oz, i.e. that the relation $\left|\partial^2\psi/\partial z^2\right| << k\left|\partial\psi/\partial z\right|$ is always satisfied, we find $\Delta_\perp\psi - i2k\partial\psi/\partial z = 0$ with $\Delta_\perp = (\partial^2/\partial x^2) + (\partial^2/\partial y^2)$. Since we need a Gaussian radial structure for the laser wave, we must look for a particular solution of the form

$\psi(r_\perp, z) = \exp[-iP(z)]\exp[-ikr_\perp^2/2q(z)]$ with $r_\perp^2 = x^2 + y^2$. P(z) is called the longitudinal wave parameter. Its real and imaginary parts represent, respectively, the variations of the phase and the variations of the amplitude of the wave along Oz. On the other hand, q(z) gives the radial variations of the field in the plane z = constant. The real and imaginary parts of 1/q(z) represent, respectively, the variations along z of the wave surface and the distribution of the field amplitude around the axis of the beam. q(z) is called the radial wave parameter. By carrying this particular solution into the differential equation written with the radial Laplace operator, we obtain

$$-i2k/q - k^2 r_\perp^2/q^2 - 2k\left[dP/dz - kr_\perp^2(dq/dz)/2q^2\right] = 0$$

This equation is satisfied for all values of $r_\perp^2$ only if we have simultaneously $dq(z)/dz = 1$ and $dP(z)/dz = -i/q(z)$.

— The first equation defines the radial wave parameter. Its integration gives $q(z) = q(0) + z$. By assuming that $R(0) = \infty$ (plane wave at z = 0), and separating the real and imaginary parts of 1/q(z) by writing $1/q(z) = 1/R(z) - i\lambda/\pi\omega^2(z)$, we find

$$R(z) = z\left\{1 + \left[\pi\omega^2(0)/\lambda z\right]^2\right\} \quad \text{and} \quad \omega^2(z) = \omega^2(0)\left\{1 + \left[\lambda z/\pi\omega^2(0)\right]^2\right\}$$

— By choosing the phase origin such that P(0) = 0, the second equation leads, after integration, to the expression $P(z) = i\mathrm{Ln}\{q(0)/[q(0) + z]\}$.

Finally we obtain

$$E(x, y, z) = E_0\left[\omega(0)/\omega(z)\right]\exp - i\left\{k\left[z + r_\perp^2/2R(z)\right] - \Phi(z)\right\}$$
$$\times \exp\left[-r_\perp^2/\omega^2(z)\right]$$

$$\text{with } \Phi(z) = \mathrm{Arctg}\left[\lambda z/\pi\omega^2(z)\right].$$

3. The discussion and the interpretation are given in Section II.2.2.3.

II.3.3. Treatment of Gaussian waves with lenses

A thin lens of focal length f, placed at the origin z = 0 of a coordinate axis Oz, induces a radial modification of the phase of an optical wave propagating along the Oz axis according to the relation $E(x,y,0+\varepsilon) = E(x,y,0-\varepsilon)\exp(ikr^2/2f)$. One will admit without demonstration that the beam parameter q and the radius of curvature R are also modified according to the relations $1/q(0+\varepsilon) = 1/q(0-\varepsilon) - 1/f$ and $1/R(0+\varepsilon) = 1/R(0-\varepsilon) - 1/f$. The radius of curvature R is defined to be positive if the concave part of the wave faces the negative values of z.

1. Let us consider a objet beam-waist (radius ω_1, beam parameter q_1) placed at a distance d_1 from the center of the lens. The lens produces an image (radius ω_2, beam parameter q_2) located at the distance d_2 from the lens. Express q_2 as a function of q_1, d_1, d_2, and f.

2. Express d_2 and ω_2 as functions of ω_1, d_1, f, and λ (the wavelength of the radiation).

1. We have the relation $1/(q_2 - d_2) = 1/(q_1 + d_1) - 1/f$ and after some elementary transformations (*cf.* Section VII.1.2) we find

$$q_2 = \left[q_1\left(1 - d_2/f\right) + d_1 + d_2 - d_1 d_2/f\right] / \left[-q_1/f + \left(1 - d_1/f\right)\right]$$

2. Remembering that $q_1 = i\pi\omega_1^2/\lambda$ and $q_2 = i\pi\omega_2^2/\lambda$, we find

$$i\pi\omega_2^2/\lambda = \left[(d_1 + d_2) f - d_1 d_2 + i\left(\pi\omega_1^2/\lambda\right)(f - d_2)\right]$$

$$\times\left[f - d_1 + i\left(\pi\omega_1^2/\lambda\right)\right] / (f - d_1)^2 + \left(\pi\omega_1^2/\lambda\right)^2$$

Setting the real part of the right member of the above equation equal to zero, we obtain the *position* relation $d_2 - f = (d_1 - f) f^2 / \left[(d_1 - f)^2 + \left(\pi\omega_1^2/\lambda\right)^2\right]$ of the image beam-waist. In the same way, setting the imaginary parts to be equal to each other, we obtain the following *magnification* relation:

$$\omega_2^2 = \omega_1^2 f^2 / \left[(d_1 - f)^2 + \left(\pi\omega_1^2/\lambda\right)^2\right]$$

II.4. Appendices

II.4.1. Table of electromagnetic waves (the numerical values are approximated by their order of magnitude)

parameter	cosmic ray	γ ray	X ray	UV	visible	IR	micro wave	radio frequency
λ(nm)	1×10^{-3}	1×10^{-1}	1×10	4×10^{2}	8×10^{2}	3×10^{6}	1×10^{9}	
ν(Hz)	3×10^{20}	3×10^{18}	3×10^{16}	8×10^{14}	4×10^{14}	1×10^{12}	3×10^{8}	
ω(rd.s^{-1})	2×10^{21}	2×10^{19}	2×10^{17}	5×10^{15}	2.5×10^{15}	6×10^{12}	2×10^{9}	
$\bar{\nu}$(cm^{-1})	1×10^{10}	1×10^{8}	1×10^{6}	2.5×10^{4}	1×10^{4}	3×10	1×10^{-2}	
E (KJ.mol^{-1})	1×10^{8}	1×10^{6}	1×10^{4}	3×10^{2}	1.5×10^{2}	4×10^{-1}	1×10^{-4}	
E (eV.mol^{-1})	7.5×10^{29}	7.5×10^{27}	7.5×10^{25}	2×10^{24}	1×10^{24}	2.5×10^{21}	7.5×10^{17}	

II.4.2. Various forms of Maxwell's equations (in SI units; in the following equations, we shall write $\rho = J = 0$ in the absence of sources,)

Vector formalism (3 D space)

— *In vacuum (real notation of the fields) :*

$$\Delta \mathbf{E} = \rho / \varepsilon_0 \,; \quad \nabla \times \mathbf{E} = -\dot{\mathbf{B}} \,; \quad \Delta \mathbf{B} = 0 \,; \quad \nabla \times \mathbf{B} = \mu_0 \mathbf{J} + \varepsilon_0 \mu_0 \dot{\mathbf{E}}$$

— *In vacuum (complex notation of the fields* $\mathbf{E} = \mathrm{Re}(\underline{\mathbf{E}})$:

For a progressive harmonic plane wave (frequency ω ; wave vector $\mathbf{k}$), the various operators used in Maxwell's equations are

$$\partial / \partial t = -i\omega \;;\; \nabla = i\mathbf{k} \;;\; \nabla\times = i\mathbf{k}\,\times$$

and Maxwell's equations take the form

$$\mathbf{k} \times \underline{\mathbf{E}} = \omega \underline{\mathbf{B}} \,;\; \mathbf{k} \times \underline{\mathbf{B}} = \mu_0 \mathbf{J} - (\omega / c^2) \underline{\mathbf{E}} \,;\; \mathbf{k} \underline{\mathbf{E}} = -i\rho / \varepsilon_0 \,;\; \mathbf{k} \underline{\mathbf{B}} = 0$$

— *In a material medium* ($\mathbf{D} = \varepsilon_0 \mathbf{E} + \mathbf{P}$; $\mathbf{B} = \mu_0 \mathbf{H} + \mathbf{M}$):

$$\Delta \mathbf{D} = \rho \,; \quad \nabla \times \mathbf{E} = -\dot{\mathbf{B}} \;;\; \Delta \mathbf{B} = 0 \,; \quad \nabla \times \left[(\mathbf{B} - \mathbf{M}) / \mu_0 \right] = \mathbf{J} + \dot{\mathbf{D}}$$

Tensor formalism (3D space, with Einstein's summing convention)

— *In vacuum:*

$$\nabla_i E_i = \rho / \varepsilon_0 \,; \quad \varepsilon_{ijk} \nabla_j E_k = -\dot{B}_i$$

$$\nabla_i B_i = 0 \,; \quad \varepsilon_{ijk} \nabla_j B_k = \mu_0 J_i + \varepsilon_0 \mu_0 \dot{E}_i$$

— *In a material medium* ($D_i = \varepsilon_0 E_i + P_i$; $B_i = \mu_0 H_i + M_i$):

$$\nabla_i D_i = \rho \,; \quad \varepsilon_{ijk} \nabla_j E_k = -\dot{B}_i \,; \quad \nabla_i B_i = 0$$

$$\varepsilon_{ijk} \nabla_j [(B_k / \mu_0) - M_k] = J_i + \dot{D}_i$$

Tensor formalism (in Minkowskian space-time coordinates; in vacuum)

The four Maxwell equations can be expressed by the relation

$$\nabla_j T_{kl} + \nabla_k T_{lj} + \nabla_l T_{jk} = 0$$

where j, k and l represent each of the four coordinates x, y, z, ict of the 4D space and T is the antisymmetrical induction tensor (rank 2) given by

$$[T]=\begin{bmatrix} 0 & B_z & B_y & -(i/c)E_x \\ -B_z & 0 & B_x & -(i/c)E_y \\ -B_y & -B_x & 0 & -(i/c)E_z \\ (i/c)E_x & (i/c)E_y & (i/c)E_z & 0 \end{bmatrix}$$

II.4.3. Three particular solutions to the wave equation

PLANE WAVE	SPHERICAL WAVE	GAUSSIAN WAVE
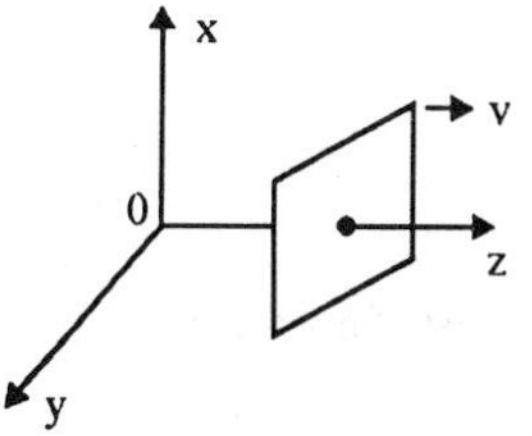	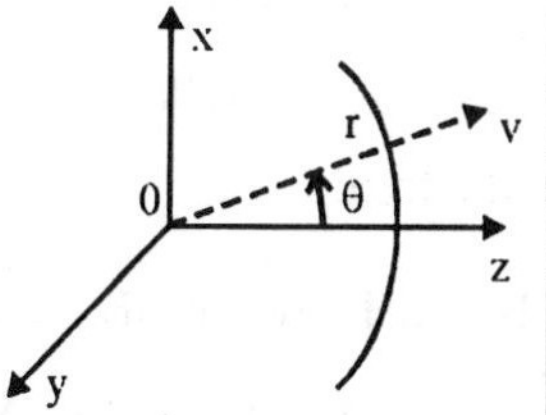	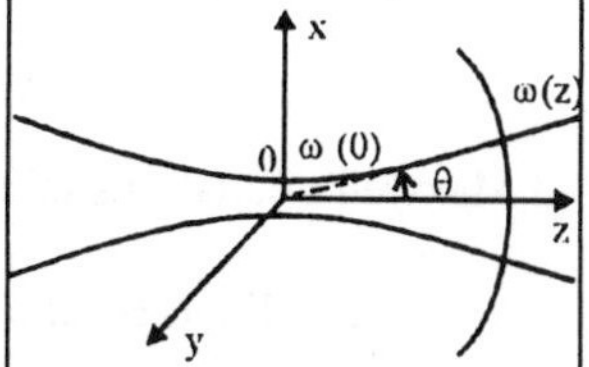
Field **E** in the wave plane	Field **E** tangent to the wave plane	Field **E** tangent to the wave plane
Progressive plane wave in the direction of positive z $E_x = f(z - vt)$	Divergent spherical wave in the direction of positive z $E_x = f(r - vt)/r$	Divergent quasi spherical wave in the direction of positive z $E_x = f(x, y, z, t)$
Plane parallel harmonic wave $\underline{E} = E_0 \exp(-i\omega t)$ $\times \exp\left[-i(\omega z / v - \varphi)\right]$	Divergent spherical harmonic wave $\underline{E} = (A_0 / r)\exp(-i\omega t)$ $\times \exp\left[-i(\omega r / v - \varphi)\right]$ and when θ is small: $\underline{E} = (A_0 / r)\exp(-i\omega t)$ $\times \exp - i(\omega / v)(z + r_\perp^2 / 2z)$ $\times \exp(i\varphi)$	Divergent Gaussian harmonic wave $\underline{E} = E_0[\omega(0)/\omega(z)]\exp(-i\omega t)$ $\times \exp - i(\omega / v)\left[z + r_\perp^2 / 2R(z)\right]$ $\times \exp(i\varphi)\exp[-r_\perp^2 / \omega^2(z)]$ far from beam-waist: $\underline{E} = (A_0 / r)\exp(-i\omega t)$ $\times \exp - i(\omega / v)(z + r_\perp^2 / 2z)$ $\times \exp(i\varphi)\exp[-r_\perp^2 / \omega^2(z)]$ spherical wave × Gaussian radial attenuation

II.5. Bibliography

2.5.1. An excellent didactic presentation of classical electromagnetism is given in:

BLUM R. and ROLLER D.E. — *Physics, vol.II : electricity, magnetism, and light,* Holden-Day, San Francisco (1982).

PEREZ J. Ph., CARLES R. and FLECKINGER R. — *Electromagnétisme*, Masson 3[sd] ed. (1997).

2.5.2. The reference book in electromagnetism is:

JACQSON J. D. — *Classical Electrodynamics*, J. Wiley, New York (1962).

2.5.3. Quantum theory of light can be found in:

COHEN-TANNOUDJI C., DUPONT-ROC J. and GRYNBERG G. — *Processus d'interactions entre photons et atomes,* InterEditions (1988).

LOUDON R. — *Quantum theory of light*, Oxford Science Publications (1983).

Chapter III

Matter and its properties

III.1. Non relativistic quantum mechanics

III.1.1. Fundamental principles and brief history

At the end of the 19th and at the beginning of the 20th century, many experimental works concerning the microscopic world struggled with major difficulties concerning their interpretation within the frame of the physics of that time. These difficulties were deep and they emerged at a most unexpected time, paradoxically, at the time corresponding to the triumph of classical mechanics, which benefited from the latest developments of Lagrange and Hamilton formalisms, and of the advent of electromagnetism due to Maxwell's theory of light vibrations. These theories were so successful that they even led to the union of electromagnetism with mechanics in 1905 by Einstein through the unifying formalism of the restricted theory of relativity.

However, some experimental facts clearly pointed to the existence of limits to the predicting and the explanatory capacities of the physics of that time. For example, starting from a joined thermodynamic and electromagnetic approach, Rayleigh and Jeans studied the spectral radiating energy density of the black body and obtained an analytic form of the total energy density which increased with frequency, and which was therefore paradoxically divergent (the UV catastrophe).

In 1887, Hertz discovered and studied the photoelectric effect consisting in the emission of electrons by a solid (metallic or oxide) submitted to the action of an incident electromagnetic wave (UV, visible or near IR). He showed that this emission takes place only if the frequency of the incident radiation is greater than a certain frequency limit ν_0, characteristic of the material and independent of the incident radiation flux. Moreover, the kinetic energy of the emitted electrons appeared to be independent of the intensity of the incident radiation, and to be related only to its frequency. This result very much contradicted the hypothesis of a continuous energy exchange between matter and a wave of frequency ν_0 during the absorption process, an idea which was widely accepted at that time.

Similarly, the line structure observed on atomic spectra suggested the existence of " privileged " electronic states in atoms associated with discontinuous series of energy levels, as for example those studied by Brackett and Balmer. This discontinuity in the atomic energy states was formalized in the combination principle of Ritz (1908), expressing that the wavenumbers of all the observed lines of the spectrum of an atom could be obtained from the difference of two terms, the so-called *spectral terms* of the atom. At that time, this principle was empirical and in no way predicted by the physics of the beginning of the century.

Last, all these experimental studies tended to demonstrate the failure of physics in accounting for phenomena concerning light-matter interaction and occurring at a microscopic level. As we shall see later, the only way to maintain a certain unity in the body of knowledge represented by physics at that time was to assume that some physical quantities (such as energy levels in atoms) could only take on a certain number of discreet values, and then to investigate the origins and consequences of these discontinuities. A clear need for a deep conceptual renewal of physics had appeared.

The first step to the birth of a new mechanics which would describe the microscopic world seems to be due to Planck who showed that in order to find an analytic expression for the spectral radiating energy density $I(\nu)$, it was necessary to admit that the exchange of energy between light and matter was discontinuous: A radiation with a frequency ν_0 can exchange energy with matter only in "lumps" of energy of magnitude $h\nu_0$, the so-called *quanta*, where h is a new universal constant (the Planck constant). With this new hypothesis, he was able to find an analytic expression for $I(\nu)$ which was in excellent agreement with the experimental data:

$$I(\nu) = (2\pi h\nu^3/c^2)\,\{1/[\exp(h\nu/kT) - 1]\} \qquad (3.1.1)$$

where c is the speed of light in vacuum, k the Boltzmann constant and T the temperature. The value of h is 6.62×10^{-34} Js.

It is fundamental to notice that at that time, the exact origin of Planck's quanta remained unknown, and that Planck himself tried (without success!) to establish his result without referring to them. The expression of $I(\nu)$ shows that the quantification effects become significant at very high frequencies and at low temperatures, as the exponential argument becomes sensitive to large values of $h\nu$. On the contrary, when $h\nu << kT$, quantification is no longer critical since as a first order approximation we have exp(x) − 1 ≈ x, so that for small ν the dependence on h disappears in the expression of $I(\nu)$ which thus coincides with the relation suggested par Rayleigh and Jeans.

Following the pioneering work of Planck on the photoelectric effect, Einstein proposed that light, described in Chapter II of this book, also exhibits discontinuous properties and he suggested the so-called radiation quantum (also named "photon"). The radiation quantum is a particle characterized by an energy E and a momentum whose modulus is $p = E/c$.

Last, in 1914, Bohr suggested that atomic and molecular edifices should be characterized by energy levels whose values form discrete series, and that radiation absorption and emission are possible only for specific frequency values ν_{if} such that $\nu_{if} = |E_i - E_f|$, where E_i and E_f are the values of the energy of the system before and after the emission ($E_i > E_f$) or the absorption ($E_i < E_f$). These hypothesis led Bohr to postulate his famous model of the hydrogen atom and to elucidate the origin of the Ritz combination principle. However, this approach is based on a planetary model of the atom, where the electron orbits around the nucleus under the action of an attractive electric force with a quantified momentum. Although these assumptions account for the atomic spectra, they contradict a fundamental result from electromagnetism stating that any accelerated electric charge radiates and thus looses energy. Consequently, in the case of a planetary model for the atom, the orbit of the electron cannot be a stable one, and the electron should eventually fall onto the nucleus. Nevertheless, the quantification of the energy exchanges between light and matter appears to offer a good frame for the interpretation of spectroscopy experiments.

In 1923, Louis de Broglie sets out a revolutionary hypothesis: In the same way that light — which, in classical physics is a vibration characterized by a wavelength λ — manifests a corpuscular nature, he suggested that it is possible to associate a wave to any particle with momentum **p** whose wave vector **k** is defined by $\mathbf{p} = (h/2\pi)\mathbf{k}$. The associated wavelength is such that $\lambda = h/p$. Through this hypothesis, wave mechanics was born. However, a basic inconsistency spoiled this description: The energy of a classical wave is distributed continuously and thus cannot consist of energy " lumps ". And also, the wave-particle duality shows that none of the two aforementioned classical descriptions can be thoroughly correct and that it would be totally artificial to use two mutually exclusive classical models (*the wave and the particle*) to account for the same microscopic system.

Little by little, a new path seemed to emerge, prefiguring a new mechanics, in which classical mechanics would be a special case, valid for macroscopic bodies, and where Planck's constant would play a key role. Thus, the heavier a body, or the larger its momentum as compared to $h/2\pi$, the less it would behave like a wave. This analysis can be compared to the development of physical optics in which geometrical optics is considered as a special case when the wavelength of the radiation becomes much smaller than the size of the bodies considered, and whose laws had been formulated well before the advent of Maxwell's theory. This is the way quantum mechanics was finally formalized, under the impulse of the outstanding physicists Dirac, Schrödinger, Pauli, De Broglie, Born, Heisenberg, etc., who invented the concept of "quantum phenomena ", which reconciled apparently incompatible standpoints. We shall briefly review the basic concepts of quantum mechanics, which reduce to classical mechanics for macroscopic systems.

III.1.1.1. Probabilistic description

For the sake of clarity, let us specify immediately that quantum mechanics does not take into account the wave concept and only consider the particle aspect. The microscopic system, whatever it may be (elementary particle, atom, molecule or photon), is a *particle*. Let us emphasize here the unity of this approach which treats radiation and matter in the same way.

Let us focus now on the example of the photon and let us consider the well known experiment on light wave interference with Young slits. A point-like light source S emits a monochromatic beam which is diffracted by two identical slits T_1 and T_2 in an opaque screen, set at an equal distance from S. Normal to the ST_1T_2 plane, rectilinear and equidistant interference fringes are observed in a plane parallel to the screen. An exclusively corpuscular interpretation seems difficult at first sight. One might be tempted to invoke the fact that the linear trajectories of the photons foreseen by classical mechanics could be mutually disturbed when the two slits are open simultaneously, and that such an effect could be canceled by reducing the light flux emitted by the source in such a way that the photons would reach T_1 and T_2 one by one. However, the interference pattern persists even in these conditions. These interference fringes show very clearly that the laws of classical mechanics do not hold for photons. Similar results have recently been obtained on other microscopic systems used in the same kind of experiments. The interference phenomenon does not appear to be a specific characteristic of radiation. In fact, it appears impossible to predict the points of impact of a microscopic system whose behavior is very different from that of usual particles which obey to Newtonian mechanics. It is by no means sufficient to specify the initial conditions in order to know its trajectory and position at any later time. We know the suggested interpretation according to we must admit that the corpuscle has probability P_1 to go through slit T_1, and probability P_2 to go through slit T_2, and that the fringes are caused by the interference between P_1 and P_2. Here is the key idea for the new description. Classical deterministic mechanics must be discarded to the benefit of a new mechanics with a pronounced probabilistic character, in which the state of a quantum particle is characterized by a complex vector $\psi(\mathbf{r}, t)$ depending both on space and time, and called the *state vector* of the system. Only its modulus $|\psi(\mathbf{r}, t)|^2$ has physical meaning and it stands for the *volume density probability of presence* of the particle at time t and at point $\mathbf{r}$. Of course, since the particle must necessarily be somewhere, the sum of all the elementary probabilities dP associated to all the elementary volumes into which space can be divided, is equal to unity.

$$\int_{espace} |\psi(\mathbf{r}, t)|^2 \, dV(\mathbf{r}) = 1 \tag{3.1.2}$$

This hypothesis marks a fundamental break with classical physics. A particle is no longer localized on its trajectory at a given time, but its presence is defined by a probability at a given time t and at every point in space. In the case of a collection of N particles, the state of the system is represented by the function $\psi(\mathbf{r}_1, \mathbf{r}_2, ..., \mathbf{r}_i, ..., \mathbf{r}_N, t)$ where $\mathbf{r}_i$ is a possible position of particle i. The square of the modulus $|\psi(\mathbf{r}_1, \mathbf{r}_2, ..., \mathbf{r}_i, ..., \mathbf{r}_N, t)|^2$ is the probability density of the presence at time t of particle 1 at $\mathbf{r}_1$, of particle 2 at $\mathbf{r}_2$, ..., of particle i at $\mathbf{r}_i$, ..., and of particle N at $\mathbf{r}_N$.

The state vector $\psi(\mathbf{r}, t)$ belongs to the vectorial space F of continuous functions everywhere defined and whose square can be summed over all space. According to a notation introduced by Dirac, we shall call $|\psi\rangle$ the *ket* vector, or ket, associated to $\psi(\mathbf{r}, t)$. This ket $|\psi\rangle$ can be expanded on a basis F according to

$$|\psi\rangle = c_i |e_i\rangle \tag{3.1.3}$$

We have used the Einstein summation convention throughout this book. The coefficients c_i are complex numbers. By analogy with the vectorial representation in Euclidian geometry, a ket can be written as a [n × 1] column matrix consisting of the components of $|\psi\rangle$ expressed on basis $|e_i\rangle$

$$|\psi\rangle = \begin{bmatrix} c_1 \\ c_2 \\ . \\ c_n \end{bmatrix} \tag{3.1.4}$$

$$|e_1\rangle = \begin{bmatrix} 1 \\ 0 \\ . \\ 0 \end{bmatrix} \quad |e_2\rangle = \begin{bmatrix} 0 \\ 1 \\ . \\ 0 \end{bmatrix} \quad |e_n\rangle = \begin{bmatrix} 0 \\ 0 \\ . \\ 1 \end{bmatrix} \tag{3.1.5}$$

To every ket $|\psi\rangle$ of vectorial space F there is an associated vector of space F^*, the dual of space F, which is called the *bra* vector, or bra, noted $\langle\psi|$ and written as a $[1\times n]$ line matrix .

$$\langle\psi| = \left[c_1^* c_2^* \ldots c_n^*\right] \tag{3.1.6}$$

For example, the bra associated to ket $\alpha|\varphi\rangle + \beta|\psi\rangle$ is $\alpha^*\langle\varphi| + \beta^*\langle\psi|$.

The product (bra $\times$ ket) defines a scalar product on F called the *hermitian product* and it defines a norm. A ket is said to be normed if its norm is equal to unity, and two kets are said to be orthogonal if their scalar product vanishes. The following properties will be frequently used in the rest of this chapter:

$$\begin{gathered} \langle\psi|\varphi\rangle = \int_V \psi^* \varphi \, dV \text{ (hermitian product)} \\ \langle\psi|\psi\rangle = \sum_i |c_i|^2 \text{ (norm)} \\ \langle e_i|e_j\rangle = \delta_{ij} \text{ (orthonormalized basis)} \end{gathered} \tag{3.1.7}$$

where δ_{ij} is the Kronecker symbol. The jth component of ket $|\psi\rangle$ in basis $|e_i\rangle$ can be obtained by taking the product $\langle e_i|\psi\rangle$

$$\begin{gathered} \langle e_j|\psi\rangle = c_i\langle e_j|e_i\rangle = c_i\delta_{ij} = c_j \\ \langle\psi|e_j\rangle = c_j^* \end{gathered} \tag{3.1.8}$$

Please note that the integral appearing in the definition of the hermitian product indeed converges because the square of the functions involved can be summed over space. Moreover, the bra and ket vectors depend explicitly on the time coordinate, but only implicitly on the spatial coordinate **r**, since they are built from vector ψ (**r**, t), keeping the same space and time dependencies.

Finally, we would like to emphasize that the new description will have to substitute a new equation, one describing the evolution of the state vector ψ (**r**, t), to the fundamental relation of dynamics which provides a way to predict trajectories in classical mechanics.

III.1.1.2. The problem of measurement and operators

The second major rupture with classical ideas, after the rejection of trajectories to the benefit of state vectors (and the notion of the probability of presence of a particle at a given time and place) came with the idea of measurement and the discrete variation (the quantification) of some physical observables such as energy, which must of course be dealt with in quantum mechanics. An explicit correspondence was suggested between the *physical observable* of a system and *linear operators*, mathematical objects which transform the state vector characterizing a system. The measurement of a given observable yields a certain number of values which constitute the *eigenvalues* of the corresponding operator. However, let us remember that the probabilistic nature of the new physics, as outlined in the foregoing section, usually leads to an *average* result which we shall have to learn to express correctly. All the strength of this new approach lies in the fact that it provides a natural way to introduce the discrete variations of experimental measurements associated to an observable, through the discontinuous spectrum of the real eigenvalues of the corresponding operator.

Last, there is an extra difficulty concerning the very important problem of measurement. In classical physics, a well conducted measurement does not disturb the system. This is no longer true in the microscopic world where experimental facts show that measurements can perturb the state of the investigated microscopic system. For instance, to shine a light on a macroscopic object *a priori* has no physical effect on it, whereas in the case of a so-called atomic object, it can induce a significant perturbation, for example in terms of electron-photon impacts. The postulates of quantum physics, which we shall discuss in Section III.1.2, will have to account for this very surprising experimental fact, typical of the microscopic world. But before we go on, let us recall some basic properties of linear operators.

We call *linear operator O* any linear application which to one ket $|\psi\rangle$ of F associates a new ket $O|\psi\rangle$, also belonging to F, such that

$$\forall(\lambda_1,\lambda_2)\in C^2, \forall(\varphi_1,\varphi_2)\in E^2, O|\lambda_1\varphi_1+\lambda_2\varphi_2\rangle=\lambda_1 O|\varphi_1\rangle+\lambda_2 O|\varphi_2\rangle$$

By expanding $|\psi\rangle$ on basis $|e_i\rangle$ of F, the action of the linear operator O on $|\psi\rangle$ reads

$$O|\psi\rangle = O(c_i|e_i\rangle) = c_i O|e_i\rangle$$

Let us call $|f_i\rangle$ the image of $|e_i\rangle$ by operator O. Ket $|f_i\rangle$, which belongs to F, can be expressed on basis $|e_i\rangle$ as

$$\mathcal{O}|e_i\rangle = f_i = o_{ij}|e_j\rangle \tag{3.1.9}$$

or equivalently, by using matrix formalism

$$\mathcal{O}|e_i\rangle = \mathcal{O}\begin{bmatrix} 0 \\ . \\ 1 \\ . \\ 0 \end{bmatrix} = \begin{bmatrix} o_{1i} \\ . \\ . \\ . \\ o_{ni} \end{bmatrix}$$

When writing the expansion of the ket $|f_i\rangle$ on basis $|e_i\rangle$ in this way, we see that the n^2 order square matrix consisting of the $n \times n$ coefficients o_{ij} ($1 \leq i \leq n$ and $1 \leq j \leq n$), is the matrix associated to operator O expressed on basis $|e_i\rangle$. The properties of the linear transformations and of their associated matrices are presented in Appendix III.5.1. We observe, and this is fundamental, that the jth column of this matrix consists of the coordinates of the image of ket $|e_j\rangle$, expressed on basis $|e_j\rangle$. This comment is essential because it provides a quick way to write the matrices associated to the operators, expressed in a given basis, starting from the images of each ket of the basis. The advantage of using the matrix form lies in the fact that the coordinates of the image of any ket expanded on basis B are easily obtained by taking the product of the matrix associated to the operator expressed on B by the column matrix of the ket in consideration, expanded on B. Let us consider, for example, the case of an operator O defined in a three dimensional space. The image by O of each of the three basis vectors can be expressed as

$$\mathcal{O}|e_1\rangle = o_{j1}|e_j\rangle \quad \mathcal{O}|e_2\rangle = o_{j2}|e_j\rangle \quad \mathcal{O}|e_3\rangle = o_{j3}|e_j\rangle$$

Let us write the matrix associated to operator O expressed on basis $|e_j\rangle$. The jth column is obtained from the image of vector $|e_j\rangle$ expressed on basis $|e_j\rangle$. So we obtain

$$(O) = \begin{pmatrix} o_{11} & o_{12} & o_{13} \\ o_{21} & o_{22} & o_{23} \\ o_{31} & o_{32} & o_{33} \end{pmatrix}$$

Using the multiplication rules of matrices, we find that the image of ket $|e_2\rangle$ is

$$O\begin{bmatrix} 0 \\ 1 \\ 0 \end{bmatrix} = \begin{bmatrix} o_{12} \\ o_{22} \\ o_{32} \end{bmatrix}$$

Let $|\varphi\rangle$ and $\langle\psi|$ be two elements of F and F*, respectively. Considering the expression $\langle\varphi|O|\psi\rangle$, where O is a given linear operator, we can write

$$|\psi\rangle = c_i|e_i\rangle \quad |\varphi\rangle = d_k|e_k\rangle$$

and

$$O|\psi\rangle = c_i O|e_i\rangle = o_{ji}c_i|e_j\rangle$$

Multiplying the left side of the above equation by $\langle\varphi|$ we obtain

$$\langle\varphi|O|\psi\rangle = d_k^* o_{ji} c_i \langle e_k|e_j\rangle = d_k^* o_{ji} c_i \delta_{jk} = d_j^* o_{ji} c_i$$

which is a number (using the convention of double "mute" condensation over indexes i and j). If $\langle\varphi|$ and $|\psi\rangle$ are the two elements $\langle e_i|$ and $|e_j\rangle$ of bases F* and F, we get $\langle e_i|O|e_j\rangle = o_{ij}$, which is the *matrix element* associated to operator O, expressed on basis $|e_i\rangle$, situated at the intersection of the ith line with the jth column. Generally, expressions of type $\langle\varphi|O|\psi\rangle$ are called *matrix elements,* which is strictly true only when $\langle\varphi|$ and $|\psi\rangle$ are basis vectors.

The expression $\langle e_i|O|e_i\rangle = o_{ii} = \mathrm{Tr}(O)$ represents the sum of the diagonal elements of the matrix representation and is called the *trace* of the matrix. For example, in the case of the matrix associated to the identity operator (or unity), which preserves the kets, the only nonzero elements are those of the principal diagonal and they are equal to 1 (Kronecker matrix).

— *Hermiticity*

An operator O is said to be *hermitian* if, for any $|\varphi_1\rangle$ and any $|\varphi_2\rangle$ the equation $\langle\varphi_1|O|\varphi_2\rangle = \langle\varphi_2|O|\varphi_1\rangle^*$ is satisfied. It is straightforward to show that the eigenvalues of a hermitian operator (which are in fact the *possible* results of a measurement of the corresponding physical property) are *real valued.* The converse assertion is also true. An operator O whose eigenvalues have real values is necessarily hermitian. For this reason, hermiticity is one of the fundamental properties of the operators of quantum mechanics. Moreover, when eigenvalues are *distinct from each other* (discontinuous), the corresponding eigenstates are *orthogonal.* An example of hermiticity is provided by an operator that we will frequently use in the rest of this chapter, the *density operator* ρ. Starting from the state vector, the density operator is defined as

$$\rho = |\psi\rangle\langle\psi| \tag{3.1.10}$$

It represents the state of the system in the same way as the ket $|\psi\rangle$ from which it is defined. Its trace is expressed by

$$\mathrm{Tr}(\rho) = \langle e_i|\psi\rangle\langle\psi|e_i\rangle = |c_i|^2 = 1 \tag{3.1.11}$$

It is equal to unity.

Let us calculate a matrix element $\langle e_j|\rho|e_i\rangle$. Using Equation 3.1.9, we obtain $\langle e_j|\rho|e_i\rangle = \rho_{ij}\langle e_j|e_j\rangle = \rho_{ij} = \langle e_j|\psi\rangle\langle\psi|e_i\rangle = \langle e_i|\psi\rangle\langle\psi|e_j\rangle^* = \rho_{ji}^*$. This result establishes the hermiticity of the density operator. For a hermitic operator, the average value calculated over any $|\varphi\rangle$ is real. The converse is also true and an operator O whose average value is real valued for all $|\varphi\rangle$ is necessarily hermitian.

— *Eigenvalue and eigenvector*

A ket vector $|u_n\rangle$ is called an *eigenvector* of operator O if it satisfies $O|u_n\rangle = \lambda_n|u_n\rangle$, where λ_n is a complex number called the *eigenvalue* associated to $|u_n\rangle$. The set of the eigenvalues of an operator forms its *spectrum*. When several linearly independent eigenvectors are associated to the same eigenvalue, there is *degeneracy*. The *degree* of degeneracy of an eigenvalue is equal to the greatest number of linearly independent eigenvectors which can be associated to this eigenvalue.

The matrix representation of an operator is always *diagonal* in the basis of its eigenvectors.

— *Commutator*

The *commutator* $[A, B]$ of two operators A and B is defined by

$$[A, B] = AB - BA$$

Two operators whose commutator vanishes are said to *commute*. When their eigenvalues are distinct — which is the most common case —, the two commuting operators have the same set of eigenvectors.

— *Average value*

The quantity $\langle\psi|O|\psi\rangle$ defines the average value of the measurement of observable O.

III.1.2. Principles of non relativistic quantum theory

They are summarized in Appendix III.5.3.

III.1.2.1. Postulates concerning the description of a system

Postulate P1. A vectorial space F is associated to any physical system. F is the so-called space of states. The state of the system is represented by a ket vector of F, called $|\psi\rangle$. $|\psi\rangle$ is formed from the state vector $\psi(\mathbf{r}, t)$ and it is defined to a multiplicative constant.

In chemistry, the state vectors which characterize the electrons of an atom or a molecule *individually* are called *atomic* or *molecular orbitals*. The exact nature of the information contained in $|\psi\rangle$ is specified by the following postulate.

Postulate P2. To any physical observable O corresponds a linear hermitian operator *O*. The set of the eigen kets $|u_n\rangle$ of *O* constitutes a basis of F.

The second part of this postulate is important. It states that every ket $|\psi\rangle$ can be linearly expanded on the orthonormal basis of the eigen kets of any operator associated to a physical observable. Let us now turn to the problem of measuring physical quantities.

III.1.2.2. Principles for the measurement of physical quantities

Postulate P3. The set of possible outcomes of the measurement of an observable O on a physical system constitutes the set of eigenvalues λ_n of the operator *O* associated to the physical observable O.

Since operator *O* is hermitian, its eigenvalues are real and they can therefore describe the result of a physical measurement.

Postulate P4.

Case of a discrete and non degenerate eigenvalue

In the case of a system described by a normalized ket $|\psi\rangle$, the probability $P(a_n)$ to obtain the value a_n as result of the measurement of the physical property O is given by

$$P(a_n) = |\langle u_n|\psi\rangle|^2 \tag{3.1.12}$$

where $|u_n\rangle$ is an eigenket of operator *O*, corresponding to observable O, associated to the eigenvalue a_n.

Case of a discrete and degenerate eigenvalue

Let $|u_n^i\rangle$ (i = 1, ..., g) be the g eigenkets of operator O associated to the physical observable O, corresponding to the g-fold degenerate eigenvalue a_n. The probability to find a_n as result of the measurement of O on the system described by state vector $|\psi\rangle$ is given by

$$P(a_n) = \sum_{i=1}^{g} \left|\langle u_n^i | \psi\rangle\right|^2$$

If the system happens to be in state $|u_i\rangle$, the eigenfunction of operator O associated to the non degenerate eigenvalue a_i, then $P(a_n) = |\langle u_n | u_i\rangle|^2 = \delta_{ni}$, so that the measurement of observable O will give the certain result a_i. Postulate P4 provides a way to define the average value — in the quantum sense — of a measurement. This average is expressed as

$$<a> = P(a_n)a_n \tag{3.1.13}$$

Using Equation 3.1.12, we get

$$<a> = a_n|\langle u_n | \psi\rangle|^2 \tag{3.1.14}$$

If we expand $|\psi\rangle$ linearly on the orthonormal basis $|u_n\rangle$ of operator O as follows

$$|\psi\rangle = \alpha_n|u_n\rangle$$

we obtain

$$<a> = a_n|\alpha_n|^2 \tag{3.1.15}$$

and therefore

$$\langle\psi|O|\psi\rangle = \alpha_n^*\alpha_n\langle u_n|O|u_m\rangle = a_n\alpha_n^*\alpha_m\langle u_n|u_m\rangle = a_n\alpha_n^*\alpha_m\delta_{nm} = a_n|\alpha_n|^2$$

so that we obtain

$$<a> = \langle\psi|O|\psi\rangle \tag{3.1.16}$$

Of course, this average value can be obtained by means of the density operator. Equation 3.1.16 reads

$$<a> = \alpha_n^*\alpha_m o_{nm} \tag{3.1.17}$$

However, $\alpha_n{}^*\alpha_m$ is the matrix elements ρ_{mn} of the matrix of the density operator expressed on basis $|u_n\rangle$, and then

$$<a>=\rho_{mn}o_{nm}=\langle u_m|\rho|u_n\rangle\langle u_n|O|u_m\rangle \qquad (3.1.18)$$

Remembering that $|u_n\rangle\langle u_n|$ is the identity operator, Equation 3.1.18 becomes

$$<a>=\rho_{mn}o_{nm}=\langle u_m|\rho\, O|u_m\rangle=\mathrm{Tr}\,(\rho\, O) \qquad (3.1.19)$$

In the *density representation* formalism, Equation 3.1.19 is the equivalent of Equation 3.1.16 in the *state representation*. It will be frequently used in the rest of this chapter.

Postulate P5. If the measurement of the physical variable O on a system in state $|\psi\rangle$ results in the non degenerate eigenvalue a_n, then, immediately after the measurement has been performed, the state of the system is no longer described by $|\psi\rangle$, but by the eigenstate $|u_n\rangle$ of O associated to the eigenvalue a_n. This postulate indicates the possible influence of the measurement on the state of the microscopic system, as was suggested in Section III.1.1.2.

III.1.2.3. Evolution principle

Postulate P6. The evolution of a quantum system described by ket $|\psi\rangle$ is given by the Schrödinger equation

$$(\mathrm{ih}/2\pi)|\dot{\psi}\rangle=H\,|\psi\rangle \qquad (3.1.20)$$

where H is the Hamiltonian operator of the system.

Knowing the Hamiltonian operator, whose construction will be detailed later, gives us the way to predict the evolution of the system described by ket $|\psi\rangle$ Of course, a similar equation corresponding to the description based on the density operator formalism also exists. In order to derive it, one just needs to take the conjugate of Equation 3.1.20 and remember the hermiticity of the Hamiltonian

$$-(\mathrm{ih}/2\pi)\langle\dot{\psi}|=\langle\psi|H \qquad (3.1.21)$$

Or

$$\dot{\rho}=|\dot{\psi}\rangle\langle\psi|+|\psi\rangle\langle\dot{\psi}| \qquad (3.1.22)$$

So that, inserting Equation 3.1.20 and Equation 3.1.21 in Equation 3.1.22, we obtain the evolution equation of the density operator, known as the *Liouville equation*

$$(\mathrm{ih}/2\pi)\dot{\rho} = [H, \rho] \tag{3.1.23}$$

This equation will be the starting point of the quantum presentation developed in Chapter IV.

A very important consequence of postulates P3 and P6 deals with *stationnary states*. For an isolated system, under the influence of a time independent potential, the Hamiltonian does not explicitly involve time and the energy remains constant. However, postulate P3 indicates that this perfectly known value can only be one of the eigenvalues of the Hamiltonian. The state of the system is described by its associated ket $|\psi\rangle$. Let us consider the eigenkets $|\psi_n\rangle$ of such a Hamiltonian, as obtained from the eigenvalue equation of H

$$H|\psi_n\rangle = \mathrm{E_n}|\psi_n\rangle$$

where $\mathrm{E_n}$ are the eigenvalues of H associated to $|\psi_n\rangle$. Inserting this in the evolution equation, we get

$$(\mathrm{ih}/2\pi)|\dot{\psi}_n\rangle = \mathrm{E_n}\,|\psi_n\rangle$$

and by integration we obtain the temporal dependence of the state vector. We can write

$$|\psi_n\rangle = |\varphi_n\rangle \exp[-(\mathrm{i}2\pi \mathrm{E_n} t/\mathrm{h})]$$

In this expression $|\varphi_n\rangle$ *does not depend on time*. The states described by these state vectors are called *stationary states*. The Hermitian product $\langle\psi_n|\psi_n\rangle$ and the volume density probability $|\psi_n(\mathbf{r}, t)|^2$ are *time independent* (although the state vector oscillates with the period $\mathrm{h/E_n}$). Let us note that the systems described in the rest of this book will always be in stationary states. In particular, the ground state will be the stationary state of lowest energy.

Until now, we mentioned two operators (the density operator and the Hamiltonian operator). We now need to explain how to build an operator A associated to a physical observable A whose expression is known from analytical mechanics in the Hamilton formalism. Observable A can be measured on a set of particles from their generalized variables of position ($\mathbf{r}_i$), momentum ($\mathbf{p}_i$) and time (t). Equating momentum and linear momentum ($\mathbf{p}_i = m_i \dot{\mathbf{r}}_i$) and in the absence of any external electromagnetic field, we can express the correspondence principle.

III.1.2.4. Correspondence principle

Postulate P7. Operator O associated to the physical observable O is obtained by substituting operators $\boldsymbol{r}_i$ and $\boldsymbol{p}_i$ into the classical expression $O(\mathbf{r}_i, \mathbf{p}_i, t)$ in place of the quantities $\mathbf{r}_i$ and $\mathbf{p}_i$ as indicated

$$\mathbf{p}_i \to \boldsymbol{p}_i = -(ih/2\pi)\,\nabla_i \qquad \mathbf{r}_i \to \boldsymbol{r}_i \tag{3.1.24}$$

In order to be able to represent physical quantities, the operators defined in this way must imperatively be Hermitian (i.e. have real eigenvalues). If this is not the case, the initial classical expression must be adequately " symmetrized" in order to obtain a Hermitian operator. In general, the product of two operators is not Hermitian. For example, operator $\boldsymbol{r}_i\boldsymbol{p}_i$ associated to the product $\mathbf{r}_i\mathbf{p}_i$, is not Hermitian. However, the symmetric combination $(1/2)(\mathbf{r}_i\mathbf{p}_i + \mathbf{p}_i\mathbf{r}_i)$ generates the Hermitian combination $(1/2)(\boldsymbol{r}_i\boldsymbol{p}_i + \boldsymbol{p}_i\boldsymbol{r}_i)$.

We shall now illustrate postulate P7 with two examples that will be used later.

First example: The angular momentum operator

Angular momentum is a fundamental quantity in physics. In classical mechanics, in particular, it is a quantity which is conserved. Turning to microscopic systems, we shall define the angular momentum of electrons and explore both the way they couple together and the effects of these couplings on the energy of the systems. In classical mechanics, the angular momentum $\mathbf{L}$ of a particle with momentum $\mathbf{p}$ situated at the extremity of vector $\mathbf{r}$ is defined by $\mathbf{L} = \mathbf{r} \times \mathbf{p}$. According to the correspondence principle (P7), the components of the corresponding operator read

$$L_x = -(ih/2\pi)\,(y\partial/\partial z - z\partial/\partial y)$$
$$L_y = -(ih/2\pi)\,(z\partial/\partial x - x\partial/\partial z)$$
$$L_z = -(ih/2\pi)\,(x\partial/\partial y - y\partial/\partial x)$$

It is straightforward to show that

$$[L_x, L_y] = (ih/2\pi)L_z \quad [L_y, L_z] = (ih/2\pi)L_x \quad [L_z, L_x] = (ih/2\pi)L_y$$

and

$$[L_y^2, L_z] = (ih/2\pi)\,(L_xL_y + L_yL_x)$$
$$[L_x^2, L_z] = -(ih/2\pi)\,(L_xL_y + L_yL_x)$$
$$[L_z^2, L_z] = 0$$

These relations can be summarized by

$$\boldsymbol{L}\times\boldsymbol{L} = (ih/2\pi)\boldsymbol{L} \qquad [\boldsymbol{L}^2, L_z] = 0$$

They define angular momentum in quantum mechanics in a very general way (for example, the orbital- or spin angular momentum). In particular, they show that an angular momentum can be specified by its norm and by one and only *one* of its projections along an axis of the laboratory!

Since the commutator of L^2 and L_z vanishes, these two operators share a common basis of eigenstates, denoted |l, m>, where l and m are two quantum numbers associated with the eigenfunctions of L^2 and L_z

$$L^2|l,m\rangle = l(l+1)(h/2\pi)^2|l,m\rangle \qquad L_z|l,m\rangle = m(h/2\pi)|l,m\rangle$$

It can be shown that

$$l \geq 0 \qquad \text{and} \qquad -l \leq m \leq l$$

The eigenfunctions of L^2 and L_z are standard functions known as *spherical harmonics* $Y_{l,m}(\theta, \varphi)$, such that

$$L^2 Y_{l,m}(\theta,\varphi) = l(l+1)(h/2\pi)^2 Y_{l,m}(\theta,\varphi) \qquad L_z Y_{l,m}(\theta,\varphi) = m(h/2\pi) Y_{l,m}(\theta,\varphi)$$

An important property of spherical harmonics is that they appear as a product of two functions, one depending on θ, the other depending on φ

$$Y_{l,m}(\theta,\varphi) = \Theta_{l,|m|}(\theta)\Phi_m(\varphi) = \Theta_{l,|m|}(\theta)(1/2\pi)\exp(im\varphi)$$

where $\Theta_{l,|m|}$ is an *associated Legendre function* whose first terms are given in Table 3.1.1.

Table 3.1.1 — **Analytical expression of a few Legendre functions.**

l	m	$\Theta_{l,\lvert m\rvert}(\theta)$
0	0	$1/\sqrt{2}$
1	0	$\sqrt{3/2}\cos\theta$
1	±1	$\sqrt{3/4}\sin\theta$
2	0	$\sqrt{5/8}(3\cos^2\theta - 1)$
2	±1	$\sqrt{15/4}\sin\theta\cos\theta$
2	±2	$\sqrt{15/16}\sin^2\theta$

Let us now focus on the problem of angular momentum combinations. Let $\mathbf{L}_1$ and $\mathbf{L}_2$ be two angular momenta (therefore verifying the two equalities defining angular momentum). We can show immediately that the operator $\boldsymbol{L}$ defined by $(\boldsymbol{L}_1 + \boldsymbol{L}_2)$ is also an angular momentum characterizing the *coupled* system. The eigenvalues of this new angular momentum are obviously $l = l_1 + l_2$ and m such that $|m_1 - m_2| \le m \le |m_1 + m_2|$. Later, we shall see the use of this notion.

Second example: the Hamiltonian operator

This operator is by far the most widely used one in quantum mechanics. How can we build it? We have to apply postulate P7 to its classical equivalent. In order to do so, let us first recall a few results of classical mechanics using Lagrange and Hamilton formalisms.

The most ancient formulation of the laws of dynamics was given by Newton. If we note $\mathbf{p}$ the linear momentum of a material point M(x, y, z) and $\mathbf{F}$(x, y, z) the force field acting at M, then the most general form of Newton's second law reads

$$\dot{\mathbf{p}} = \mathbf{F}$$

If the mass is time independent, then dp/dt = mdv/dt and the above equations provide a way to give the explicit forms of functions x(t), y(t) and z(t), that is to say the corpuscle trajectory, as soon as the initial conditions are specified. In carthesian coordinates, the kinetic energy T of a particle takes the form

$$T(\dot{x}, \dot{y}, \dot{z}) = (m/2)(\dot{x}^2 + \dot{y}^2 + \dot{z}^2)$$

If V(x, y, z) is the potential energy, which depends only on the position of the particle, we have, for example, along the x axis

$$m\ddot{x} = -\partial V / \partial x = F_x$$

Let L be the *Lagrangian* of the system, defined from T and V by

$$L(x, y, z, \dot{x}, \dot{y}, \dot{z}) = T(\dot{x}, \dot{y}, \dot{z}) - V(x, y, z)$$

We have

$$m\dot{x} = \partial T / \partial \dot{x} = \partial L / \partial \dot{x} \qquad -\partial V / \partial x = \partial L / \partial x$$

and Newton's equations take the form

$$(d/dt)(\partial L / \partial \dot{x}) = \partial L / \partial x\,; (d/dt)(\partial L / \partial \dot{y}) = \partial L / \partial y; (d/dt)(\partial L / \partial \dot{z}) = \partial L / \partial z$$

These relations make up the *Lagrange equations*. The value of the Lagrange formalism is two fold

— First, Lagrange equations always have the same form, whatever the coordinate system chosen to express them.

— Second, they are more convenient to handle than Newton's equations, as it is usually easier to know the potential at a given point than to identify precisely the three components of the force field acting on the system. The resolution of the three second order Lagrange equations provides a way to characterize the motion of a material point completely as soon as the six initial conditions are known [**v** (t = 0), **r** (t = 0)]. In order to characterize the evolution of a set of N material points, 3N Lagrange equations endowed with 6N initial conditions must be solved.

Another formulation of the equations of classical mechanics, using only first order differential equations, exists. Indeed, it can be shown in mathematics that any system of N second order differential equations, involving N unknowns, can be replaced by an equivalent system of 2N first order differential equations. This is the transformation which Hamilton operated on the Lagrange system of equations. For a set of N points, if p_k denotes the *generalized conjugate linear momentum* of the *generalized space coordinate* q_k

$$p_k = \partial L / \partial \dot{q}_k \qquad k = 1, 2, \ldots, 3N$$

Starting from L, the *Hamiltonian* H of the system is defined by

$$H = \sum_{k=1}^{3N} \left[p_k \dot{q}_k - L(q_k, \dot{q}_k) \right]$$

Although this definition of H contains the expression of velocities, it is easy to show that H depends only on the generalized coordinates and momenta.

Let us now focus on the physical meaning of H. We have

$$\frac{dH}{dt} = \frac{d}{dt}\left(\sum_{k=1}^{3N} \left[p_k \dot{q}_k - L(q_k, \dot{q}_k) \right] \right) = \sum_{k=1}^{3N} \left(p_k \ddot{q}_k + \dot{p}_k \dot{q}_k - \frac{\partial L}{\partial q_k} \dot{q}_k - \frac{\partial L}{\partial \dot{q}_k} \ddot{q}_k \right)$$

Moreover, since $\dot{p}_k = \partial L / \partial q_k$, dH/dt = 0. Thus, for conservative systems, H appears as a constant of the motion. In this case, it is straightforward to shown that H is nothing but the total mechanical energy of the system

$$H = T + V \qquad (3.1.25)$$

In conclusion, it is worth noting that we have introduced the Lagrangian and Hamiltonian formalisms starting from Newton's equations. However, they are more general and can also be deduced directly from the least action principle.

The use of postulate P7 is simple. For example, in the case of a system with mass m, if we write the Hamilton function as

$$H = p^2/2m + V(q)$$

the corresponding Hamiltonian will write as

$$H = -(h^2/8\pi^2 m)\,\Delta + V(q)$$

A second example is discussed in Section IV.3.2.1.

Postulates P6 and P7 therefore offer a solution to all problems of quantum mechanics since they provide, in principle, the explicit determination of the state vectors which take part in the evaluation of the average quantities related to the physical observables we are interested in. The basic problem is to solve differential equation 3.1.20, or its equivalent form (Equation 3.1.23), and all the difficulty lies in the analytic expression of the Hamiltonian build from Equation 3.1.24. Of course, the simpler the Hamiltonian expression is, the easier it will be to find the eigenvectors. Two concepts (*symmetry* and *perturbation*) will prove to be extremely fruitful to deal with this question.

The importance of the concept of symmetry will be discussed throughout all the rest of this book. Let us simply say that the very existence of some symmetry elements for a given molecule, due to the structure of the considered object (atom, molecule, complex species, ...) will induce constraints on the state vectors that may simplify the resolution of the eigenvalue equation. Indeed, the existence of a symmetry element in a molecule inevitably has consequences on the shape of its electronic cloud, whose description is given by the square of the modulus of the corresponding state vector.

The concept of "perturbation" appears when trying to simplify the expression of a Hamiltonian. Indeed, it is possible to view the Hamiltonian as the sum of several terms, some of them being "leading terms", characterizing the most important physical effects, and the others, denoted as "perturbative terms", with very weak amplitudes as compared to the leading ones, but which need to be taken into consideration to describe the states of the system accurately. This is the essence of the perturbation theory, which is in fact a mathematical technique providing a two step method to solve the eigenvalue equation:

— In the first step, only the eigenvalue equation obtained from the truncated Hamiltonian, involving just the "leading terms" (thus, simpler that the total Hamiltonian) is considered and solved

— Then, as a second step, the results of perturbation theory are applied to find an approximate solution to the complete problem, built from the solutions of the simplified problem (involving the truncated Hamiltonian). Such an approach assumes that we can classify each term of the Hamiltonian according to their magnitude, in order to detect more specifically the smaller ones. Practically, we distinguish time dependent perturbations theories (for example, the one describing the effect of an oscillating electric field of low intensity) from time independent ones.

Let us come now to the last two postulates of quantum mechanics. The first seven ones used a non relativistic formulation of the energy (Hamiltonian), and the evolution equation was not invariant under the Lorentz transformation. A relativistic evolution equation for $|\psi\rangle$, which is invariant under the Lorentz transformation, has been suggested by Dirac who assumed that the electron is not only characterized by its rest mass m_e and its charge $-e$, but also by an intrinsic angular momentum called *spin*, a vectorial quantity without any classical equivalent image. It is possible however, as was shown by Pauli, to keep the non relativistic Schrödinger equation and to introduce the existence of spin through a postulate. Within this frame, the state vector describing a system will no longer be related only to the space variables but will also have to specify the spin variables. Since spin is an angular momentum, the associated operators $\boldsymbol{S}$ and S_z verify the general properties of angular momenta, properties which we recalled earlier in this chapter (in particular, those concerning the rules of their combination and commutation).

III.1.2.5. Spin creation principle

Postulate P8. The electron is a particle with an angular momentum of spin S. The two eigenfunctions $|\alpha\rangle$ and $|\beta\rangle$ of operator S_z define a two dimensional space F_S of the spin states of the electron, and they are such that

$$S_z|\alpha\rangle = (1/2)(h/2\pi)|\alpha\rangle \qquad S_z|\beta\rangle = -(1/2)(h/2\pi)|\beta\rangle \qquad (3.1.26)$$

The space of states F of the electron is defined by the product space $F = F \times F_S$ of two spaces, F concerning the particle and F_S describing its spin states. The basis of space F is the set of functions obtained by multiplying a basis function of F by a basis function of F_S in all possible ways.

The eigenvalues $\pm(1/2)$ (in units of $h/2\pi$) were deduced from Stern and Gerlach's experiments. On the other hand, it must be said that the analytical form of spin vectors $|\alpha\rangle$ and $|\beta\rangle$ are not explicitly known. In fact, it is sufficient to know that $|\alpha\rangle$ and $|\beta\rangle$ are associated to the eigenvalues $+(1/2)$ and $-(1/2)$ (in units of $h/2\pi$).

These eigenvalues are defined from the eigenvalues of the two operators $\boldsymbol{S}^2$ and S_z, whose values are, respectively, $s(s+1)(h/2\pi)^2$ and $m_s(h/2\pi)$ [with $m_s = \pm(1/2)$; *cf.* Equation 3.1.26].

III.1.2.6. Antisymmetrization principle

Postulate P9. When a system contains n electrons, only those state vectors of space $F = F^{(1)} \times F^{(2)} \times \ldots \times F^{(n)}$ which are antisymmetrical with respect to the exchange of any two electrons can describe physical states.

Let us consider, for example, a system consisting of two electrons, and let $|\psi(1,2)\rangle$ be the state vector representing the system in space $F = F^{(1)} \times F^{(2)}$, the product of the spaces describing the states of each of the two electrons (1) and (2). Assuming that the state vector $|\psi(1,2)\rangle$ can be expressed as the product of states vectors relative to each of the electrons, we can write

$$|\psi(1,2)\rangle = |\psi_1(1)\rangle|\psi_2(2)\rangle$$

However, this vector does not satisfy postulate P9 because

$$|\psi(1,2)\rangle \neq -|\psi(2,1)\rangle$$

but the linear combination

$$|\psi(1,2)\rangle = (1/\sqrt{2})\left[|\psi_1(1)\rangle|\psi_2(2)\rangle - |\psi_1(2)\rangle|\psi_2(1)\rangle\right]$$

is normed and antisymmetrical with respect to exchange. Noticing we can write $|\psi(1,2)\rangle$ as the determinant

$$|\psi(1,2)\rangle = \frac{1}{\sqrt{2}}\begin{vmatrix} |\psi_1(1)\rangle & |\psi_1(2)\rangle \\ |\psi_2(1)\rangle & |\psi_2(2)\rangle \end{vmatrix},$$

this result can be generalized immediately to the construction of a state vector describing a set of N electrons from the N monoelectronic state vectors $|\psi_1(1)\rangle, \ldots, |\psi_N(N)\rangle$:

$$|\psi(1, \ldots, N)\rangle = \frac{1}{\sqrt{N!}}\begin{vmatrix} |\psi_1(1)\rangle & |\psi_1(2)\rangle & . & |\psi_1(N)\rangle \\ |\psi_2(1)\rangle & |\psi_2(2)\rangle & . & |\psi_2(N)\rangle \\ . & . & . & . \\ |\psi_N(1)\rangle & |\psi_N(2)\rangle & . & |\psi_N(N)\rangle \end{vmatrix} \quad (3.1.27)$$

The state vector obtained in this way (also named " Slater determinant ") is antisymmetrical by construction, because the value of a determinant changes sign when any two of its columns are permuted, that is, when any two electrons are exchanged.

A fundamental consequence of the ninth postulate, which appears naturally when the state vector is considered as a Slater determinant, is that in a many electron system, no two of the electrons can be in the same individual state simultaneously. This way, we rediscovered the *exclusion principle* stated by Pauli, which appears as a consequence of postulate P9. Indeed, if two electrons were in the same state, two columns of the determinant would be identical and the determinant would vanish.

We shall now recall some elements of group theory, applied to the study of the symmetry of microscopic systems.

III.2. Symmetry and group theory

III.2.1. Symmetry elements and symmetry operations

A *symmetry operation* is a geometric transformation which transforms a geometric object (here a microscopic system) into an image which cannot be distinguished from the original object. A symmetry operation is generally associated to a *symmetry element*, which is a geometrical object (plane, axis, point) with respect to which the symmetry is taken; note that in the following, we will not distinguish between the symmetry elements and the symmetry operations. The basic symmetry operations are the following:

Rotation: If a clockwise rotation by an angle $\theta = 2\pi/n$ (n being an integer) brings the nuclear framework of a molecule into coincidence with itself, the molecule is said to have a n-fold symmetry axis denoted C_n. Of course, $C_n^n = I$. The axis with the largest value of n is called the *principal axis*.

Reflection: If reflection on a plane brings the nuclear framework into coincidence with itself, the molecule is said to have a symmetry plane denoted σ. Of course, $\sigma^2 = I$. If the plane contains the principal axis, it is noted σ_v (v for vertical). It is named σ_h if it is perpendicular to the principal axis (h for horizontal). It is labeled σ_d (d for dihedral) if it bisects the angle between two two-fold axes of symmetry which are perpendicular to the principal axis.

Rotation-reflection (*cf.* Figure 3.2.1): If the nuclear framework of a molecule is brought to coincidence after a clockwise rotation around the C_n axis followed by a reflection on a plane perpendicular to that axis, the molecule is said to have a n-fold improper axis labeled S_n. Consequently, if a molecule possesses simultaneously a C_n axis and a plane of symmetry σ_h, the C_n axis is also a S_n axis. It is easily seen that $S_k^k = E$; $S_n^2 = C_n^2$; $S_n^k = \sigma_h C_n^k$ if k is uneven and $S_n^k = C_n^k$ if k is even.

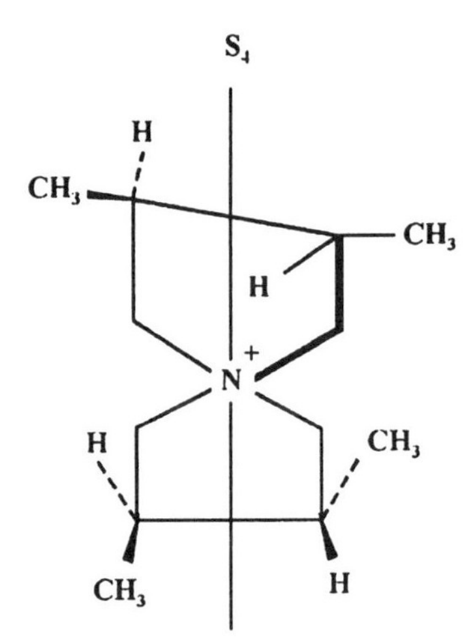

Figure 3.2.1: Example of an improper rotation axis S_4.

The inverse operation: Let I be a point that we call the symmetry center. The image of any point P through the symmetry center I is the point P' such that I lies in the middle of PP'. A molecule will be said to have a symmetry center if there exists such a point I which brings the nuclear framework into coincidence with itself. This operation is symbolized by i. It is easily shown that $i^2 = E$ and that $i = S_2$.

III.2.2. Representation of an operation by an operator

A symmetry operation, associated to a symmetry element, can be represented by an *operator* which explicitly gives the correspondence between the original object and its transformation by the considered symmetry operation. By choosing a basis, it is possible to represent the operator by a matrix associated with the symmetry operation in question, as was shown in Section 3.1. In this way, several symmetry operations can be applied successively, thus defining a transformation whose associated matrix is obtained by taking the product of the matrices representing each individual symmetry operation. However, an operator is thoroughly specified as soon as the images of each of the basis vectors in which it is defined are known (*cf.* Section III.1.1.2). For example, the matrix R_θ associated to a rotation by angle θ in the plane can be obtained from the images of the basis vectors (**i**, **j**) of the plane and reads

$$\left(R_\theta\right) = \begin{pmatrix} \cos\theta & -\sin\theta \\ \sin\theta & \cos\theta \end{pmatrix}$$

III.2.3. Group structure and classification

A *group* is a set F of entities endowed with a composition law (labeled for example $*$) verifying the four following properties:

— The law $*$ is *internal*: The *products* $A*B$ and $B*A$ of any two elements A and B of the group must also belong to the group.

— The law $*$ is *associative*: For any elements A, B and C of the group, $A*(B*C) = (A*B)*C$.

— There is an *identity element* E in the group, such that for any element A of the group, $A*E = E*A = A$.

— Every element A in the group has an *inverse*, noted A^{-1}, which belongs to the group and which is such that $A*A^{-1} = A^{-1}*A = E$.

If, moreover, the group elements commute with each other, *i.e.* $A*B = B*A$ for any elements A and B of the group, the group is called an Abelian group.

As an example, Z, the set of positive and negative integers, endowed with addition form an Abelian group (the identity element is 0 and the inverse of an element z of Z is $-$ z, which belongs to Z). However, Z is not a group when endowed with multiplication. There is indeed an identity element in Z ($z = 1$), but not every element has an inverse (the inverse of $z = 4$ is $z^{-1} = 0.25$, which is not an integer). Similarly, the set $F = \{-2, -1, 0, 1, 2, 3\}$ endowed with addition is not a group since the rule is not internal (the sum $2 + 3 = 5$ is not included in F, although an identity element does exist and every element has its inverse in F). In general, a group is thoroughly specified as soon as its multiplication table (i.e. its composition law) is known.

It is easy to show that the set of all the *symmetry operations* which leave a molecule unchanged forms indeed a group. If the number elements of the set is h, h is called the *order* of the group. We emphasize that this group, whose identity element will be the *identity operation*, is the set of *symmetry operations*, and not the set of *symmetry elements*. The group is characterized by a multiplication table which gives the result of the composition of any two symmetry operations. Indeed, it is possible to classify molecules according to the multiplication table that shows the symmetry elements which leave them unchanged. This multiplication table translates the geometric relations existing between the various symmetry elements to which these operations are associated, and the relative arrangement of atoms within the molecule. This type of classification is very powerful since it does not depend on the intrinsic complexity of the molecule (nature of the atoms and bonding, spatial extension of molecular groups, ...) which may *a priori* be infinite, and relies only on the nature of the possible arrangement of the symmetry elements of an object in space, the number of which has a natural limit because of geometric constraints.

We shall now describe and classify those groups consisting of symmetry transforms which leave objects of finite sizes (such as molecules) unchanged according to a nomenclature due to Schoenflies. These groups are named *point groups*. They necessarily have one fixed point (the center of mass of the object considered), because in the absence of any fixed point, the object could at some point undergo a translation and thereafter would no longer superimpose on itself after being transformed by a symmetry operation.

— C_n *groups*: They contain one n-fold axis of symmetry. The C_1 group consists only of the identity operation and therefore corresponds to a total lack of symmetry.

— S_{2n} *group*: It is the rotation group about an improper axis with even order 2n.

— C_{nh} *group*: These groups are obtained by associating a horizontal plane (h for horizontal) with a n-fold axis of symmetry.

— C_{nv} *group*: These groups are obtained by associating a vertical plane (v for vertical) with an n-fold axis of symmetry. These two symmetry elements imply the existence of (n – 1) axial planes forming the dihedral angle $\alpha = \pi/n$ between them.

— D_n groups: These groups are obtained by associating an orthogonal two-fold axis of symmetry with an n-fold axis of symmetry. In this case too, the conjunction of these two symmetry elements implies the existence of (n – 1) further two-fold axes of symmetry, so that finally there are n two-fold symmetry axes, forming a dihedral angle $\alpha = \pi/n$ between them.

— D_{nh} groups: These groups are obtained by adding a horizontal plane containing n two-fold axes of symmetry to the symmetry elements of the D_n group. The conjunction of these symmetry elements results in n vertical planes of symmetry, each specified by the principal axis of rotation together with one of the n secondary rotation axes.

— D_{nd} *groups*: These groups (d for dihedral) are obtained by adding vertical symmetry planes, bisecting two neighboring secondary rotation axes to the set of symmetry elements of the D_n group.

— *Cubic groups*: These groups collect all the symmetry transformations that leave the octahedron or the cube (O, O_h) or the tetrahedron (T, T_h, T_d) unchanged.

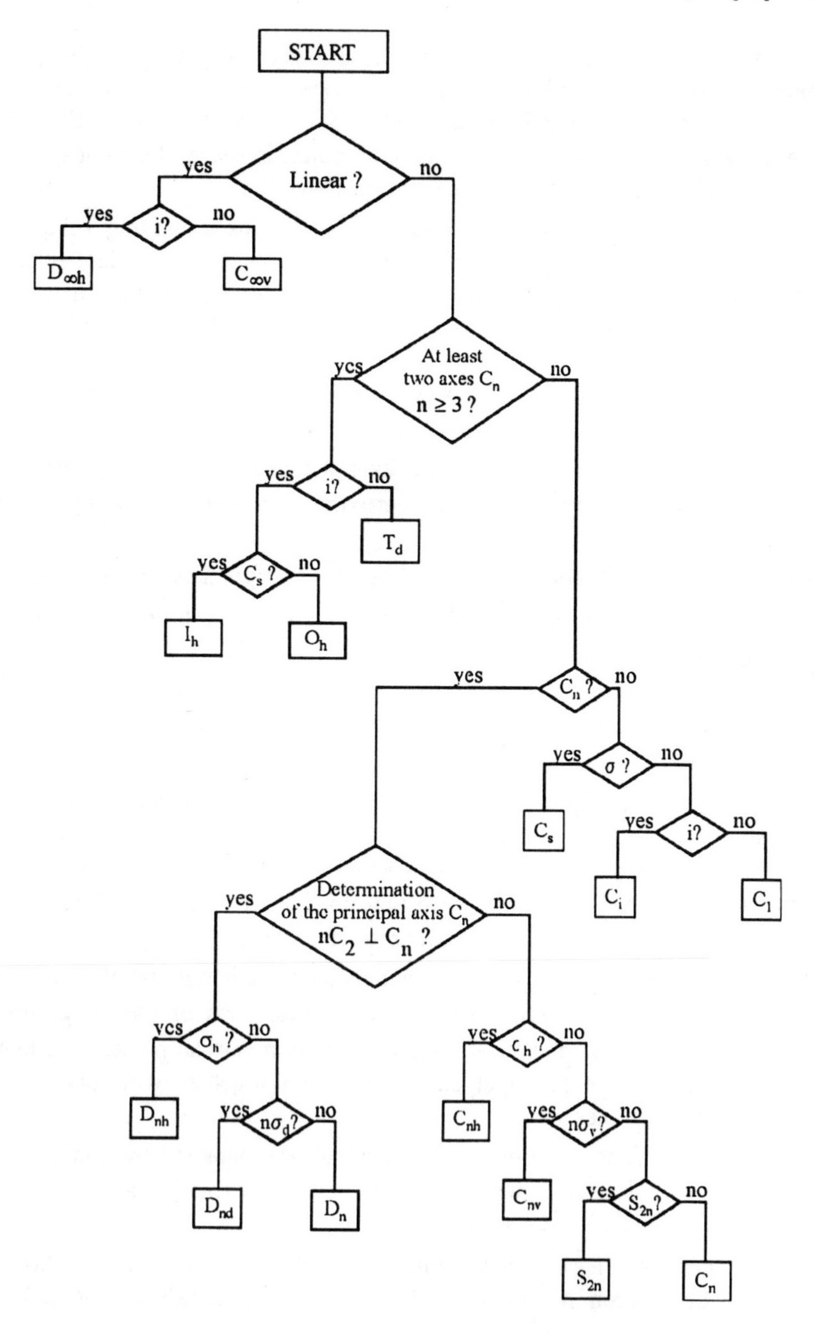

Figure 3.2.2: Flowchart for the determination of molecular symmetry.

It is clear that for a molecule to belong to any of the above groups depends on whatever or not it possesses typical symmetry elements. The group of a molecule can be determined by using the flowchart shown on Figure 3.2.2.

For example, it is easy to show that PCl_5 belongs to the point group D_{3h} (*cf.* Figure 3.2.3).

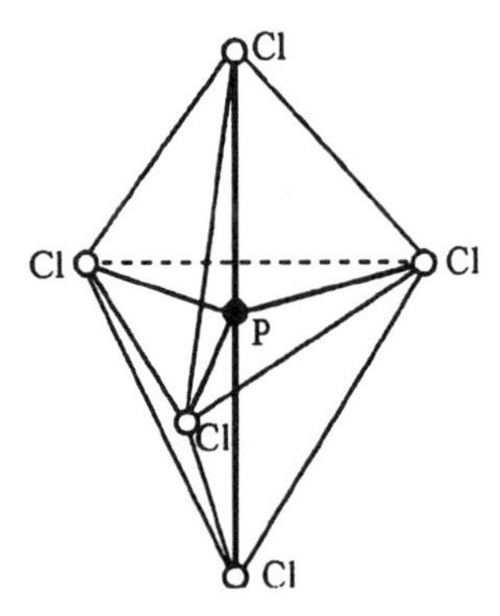

Figure 3.2.3: The bipyramidal structure with triangular basis of the PCl_5 molecule.

Very convenient visual methods also exist (*cf.* Bibliographic reference 1.2.1) as do computer aided research strategies.

III.2.4. Group representation

III.2.4.1. Introduction to the notion of representation

Let us consider, for example, an equilateral triangle. The flowchart of Figure 3.2.2 tells us that the object belongs to the C_{3v} group which contains the identity operation, two rotation operations and three reflection operations.

We will use the coordinate system displayed on Figure 3.2.4.

We shall note E the identity operation; A, B and C the three reflection operations about the yOz, zOb and zOc planes, respectively, and D and F the two rotation operations of angle $2\pi/3$ about the Oz axis, in inverse (–) and direct (+) directions, respectively.

In Table 3.2.1 we established the multiplication table of the group, obtained by applying first the symmetry operation indicated above the column and then the one indicated in front of the line. For example, transforming successively the above triangle first by operation B and then by A is the same as to transform the initial triangle directly by F. Consequently, we can write $A*B = F$.

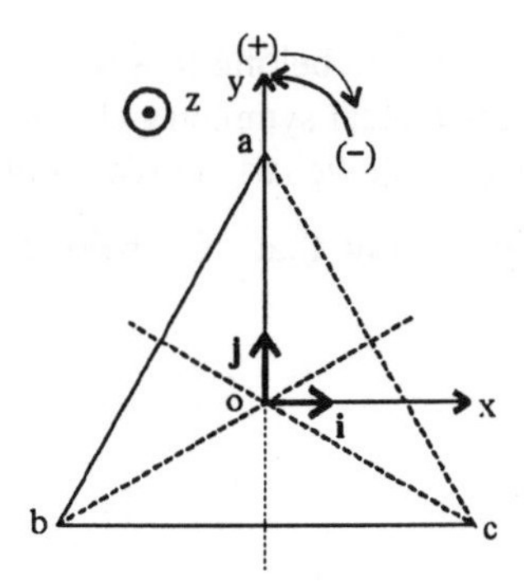

Figure 3.2.4: Symmetry elements in an equilateral triangle (point group C_{3v}).

Table 3.2.1 — **Multiplication table of C_{3V} group.**

*	E	A	B	C	D	F
E	E	A	B	C	D	F
A	A	E	D	F	B	C
B	B	F	E	D	C	A
C	C	D	F	E	A	B
D	D	C	A	B	F	F
F	F	B	C	A	E	D

A glance at the table shows that the set of symmetry transformations combined with the composition rule defined above definitely has a group structure. The rule is internal, associative, a neutral element exists and the elements can be inversed (for example, A is its own inverse because A∗A = E). It should be clear that the multiplication table is in fact determined by the relative orientation of the symmetry elements of the C_{3V} group. More precisely, the nature of a group is more closely related to the formal structure of its composition rule (and thus to its table) than to the elements it is made up of. Let us consider for example, vectors **i** and **j** which form a basis of the plane, and let us focus on the effect of the symmetry operations of the C_{3V} group on this basis. The action of the symmetry operations on each of the basis vectors provides a way to describe the set of the 2×2 matrices, called Γ_3, associated with each of the symmetry operations. Finally, we obtain six 2nd order matrices, expressed on the **i**, **j** basis and representing the six symmetry operators of the group. They are given on the first line of Table 3.2.2.

Γ_3, the set of matrices associated to all the symmetry elements, endowed with the matricial product rule, forms a group *representation* whose table is identical to that of the C_{3v} group. This example brings out the concept of group *representation*.

Table 3.2.2 — **Matrices and characters associated to the symmetry operations of the C_{3v} group, expressed in the (i, j) basis.**

	E	A	B	C	D	E
Γ_3	$\begin{pmatrix}1 & 0\\0 & 1\end{pmatrix}$	$\begin{pmatrix}-1 & 0\\0 & 1\end{pmatrix}$	$\begin{pmatrix}\frac{1}{2} & \frac{\sqrt{3}}{2}\\\frac{\sqrt{3}}{2} & -\frac{1}{2}\end{pmatrix}$	$\begin{pmatrix}\frac{1}{2} & -\frac{\sqrt{3}}{2}\\-\frac{\sqrt{3}}{2} & -\frac{1}{2}\end{pmatrix}$	$\begin{pmatrix}-\frac{1}{2} & \frac{\sqrt{3}}{2}\\-\frac{\sqrt{3}}{2} & -\frac{1}{2}\end{pmatrix}$	$\begin{pmatrix}-\frac{1}{2} & -\frac{\sqrt{3}}{2}\\\frac{\sqrt{3}}{2} & -\frac{1}{2}\end{pmatrix}$
χ	2	0	0	0	−1	−1

A point group is first of all defined by a composition rule which finally conveys the distribution and the relative orientation of the symmetry elements which leave the considered molecule unchanged. We call group *representation* a set of matrices associated to the symmetry operations of the group, expressed in a given basis.

The concepts of *basis* and that of *representation* are thus closely related. In the previous example, we chose the 2 dimensional basis (**i**, **j**) in order to study the effect of all the symmetry operations of the group, and we ended up with a representation of 2nd order matrices. Similarly, we could choose a 6 dimensional basis, represented by the six vectors $\mathbf{e}_i$, $1 \leq i \leq 6$ (an orhonormed couple of vectors $\mathbf{e}_i$ at each of the triangle summits, for instance) and in this basis, express the effect of the six symmetry operations of the C_{3v} group on each of the six vectors of the basis. We would obtain a set of six 6-ordered matrices, yielding another *representation* of the C_{3v} group. The multiplication table of these matrices is identical to that of the first representation. It is thus clear that *a priori* an infinity of group *representations* exist, depending on the choice of basis of given dimension, in the same way as an infinity of matricial representations exist for any operator.

Let us now focus on a representation Γ of dimension n, of a group G. Let us assume that the n basis vectors have the following specific property: The basis vectors form two distinct sets, the first one consisting of the first p basis vectors, the second of the (n – p) remaining ones, and for all the symmetry operations of the group considered, the vectors of each set transform exclusively among each other, independently of the other set of basis vectors. In other words, whatever the operation R of group G, we have

$$\text{for } 1 \leq i \leq p: \quad R(v_i) = \sum_{j=1}^{p} c_{ji} v_j$$

$$\text{for } -1 \leq i \leq n: \quad R(v_i) = \sum_{j=p+1}^{n} c'_{ji} v_j$$

In this way, we express that the two sets $\{v_i\}_{1 \leq i \leq p}$ and $\{v_i\}_{p+1 \leq i \leq n}$ never "mix" together during the action of any of the symmetry operations of the group. The $n \times n$ matrices associated to each of the symmetry operations of the group are *block-diagonal* and written as

$$(R) = \begin{pmatrix} \boxed{A} & 0 \\ 0 & \boxed{B} \end{pmatrix}$$

where A and B are matrices of order $p \times p$ and $(n - p) \times (n - p)$, respectively. According to the rules of matrix calculus, such matrices multiply block by block. Consequently, each of the matrices associated with the symmetry operations contains two sub-matrices also satisfying the group multiplication table when considered individually. Therefore, these sub-matrices form a couple of new representations of the group, of smaller dimension (of dimensions p and $n - p$), and representation Γ is said to be *reducible* in (at least) two representations, of dimensions p and $n - p$.

A representation is said to be *irreducible* if it is impossible to divide its basis into smaller stable subsets for all the symmetry operations of the group (that is to say, which do not mix with other elements of the basis during the application of any symmetry operation of the group). The existence and the number of irreducible representations of a group constitute one of its fundamental properties.

A representation Γ, reducible into two representations labeled Γ_1 and Γ_2, will be said to be the *direct sum* of Γ_1 and Γ_2, and we shall write $\Gamma = \Gamma_1 \oplus \Gamma_2$. Thus, it is the custom to say that an *irreducible representation occurs as many times in a reducible representation as one of its bases appears in the basis of the reducible representation considered.* For example, if Γ, Γ_1 and Γ_2 are three representations of respective dimensions d, d_1 and d_2, and if $\Gamma = a_1\Gamma_1 \oplus \Gamma_2$ (a_1 being an integer), then we have $d = a_1 d_1 + d_2$.

It is clear that the quest for the reducibility of a representation assumes finding a basis in which all the matrices associated with the symmetry elements of the group are block-diagonal. *A priori*, the choice of possible bases for a representation is infinite. As to the explicit construction of such specific bases which yield a partial (or complete) diagonalization of the set of the matrices associated to the operations of the group, this raises the difficulty of their existence itself. Without going into the details of the justifications, we shall now present some fundamental results concerning reducible and irreducible representations of a group G.

III.2.4.2. Representation properties

Due to the fact that many bases are possible for a representation of given dimension, matrices associated with symmetry operations are not themselves representatives of the group, although they verify the multiplication table. However, matrices have a property which is basis independent: The *trace*, which is the algebraic sum of the diagonal elements. In the following, the trace will be called *character* and labeled χ. In general, χ is a complex number.

Two elements A and B of a group are said to be each others *conjugate* if there exists an element C of the group such that $A = C*B*C^{-1}$. If A is conjugate to B and B is conjugate to C, then A is conjugate to C. A set of mutually conjugate elements forms a class. It is possible to gather the elements of a group into classes, a class being completely specified if one of its elements is known, since the other ones are obtained by forming the products $X*B*X^{-1}$ (X varying amongst all the elements of the group). Of course, every element can belong to only one class whose elements obviously have the same character. Considering again our example of the C_{3v} group, the three reflection operations A, B and C form a class, which we shall denote $3\sigma_v$ (as it contains three elements), and the two rotation operations D and F form another class labeled $2C_3$ (for the same reason). It is then possible to describe the representation Γ_3 of the C_{3v} group from its *character table,* which is a double entrance table. The upper line mentions all the classes of the group, and the character of the corresponding symmetry operation is given in every column for the representation considered. Finally, the character table of representation Γ_3 takes the form:

Table 3.2.3 — **Character table of representation Γ_3 of the C_{3v} group.**

	E	$3\sigma_v$	$2C_3$
χ_{Γ_3}	2	0	–1

Let us immediately call attention to a point which is essential and general. The *dimension* of a representation (reducible or irreducible) can be obtained by inspecting the character of the representative matrix of the identity operation (E) of this representation.

We shall now state five fundamental properties of the irreducible representations of a given group G:

— The number r of irreducible representations of a group is equal to the number of its classes.

This is a fundamental result expressing that, contrary to the number of reducible representations, the number of irreducible representations of a point group is finite, thereby justifying the important part played by the irreducible representations in the characterization of a group.

— The sum of the square of the *dimensions* l_i of the *irreducible representations* of a group G is equal to the order h of the group:

$$\sum_i l_i^2 = h$$

— The vectors whose components are formed by the characters of the irreducible representations of a group G are orthogonal:

$$\sum_R \left[\chi_i(R)\chi_j^*(R)\right] = h\delta_{ij}$$

— The sum of the squares of the characters of all the symmetry operations of each of the irreducible representations of a group is equal to the order h of the group:

$$\sum_R \left[\chi_i^*(R)\right]^2 = h$$

These sums extend over *all* elements of each class. This property is just a special case (i = j) of the previous one.

— In every group G, there exists an unique irreducible representation which is totally symmetrical and whose characters are all equal to +1.

These five properties provide a way to build the character tables of the irreducible representations associated to every point group. For example, in the case of the C_{3v} group which we already studied, the first property tells us that C_{3v} is characterized by three irreducible representations whose dimensions l_i are such that $l_1^2 + l_2^2 + l_3^2 = 6$ (second property). The fifth property gives $l_2^2 + l_3^2 = 5$, establishing the existence of two irreducible representations of dimension 2 and 1 whose characters may be obtained from the third and fourth properties.

Some typical features of these irreducible representations are outlined in a notation convention suggested by Mulliken, and presented in Table 3.2.4.

Table 3.2.4 — **Mulliken notations of the irreducible representations of a group.**

		DEFINITION
CAPITAL LETTER	A	Dimension 1 and symmetrical for the rotation about the C_n principal axis
	B	Dimension 1 and antisymmetrical for the rotation about the C_n principal axis
	E	Dimension 2
	T or F	Dimension 3
	G	Dimension 4
	H	Dimension 5
NUMERICAL INDEX	1	Symmetrical for the rotation about the secondary C_2 axis
	2	Antisymmetrical for the rotation about the secondary C_2 axis (without a secondary C_2 axis, use a σ plane).
EXPONENT	'	Symmetrical about σ_h
	"	Antisymmetrical about σ_h
LETTER INDEX	g	Symmetric for inversion
	u	Antisymmetrical for inversion

Irreducible representations (and their corresponding character tables) of all the point groups are given in Appendix III.5.4. Among others, you can see that to the right of the column containing the characters, one (possibly two) columns can appear containing the abbreviated forms of the atomic orbitals (s, p, ...) or/and a few letters (x, y, xy, ...).

These symbols indicate that the atomic orbitals or the corresponding literal functions (s or $x^2 + y^2 + z^2$, p_x or x, d_{xy} or xy) have the same symmetry as the irreducible representations next to which they are placed and thus can constitute bases for them. Similarly, a symbol of type R_x signifies that a rotation motion about the Ox axis of the reference frame has the same symmetry as the irreducible representation considered, which property may have important applications in spectroscopy.

We shall conclude our study of the properties of irreducible representations of a group by stating a fundamental result which permits to know if, and how many times, an irreducible representation occurs in a reducible representation of a group by using only the character table of the group considered.

— Let Γ be a reducible representation of a group G. The number of times n_i that an irreducible representation Γ_i occurs in Γ is given by:

$$n_i = \frac{1}{h}\sum_R \left[\chi_i^*(R)\chi(R)\right] \tag{3.2.4}$$

You will have noticed the analogy between the above formula and the definition of the Euclidian scalar product. Considering again the above example of the C_{3v} group, let us investigate the irreducible representations that occur in Γ_3. We have

$$n(A_1) = (1/6)\ [1(1\times2) + 3(1\times0) + 2(1\times-1)] = 0$$
$$n(A_2) = (1/6)\ [1(1\times2) + 3(-1\times0) + 2(1\times-1)] = 0$$
$$n(E) = (1/6)\ [1(2\times2) + 3(0\times0) + 2(-1\times-1)] = 1$$

and finally $\Gamma_3 = E$, which was *a priori* an obvious result from the inspection of the character table of the C_{3v} group.

III.2.4.3. Direct product of two representations

Let us consider two representations Γ_1 and Γ_2 of the *same* group G. Their respective dimensions are given by the characters of the identity operation. Let n_1 and n_2 be these dimensions. We note $\{f_1, f_2, \ldots, f_{n_2}\}$ the n_1 basis vectors of Γ_1 and $\{g_1, g_2, \ldots, g_{n_2}\}$ the n_2 basis vectors of Γ_2.

The question we ask ourselves is what information can we obtain about the symmetry of the $n_1 \times n_2$ products of type $f_i g_j$, knowing that the sets $\{f_i\}$ and $\{g_j\}$ individually constitute the bases of the two representations of group G. The answer to this question is given by the group theory which tells us that the set of $n_1 \times n_2$ products of type $f_i g_j$ forms the basis of a *new* representation of group G, labeled $\Gamma_1 \otimes \Gamma_2$, whose dimension is $n_1 \times n_2$ and which is called the *direct product* of the two representations Γ_1 and Γ_2.

Let us take an example. Let Γ_1 and Γ_2 be two representations, of dimension 2 and 3, respectively, of the same group G. Keeping the previous notations, the basis of the representation $\Gamma_1 \otimes \Gamma_2$ will be formed by the six products $\{f_1g_1, f_1g_2, f_1g_3, f_2g_1, f_2g_2, f_2g_3\}$. Since the direct product $\Gamma_1 \otimes \Gamma_2$ is a representation of group G, we can calculate the character associated to the symmetry operation R for this representation. Noting $M_{\Gamma i}(R)$ the matrix associated to the symmetry operation R expressed in the basis of the Γ_i representation, we have

$$(M_{\Gamma_1}[R]) = \begin{pmatrix} a_{11} & a_{12} \\ a_{21} & a_{22} \end{pmatrix} \qquad (M_{\Gamma_2}[R]) = \begin{pmatrix} b_{11} & b_{12} & b_{13} \\ b_{21} & b_{22} & b_{23} \\ b_{31} & b_{32} & b_{33} \end{pmatrix}$$

with $\chi_{\Gamma 1}(R) = a_{11} + a_{22}$ and $\chi_{\Gamma 2}(R) = b_{11} + b_{22} + b_{33}$.

The matrix associated to R, expressed on basis $\{f_i g_j\}$, will be obtained from the image given by R of each of the six products $f_i g_j$. Remembering that the result of the action of R on a product of functions is the product of the action of R on each of the functions separately, we shall have for example

$$R(f_1 g_1) = R(f_1) \times R(g_1) = (a_{11} f_1 + a_{21} f_2) \times (b_{11} g_1 + b_{21} g_2 + b_{31} g_3)$$

then, expressing the result on basis $\{f_i g_j\}$, we obtain the first column of the desired matrix. Finally, we obtain

$$(M_{\Gamma_1 x \Gamma_2}[R]) = \begin{pmatrix} a_{11}b_{11} & a_{11}b_{12} & a_{11}b_{13} & a_{12}b_{11} & a_{12}b_{12} & a_{12}b_{13} \\ a_{11}b_{21} & a_{11}b_{22} & a_{11}b_{23} & a_{12}b_{21} & a_{12}b_{22} & a_{12}b_{23} \\ a_{11}b_{31} & a_{11}b_{32} & a_{11}b_{33} & a_{12}b_{31} & a_{12}b_{32} & a_{12}b_{33} \\ a_{21}b_{11} & a_{21}b_{12} & a_{21}b_{13} & a_{22}b_{11} & a_{22}b_{12} & a_{22}b_{13} \\ a_{21}b_{21} & a_{21}b_{22} & a_{21}b_{23} & a_{22}b_{21} & a_{22}b_{22} & a_{22}b_{23} \\ a_{21}b_{31} & a_{21}b_{32} & a_{21}b_{33} & a_{22}b_{31} & a_{22}b_{32} & a_{22}b_{33} \end{pmatrix}$$

The character can be obtained immediately. We note in particular that

$$\chi_{\Gamma 1 \otimes \Gamma 2}(R) = (a_{11} + a_{22}) \times (b_{11} + b_{22} + b_{33}) = \chi_{\Gamma 1}(R) \times \chi_{\Gamma 2}(R)$$

This important result is very general and it allows the evaluation of the characters of the direct product of two known representations. Thus, for every operation R of the group, we have $\chi_{\Gamma 1 \otimes \Gamma 2}(R) = \chi_{\Gamma 1}(R) \times \chi_{\Gamma 2}(R)$ and all the properties of the representation of a group (reducibility, ...) still hold for representations of the direct product.

The direct product has a very powerful application in the evaluation of integrals in quantum mechanics. In mathematics the simple use of symmetry arguments frequently makes it possible to simplify the calculation of integrals over symmetric domains, delivering us from the — often difficult — search for an explicit primitive. For example, the integral of an even function defined over a symmetrical domain *never* vanishes, whereas, on the contrary, the integral of an uneven function, defined over a symmetrical domain, *always* vanishes.

Generalizing this approach, let us assume that we are looking for the value of the integral $I = \langle f|O|h\rangle$. $\langle f|$ and $|h\rangle$ are two vectors and *O* is an operator, belonging, respectively, to the three representations F, H and O of the *same* group G. If the representation F⊗O⊗H contains the totally symmetrical representation (in which case it is reducible) or if it is equal to the totally symmetrical representation (in this case, it is irreducible), then I does not vanish (by analogy with the property of integrals of even functions).

Thus, group theory sometimes permits to detect very simply the non vanishing property of integrals, in fact, as soon as the symmetry of the integrand is known, and this can be essential in the study of some physical properties (such as, for example, the selection rules discussed in Chapter IV).

Observation: The projection operator

A priori, an infinity of representations and bases exist for a given group. As said, it is possible to simplify the study of these representations by reducing them to a few " canonical representations " of the group, namely its irreducible representations. Then the problem is, starting from the known basis functions of the initial representation, to express new functions that may serve as a basis for the irreducible representations of the investigated group. This problem is far from simple because these irreducible representations are, *a priori*, known to us only by their character. Some cases can be extremely simple, as the representation of dimension 1 of type A. Here, we just need to find an initial function which is symmetrical with respect to the rotation around the principal axis. It can be more difficult in the case of representations with larger dimensions. Using group theory, we need to built the operator $P_{\Gamma i}$ which is defined for the representation Γ_i by

$$P_{\Gamma_i} = \sum_{R} \chi_{\Gamma_i}(R) R \tag{3.2.5}$$

This operator, which can act on any vector of the basis chosen for the problem in question, can be considered as a *generator of functions* (not necessarily normalized ones) which have the symmetry of the representation from which this operator is built (Γ_i). If Γ_i is one dimensional, the application of the operator to any element of the basis will of course yield a basis function of Γ_i. However, if Γ_i is of dimension $n \geq 1$, the functions generated using the projection operator are not necessarily orthogonal to each other, so this method to find a basis should be conducted with care. Two examples will be discussed in Section III.3. A few simple symmetry questions are presented in Problem III.4.1.

III.3. Application of quantum mechanics and group theory to the description of stationary electronic states in atoms and molecules

III.3.1. Description of the electronic structure of the hydrogen atom and of the hydrogenic atoms

The hydrogen atom (*cf.* Figure 3.3.1) is formed of a nucleus (mass $M = 1.67 \times 10^{-27}$ kg) and an electron ($m = 9.1 \times 10^{-31}$ kg) which have opposite charges with the same modulus $e = 1.6021 \times 10^{-19}$ C, located at a distance r from each other. More generally, we will focus on " hydrogenoïd " atoms, consisting of a nucleus of charge (+Ze) and of a single electron (the atom is thus ionized Z −1 times).

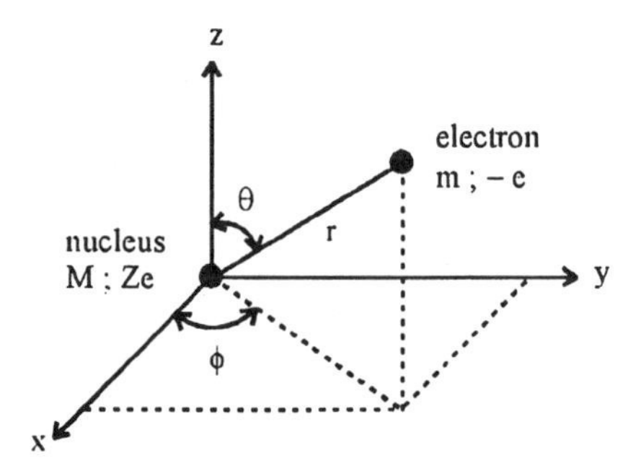

Figure 3.3.1: Schematic description of the variables describing a monoelectronic atom.

III.3.1.1. Solution of the eigenvalue equation for the Hamiltonian operator

The energy E and the state vector $|\psi\rangle$ of these hydrogen like systems can be found by solving the eigenvalue equation of the Hamiltonian operator obtained by applying postulate P7 to the Hamilton function (*cf.* Section III.1.2.4)

$$-\frac{h^2}{8\pi^2 m}\Delta|\psi\rangle + V(\boldsymbol{r})|\psi\rangle = -\frac{h^2}{8\pi^2 m}\Delta|\psi\rangle - \frac{Ze^2|\psi\rangle}{4\pi\varepsilon_0 \boldsymbol{r}} = E|\psi\rangle \quad (3.3.1)$$

By using spherical coordinates, a simplification emerges because we can use the angular momentum $\boldsymbol{L}^2$:

$$-[(h^2/8\pi^2 m)(1/r)(\partial^2/\partial \boldsymbol{r}^2) + (\boldsymbol{L}^2/2mr^2) + V(\boldsymbol{r})]|\psi\rangle = E|\psi\rangle$$

Since the eigenfunctions of the angular momentum are known (spherical harmonics) and since L^2 commutes with H, we set

$$\psi(r,\theta,\varphi) = R(r)Y_{l,m}(\theta,\varphi)$$

R(r) is called the *radial function*. After substitution, we finally obtain

$$\frac{d^2R(r)}{dr^2} + (2/r)\frac{dR(r)}{dr} - \left\{\frac{l(l+1)}{r^2} + 2m(2\pi/h)^2[E - V(r)]\right\}R(r) = 0 \qquad (3.3.2)$$

which is the *radial equation* It depends on parameter l which comes from the quantification of the angular momentum. The normalization of the wave function

$$\langle\psi|\psi\rangle = \int_V |\psi(r,\theta,\varphi)|^2 \, dV = 1$$

splits into parts and finally gives the relation

$$\int_0^\infty |R(r)|^2 \, r^2 dr = 1 \qquad (3.3.3)$$

The solution of differential equation 3.3.2, together with condition 3.3.3, provides the expressions of R(r) and E. Thus, we can express the state (eigenvector $|\psi\rangle$) and the energy (E) of the hydrogen atom.

III.3.1.2. Energy and shell model

The energy E appears in fact as a *discrete series* of values E_n given by

$$E_n = -(2\pi^2 me^4/h^2)(1/n^2) \qquad (3.3.4)$$

where n is the principal quantum number and takes on the values n = 1, 2, 3, ... If we wish to take into account the motion of the nucleus, instead of the mass m of the electron we must substitute the reduced mass $\mu = mM/(m+M)$ where M is the mass of the nucleus, which is very much heavier than the electron.

Every energy level is associated with a value of n. We shall speak of *energy shells* characterized by a value of n. We denote these shells by the letters K, L, M, N, ..., for n = 1, 2, 3, 4, ..., respectively. The shell model only refers to the energy and does not concern the electronic structure of the atom. *Talking of " electronic shells " makes no sense.*

III.3.1.3. State vector and atomic orbital multiplicity

The radial wave function is defined from the associated *Laguerre polynomials* L_{n+l}^{2l+1} (Table 3.3.1) by

$$R(r) = R_{n,l}(r) = \rho^l \exp(-\rho/2) L_{n+l}^{2l+1}(\rho) \quad \rho = (2me^2/n)(2\pi/h)^2 r \tag{3.3.5}$$

The eigenfunctions of the Hamiltonian can thus be expressed as

$$\psi_{n,l,m}(r,\theta,\varphi) = R_{n,l}(r)\Theta_{l,m}(\theta)\Phi_m(\varphi) \tag{3.3.6}$$

and they depend on three quantum numbers n, l, m, which form an orthonormal basis. States characterized by the same value of n, but with different values of l and m, have the same energy. We can calculate the degeneracy g_n of a state associated to a given value of n. There are n different values of l for each value of n, and m can takes (2l + 1) values for each value of l. The degeneracy is thus

$$g_n = \sum_{l=0}^{n-1}(2l+1) = n^2$$

Here we talk of *accidental degeneracy*, as it is caused by the special form of the potential and not by its spherical symmetry. When the degeneracy is a consequence of the symmetry of the problem (which is most frequent), we talk of *essential degeneracy*.

Table 3.3.1 — **Analytical expressions of a few radial functions $R_{n,l}$.**
($a = \varepsilon_0 h^2/\pi m e^2 = 0.0529177725$ nm).

n	l	$R_{n,l}$
1	0	$(Z/a)^{3/2}\, 2\exp(-Zr/a)$
2	0	$(Z/2a)^{3/2}\, 2\,(1 - Zr/2a)\exp(-Zr-/2a)$
2	1	$(Z/2a)^{3/2}\, 2/\sqrt{3}\,(Zr/2a)\exp(-Zr/2a)$
3	0	$(Z/3a)^{3/2}\, 2[1-2(Zr/3a)+2/3(Zr/3a)^2]\exp(-Zr/3a)$
3	1	$(Z/3a)^{3/2} 4\sqrt{2}/3\,(Zr/3a)[1-1/2(Zr/3a)]\exp(-Zr/3a)$
3	2	$(Z/3a)^{3/2}\, 2\sqrt{2}/3\sqrt{5}\,(Zr/3a)^2]\exp(-Zr/3a)$

Finally, considering the electron spin and in accordance with postulate P9, we shall write the state vector as $\psi_{n,l,m}(r,\theta,\varphi)\sigma(m_s)$ (the product of a space function by a spin function) where $\sigma(m_s)$ is one of the functions $|\alpha\rangle$ or $|\beta\rangle$ (or more generally a linear combination of the two functions). So, the degeneracy of a state n is equal to $2n^2$ and, conventionally, the state vector will be noted according to the Dirac notation as $|n, l, m, m_s\rangle$.

Practically, the number l is not explicitly specified and it is replaced by a letter whose meaning is given on Table 3.3.2.

Table 3.3.2 — **l value notation.**

value of l	0	1	2	3	4
Corresponding letter	s	p	d	f	g

III.3.1.4. Various representations of probability densities in real space

Only the modulus of the state vector $|\psi\rangle$ has a physical meaning and it stands for the volume density of the probability of presence. It is therefore better to be able to sketch some of its geometric representations. For example, in the case of the hydrogen atom, the normalization condition expressed in spherical coordinates takes the form

$$\int_0^{\infty}\int_0^{\pi}\int_0^{2\pi} R\Theta\Phi^2 r^2 \sin\theta d\theta d\varphi dr = 1$$

This equality suggests that we introduction two types of probability densities: One *radial*, defined for a given orientation (θ, φ) by the graph of R(r) as a function of r, and the other one *angular*, defined for a given value of r by the graph of $Y(\theta)$ or of $|Y(\theta)|^2$ as a function of θ.

The representation of R(r), proportional to the radial density of the probability of presence, gives a feeling of the extension in space of the orbitals according to the values of the quantum numbers (*cf.* Figure 3.3.2).

Similarly, the angular density of the probability of presence gives information about the shape of the orbitals for a given value of r and provides hints about the favored directions in space where the electron probabilities of presence are largest, and this is a fundamental issue in the study of the chemical bond. A few examples are given in Figure 3.3.3.

Orbital d_{z^2} corresponds to m = 0; orbitals p_x, p_y, d_{yz} and d_{zx} are linear combinations of the spherical harmonics m = + 1 and m = – 1; orbitals d_{xy} and $d_{x^2-y^2}$ are linear combinations of the spherical harmonics m = + 2 and m = – 2.

It is possible to think of other ways of picturing the orbitals, such as, for example, sketching equal electronic volume density surfaces (isodensity curves).

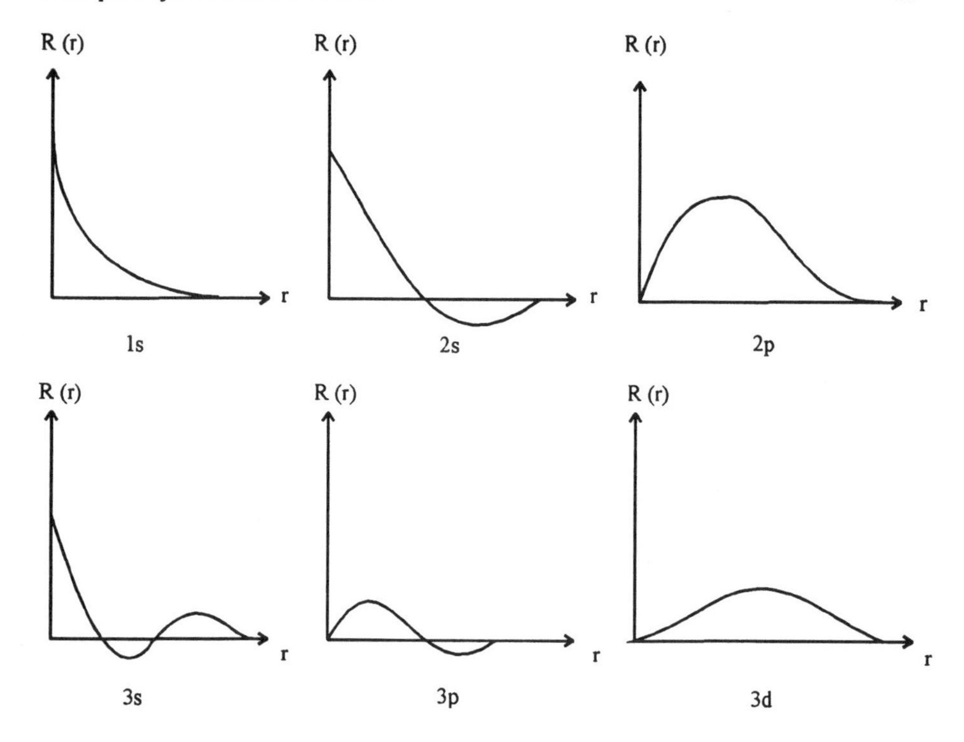

Figure 3.3.2: R(r) representations as a function of r for n = 1, 2 and 3.

III.3.1.5. Spin orbit coupling

The electron, as a charged particle, generates a current loop due to its orbital motion and, consequently, a magnetic induction **B**. The intensity of **B** will of course be proportional to the value of the quantum number l quantifying the orbital momentum. Moreover, the electron also has an intrinsic magnetic momentum μ, proportional to its spin angular momentum **s**

$$\mu = -g_e \mu_B \mathbf{s}$$

where g_e and μ_B are, respectively, the Landé factor and the Bohr magneton.

The existence of **B**, due to the orbital motion of the electron, results in a modification of the orientation of μ. This is called the *spin-orbit coupling*. The interaction energy E between a magnetic moment μ and the induction **B** is given by

$$E = -\mu \mathbf{B}$$

and the spin-orbit coupling Hamiltonian H_{SO} for a monoelectronic system takes the form $H_{SO} = \xi(r)\boldsymbol{ls}$

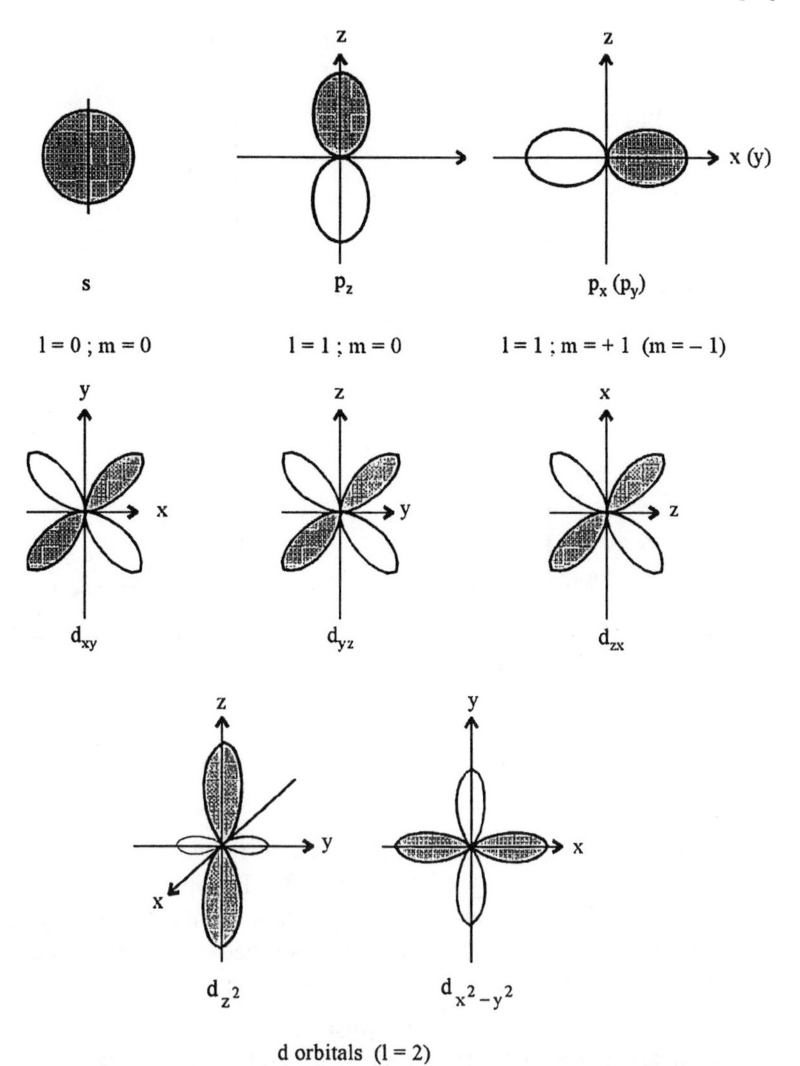

Figure 3.3.3: Angular parts of a few real orbitals for l ranging from 0 to 2.

Here, ***ls*** denotes the operator built from the scalar product **ls**. The analytic expression of $\xi(r)$ can be obtained from relativistic quantum electrodynamic arguments which lie out of the scope of this book. In the case of hydrogen like atoms, it is given by

$$\xi(r) \approx (1/2m_e^2c^2)[(1/r)(\partial v/\partial r)]$$

V is the electrostatic potential energy of the electron in the field of the nucleus with charge (Ze). It can be shown that the expression for H_{SO} still holds in the case of many-electron atoms. Taking into consideration the influence of the spin-orbit coupling, the eigenvalue equation of the total Hamiltonian takes the form

$$\left(H_0 + H_{SO}\right)|\psi\rangle = \mathrm{E}|\psi\rangle$$

where H_0 represents the Hamiltonian without coupling.

The expression of $\xi(r)$ shows that the spin orbit coupling becomes more important as the nucleus charge increases (large Z). However, it remains weak with respect to the leading effect (differences between the energy levels associated to the eigenvalues of H) and the perturbation theory is sufficient to account for it. One may notice that the operators l_z and s_z no longer commute with the total Hamiltonian, because of the coupling term. As a consequence, the quantum numbers l and s are not completely appropriate for the description of the states of a monovalent atom, since the eigenfunctions of l_z and s_z are not shared by the Hamiltonian. However, the total angular momentum j, defined by $\boldsymbol{j} = \boldsymbol{l} + \boldsymbol{s}$, is such that j^2 and j_z commute with the full Hamiltonian. Consequently, these operators share a common basis of eigenfunctions. We note j (j + 1) and $m_j = m_l + m_s$ the eigenvalues (in units of $[h/2\pi]^2$ and $[h/2\pi]$, respectively,) associated to j^2 and j_z and we can say that j and m_j are *" good " quantum numbers* of the problem.

The splitting of the main lines in the optical spectra of monovalent atoms (fine structure) has its origin in the spin-orbit coupling. This type of splitting has been observed especially in hydrogen and sodium, whose yellow emission line (D line: 3p $\rightarrow$ 3s transition), is split into two lines D_1 and D_2, of respective wavelengths $\lambda_1 = 589.0$ and $\lambda_2 = 589.6$ nm.

III.3.2. The description of the electronic structure of many-electron atoms

The case of the hydrogen atom and of hydrogen like atoms was especially simple, since it dealt with a single electron facing a potential created by a nucleus assumed to be fixed. Let us now discuss a much more difficult, but more general case, dealing with a nucleus of charge (+Ze), with N electrons (if N = Z, the atom is neutral). Assuming the nucleus to have infinite mass (thus it is fixed), the Hamiltonian takes the form (neglecting spin effects)

$$H = \sum_{i=1}^{N}(-\mathrm{h}^2/8\pi^2\mathrm{m})\Delta_i - \sum_{i=1}^{N}(\mathrm{Ze}^2/4\pi\varepsilon_0 r_i) + \sum_{i=1}^{N}\sum_{j=i+1}^{N}(\mathrm{e}^2/4\pi\varepsilon_0 r_{ij}) \quad (3.3.7)$$

where $\mathbf{r_i}$ is the distance between electron i and the nucleus, and $\mathbf{r_{ij}}$ the distance between electrons i and j. The first term in H accounts for the kinetic energy of the electrons, the second one for their electrostatic energy in the field of the nucleus, and the third one describes the electron pair repulsion. This last term makes it impossible to find an exact solution to Equation 3.3.7 (note that this last term cannot be neglected because there are $Z[Z-1]/2$ terms of type e^2/r_{ij} and their summation leads to a value of the same order of magnitude as that of the summation of the terms of type Ze^2/r_i). The inter-electronic repulsion term makes on the one hand that a given electron acts as if it were interacting with a diffuse charged cloud (because of the presence of the other electrons) which, *a priori*, has no privileged direction and thus exhibits spherical symmetry, and, on the other hand, it interacts with the (+Ze) charged nucleus. So, we are forced to look for an approximate solution.

III.3.2.1. Central field approximation: The Hartree-Fock method and the self consistent field

Considering the above comments, we can rewrite the Hamiltonian as

$$H=\sum_{i=1}^{N}-\frac{h^2}{8\pi^2 m}\Delta_i+\sum_{i=1}^{N}\left[\frac{-Ze^2}{4\pi\varepsilon_0 r_i}+V(r_i)\right]+\sum_{i=1}^{N}\sum_{j=i+1}^{N}\left[\frac{e^2}{4\pi\varepsilon_0 r_{ij}}-V(r_i)\right]$$

where $V(r_i)$ is a function depending only on r_i. $V(r_i)$ is obtained by an iterative method that we shall discuss later, in such a way that the summation

$$H_\varepsilon=\sum_{i=1}^{N}\sum_{j=i+1}^{N}\left[\frac{e^2}{4\pi\varepsilon_0 r_{ij}}-V(r_i)\right]$$

is sufficiently small so as to be negligible, in a first approximation, with respect to the other terms of H. Finally, setting

$$\sum_{i=1}^{N}U(r_i)=\sum_{i=1}^{N}\left[\frac{-Ze^2}{4\pi\varepsilon_0 r_i}+V(r_i)\right]$$

H can be written according to

$$H=\sum_{i=1}^{N}-\frac{h^2}{8\pi^2 m}\Delta_i+\sum_{i=1}^{N}U(r_i)+H_\varepsilon\approx\sum_{i=1}^{N}-\frac{h^2}{8\pi^2 m}\Delta_i+\sum_{i=1}^{N}U(r_i)\approx\sum_{i=1}^{N}h_i \quad (3.3.8)$$

i.e. as the sum of monoelectronic type Hamiltonians h_i, each identical to that of the hydrogen atom, and of potential terms $U(\mathbf{r}_i)$.

Of course, the total potential term depends on the number of electrons, and thus on the very nature of the atom under consideration. Within this frame, an eigenfunction of H can be expanded as a product of N monoelectronic functions $\varphi_i(\mathbf{r}_i)$, each an eigenfunction of the monoelectronic operator h_i and depending only on the position variables of electron i. The choice of $V(\mathbf{r})$ must be such that H_ε is a negligible term, although neither $V(\mathbf{r})$ nor the interelectronic repulsive potential are small individually. In fact, $V(\mathbf{r})$ is nothing but an *average central potential* which for each electron conveniently accounts for the Coulomb repulsive potential created by all the remaining electrons seen as a whole.

We write

$$H|\psi\rangle = E|\psi\rangle$$

thus

$$(h_1 + h_2 + ... + h_N)|\varphi_1\varphi_2...\varphi_N\rangle = (E_1 + E_2 + ... + E_N)|\varphi_1\varphi_2...\varphi_N\rangle$$

where φ_1 depends only on $\mathbf{r}_1$, φ_2 on $\mathbf{r}_2$, ..., φ_N on $\mathbf{r}_N$, with $h_i|\varphi_i\rangle = E_i|\varphi_i\rangle$.

The eigenfunctions $\varphi_i(\mathbf{r}_i)$ do not totally coincide with those of the hydrogen atom, since the potential $U(\mathbf{r}_i)$ does not strictly vary as $1/\mathbf{r}$. However, because $U(\mathbf{r}_i)$ displays spherical symmetry, a separation of variables (r_i, θ_i, ϕ_i) of electron i can be performed

$$\varphi(\mathbf{r}_i) = R'_{n_i,l_i}(r_i)Y_{l_i,m_i}(\theta_i,\varphi_i)$$

where $Y_{li,mi}$ denote spherical harmonics and $R'_{ni,li}$ a radial function, differing from that of the hydrogen atom and whose empirical expression was suggested by Slater (*cf.* Bibliographical reference 1.2.1). This time, for each monoelectronic orbital $\varphi_i(\mathbf{r}_i)$, the energy E_i depend both on n_i and on l_i because of the form of the potential $U(\mathbf{r}_i)$ (contrary to the case of the hydrogen atom, where it depended exclusively on n). *The accidental degeneracy is raised.* We notice that the energy does not depend on the quantum number m_i. Of course, the overall statevector is written as a Slater determinant (*cf.* Equation 3.1.27), including the spin vectors and with $|\psi\rangle = |\varphi_1\varphi_2...\varphi_N\rangle$. In the central field approximation, the energy of a many-electron atom is the sum of the individual electronic energies

$$E_{n,l} = \sum_{i=1}^{N} E(n_i, l_i)$$

Again, we find a shell type energy distribution. The energy $E_{n,l}$ is independent of m and can take $(2l + 1)$ values. Taking into account the spin, the degeneracy of level $E_{n,l}$ is $2(2l + 1)$, and it does not depend on n.

The *Klechkowsky rule* (*cf.* Bibliographic reference 1.2.2) stands as a simple mnemonic method to classify the levels E(n, l) by order of energy. The energy associated to a state characterized by a couple (n, l) is an increasing function of the sum (n + l). For equal values of (n + l), the energy increases with n. Thus, the energy of level 4s is greater than the energy of level 3d.

An *electronic configuration* is the description of the state of N electrons which are in energy state E(n, l). It is conventionally written by associating a number (the n value) to a letter (the l value) and specifying as a superscript the number of electrons concerned. Finally, the overall electronic configuration of the atom is obtained by writing all the electronic configurations from left to right, in order of increasing energy levels.

For example, the electronic configuration of the potassium atom in its fundamental state is written $1s^2 2s^2 2p^6 3s^2 3p^6 4s^1$.

As was previously discussed, V(**r**) is evaluated using an iterative method developed in 1928 by Hartree. The reader will find a detailed outline of the method — based on the *variation theorem* — in the Bibliographic reference 1.2.2.

III.3.2.2. Electrostatic and magnetic interactions: Russel-Saunders and spin-orbit couplings

We already mentioned (*cf.* Section III.3.1.5) the spin-orbit coupling, essentially of magnetic origin, which occurs in the hydrogen atom. For many-electron atoms, the case is more complicated, because of the existence of an additional interaction, of electrostatic nature, due to inter-electron repulsions.

— Electrostatic type interactions: The Russel-Saunders coupling

In a many electron atom, the electrons have correlated movements, among others because they experience mutual repulsions. Consequently, the angular momentum, which is constant for a single electron, varies as a function of time for a collection of interacting electrons. Finally, the angular momenta of the individual electrons, together with their associated quantum numbers (l_i, s_i), are no longer relevant to describe the state of the system, because the associated operators do not commute with hamiltonian H_{RS} describing the Russel-Saunders coupling. At each instant, we can define the *total angular momentum* of a many electron atom as the sum of the individual angular momenta of each electron i. The angular and spin momentum operators take the form

$$\boldsymbol{L} = \sum_i \boldsymbol{l}_i \quad \boldsymbol{S} = \sum_i \boldsymbol{s}_i \tag{3.3.9}$$

This way to combine the angular momenta is the *Russel-Saunders coupling* (or L-S coupling), and now a total angular momentum can be defined as

$$J = L + S \tag{3.3.10}$$

The quantum number J associated to the operator $\boldsymbol{J}$ ranges between L – S and L + S by steps of one unit. The introduction of this coupling scheme is very powerful, as it is possible to show that both operator $\boldsymbol{L}^2$ and operator $\boldsymbol{S}^2$ commute with the Hamiltonian H_{RS} and thus share a common set of eigenvectors. In other words, eigenfunctions of $H_0 + H_{RS}$ (H_0 being the unperturbed Hamiltonian) can be obtained by seeking those of $\boldsymbol{L}^2$ and $\boldsymbol{S}^2$. Thus, the quantum numbers L and S indeed characterize each energy level of a many-electron atom undergoing an electrostatic repulsion perturbation.

— ***Magnetic type interactions: Spin-orbit coupling***

The spin-orbit coupling — a relativistic effect — emerges through an additional term H_{SO} of the Hamiltonian. For a many-electron system, the term H_{SO} can be deduced straightforwardly from the monoelectronic case studied above

$$H_{SO} = \sum_i \xi(\mathbf{r}_i) \boldsymbol{l}_i \boldsymbol{s}_i$$

The spin-orbit coupling is usually weak, so that the Hamilton functions are such that $H_{SO} << H_0 + H_{RS}$, and the corresponding effect may be calculated by using perturbation theory. Although $\boldsymbol{L}^2$ and $\boldsymbol{S}^2$ do not *strictly* commute with the total Hamiltonian, L, S and J are still considered as " good quantum numbers " providing a correct description of the system (since term H_{SO} is small with respect to H, $\boldsymbol{L}^2$ and $\boldsymbol{S}^2$ " almost " commute with H).

When the atomic number of the atom increases, the velocity of the electrons also increases, so that relativistic effects (as the spin-orbit coupling) become more important. Consequently, $H_0 + H_{SO}$ become the leading term in the Hamiltonian and H_{RS} can be treated as a perturbation. In this case, $\boldsymbol{L}^2$ and $\boldsymbol{S}^2$ do not commute with $H_0 + H_{SO}$ and L and S are no longer relevant to describe the energy states of the atom. More precisely, since the leading term is $H_0 + H_{SO}$, an analogy with the case of the spin-orbit coupling of a monoelectronic system shows that a " good " quantum number (*i.e.* characteristic of the system) is obtained by combining the angular momentum operators of every electron individually ($j_i = j_i + s_i$), in order to define the overall angular momentum operator by

$$J = \sum_i j_i \tag{3.3.11}$$

This type of combination of the angular momentum is the so-called *j-j coupling*. Of course, this approach does call for H_{RS} to be very small as compared to $H_0 + H_{SO}$. Unfortunately, intermediates situations, where $H_{RS} \approx H_{SO}$, are frequent and require tedious calculations.

III.3.2.3. Description of configurations

We just outlined several types of interactions which exist within a many-electron atom. According to their relative amplitudes, the description of the corresponding states of the atom is different. For instance, when the electrostatic repulsion is greater than the spin-orbit coupling, L and S are relevant quantum numbers to describe the system (L-S coupling). A *configuration* is the specification of the orbitals characterizing the electrons of an atom. The *terms,* arising from a configuration, represent all the energy levels associated to an atom when the electrostatic repulsion interaction is considered. A term is specified by the knowledge of L and S and its degeneracy is (2L + 1)(2S + 1). A term is denoted ^{2S+1}L, the value of L being indicated by the letter S, P, D, F, G, H, ..., when L = 0, 1, 2, 3, 4, 5, ... Finally, the term degeneracy is removed by incorporating the spin-orbit coupling, meaning that quantum number J is known and we obtain *levels* labeled $^{2S+1}L_J$. These levels have a degeneracy equal to (2J+1), which can be removed by applying an external field to induce *states*. Thus, a hierarchy emerges for the definitions of configuration, term, level and state. It corresponds to putting under one name the state vectors associated to the same energy as first the electrostatic repulsion, then the spin-orbit coupling and last the influence of an external field are considered. Figure 3.3.4 sketches the progression when R-S coupling is dominant.

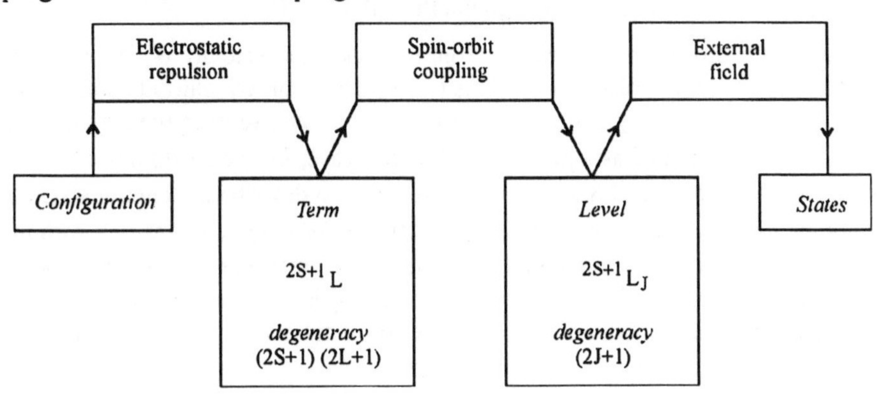

Figure 3.3.4: Outline of a progression in the case of R-S coupling.

A *correlation diagram* (Figure 3.3.5) can be drawn, in an L-S or in a j-j scheme, which shows the successive degeneracy removals of a given

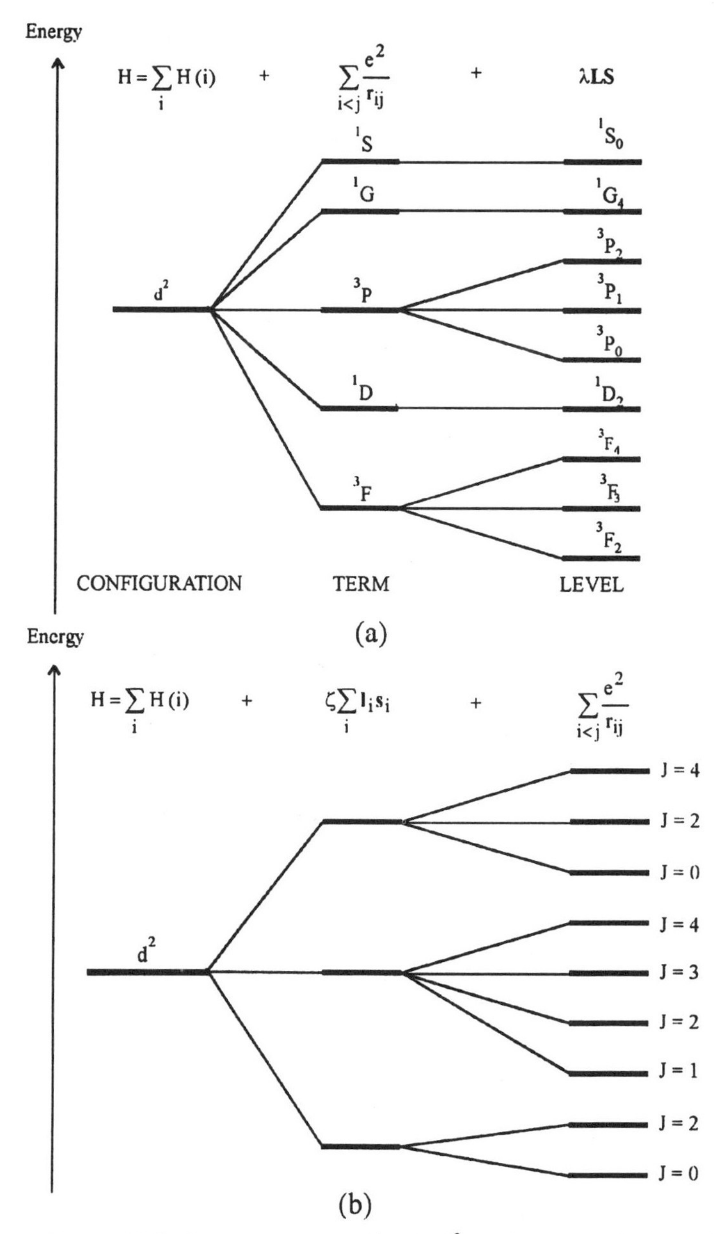

Figure 3.3.5: Degeneracy removal of a d^2 configuration. (a) Russel-Saunders coupling case where the interelectronic repulsion interaction is greater than the spin-orbit coupling interaction; (b) j-j coupling case, where the spin-orbit coupling energy is greater than the interelectronic repulsion energy.

configuration when the electronic repulsion and the spin orbit coupling are considered.

Such diagrams are obtained from the eigenstates of many-electron Hamiltonians and rely on tedious numerical calculations. Thus, it might be quite interesting to derive some simple criteria providing, for example, a classification by increasing energy of the terms arising from a given configuration, in the same way as the Klechkowsky rule offers an ordering of the electronic configurations. As far as the terms are concerned, the correlations are too complex and such a set of rules does not exist. However, the fundamental term can be obtained from a given configuration using a set of rules, namely the *empirical Hund's rules*. Let us emphasize again that they predict *only* the fundamental term and do not give any information about the nature of the excited terms.

Hund's rule states that the term with the largest spin multiplicity has the smallest energy among the terms arising from the same electronic configuration. If several terms correspond to the largest spin multiplicity, it is the term with the largest orbital multiplicity which is more stable. For instance, the fundamental term is ^3F for the d^2 configuration.

Great similarities exist between terms arising from so-called *reciprocal configurations*: p^n and p^{6-n}, d^n and d^{10-n}, f^n and f^{14-n}, ... For example, the total degeneracy (orbital and spin degeneracies) of a d^2 configuration is $C_{10}^2 = 45$ and it is identical to that of a d^8 configuration (C_{10}^8). More precisely, let us focus on the case of the spin-orbit coupling for the reciprocal configurations d^1 and d^9. If we consider configuration d^9 as a positive hole in a completely filled d shell, then it is clearly understood why the spin-orbit coupling constant for a d^9 configuration (case of a single positive charge) has a sign opposite to that of a d^1 configuration (case of a single negative charge). This result matches that concerning the sign of ζ for atoms whose shells are more or less that half filled, according to the L-S coupling scheme. More generally, such an effect (sign change of the coupling constant) is expected for two reciprocal configurations in which the effect of a so-called *monoelectronic* type operator, i.e. involving a *single electronic charge*, is considered. This is the case of the spin-orbit coupling. On the contrary, the operator related to the electronic repulsion depends on the *product of two electrical charges* e^2/r_{ij}, and thus it does not depend on the sign of these charges, whether positive for holes (+e)(+e), or negative for electrons (− e)(− e). Consequently, all the terms arising from the same configuration are identical for two reciprocal configurations which have, in particular, the same fundamental term.

Let us also mention the case of atoms with *filled shells* or *closed shells*: s^2, p^6, d^{10}, f^{14}, ... These atoms have a vanishing resulting orbital and spin angular momentum so they do not contribute to the values of L, S and J. For example, all rare gases have $^1S_0$ as fundamental term.

The multiplicity of the term $^{2S+1}L_J$ is given by the value of the quantity 2S + 1, where S is the total spin angular momentum. Thus, the term is said to be *triplet* when 2S + 1 = 3, *doublet* when 2S + 1 = 2 and *singlet* when 2S + 1 = 1. Please notice that when L ≥ S, the *multiplicity* term is equal to its *level* number, whereas when S ≥ L, only (2L + 1) J values can be defined and the term exhibits 2L + 1 levels.

III.3.3. Description of electronic structure of molecules

III.3.3.1. General presentation

During our description of matter, we first discussed the case of the atom, i.e. a system consisting of a single nucleus surrounded by one or several electrons. We shall now consider molecules, i.e. a system formed by a group of nuclei (N) surrounded by electrons (e). The first step will consist in defining the Hamiltonian of the system within the stationary states approximation (*cf.* Section III.1.2.3), and then to solve the eigenvalue equation associated to this Hamiltonian in order to get the eigenstates $(E, |\psi\rangle)$ of the molecule. In its general form, the non relativistic Hamiltonian characterizing a molecule can be written as

$$H = T_N + T_e + V_{NN} + V_{eN} + V_{ee}$$

Using Greek letters for the variables related to the nuclei and Roman ones for those concerning electrons, $T_N = -(h^2/8\pi^2)\sum_\alpha(\Delta_\alpha/M_\alpha)$ represents the operator associated to the kinetic energy of the nuclei $T_e = -(h^2/8\pi^2 m_e)\sum_i \Delta_i$ corresponds to the kinetic energy of the electrons, $V_{NN} = \sum_\alpha\sum_{\beta>\alpha}(Z_\alpha Z_\beta e^2/4\pi\varepsilon_0 r_{\alpha\beta})$ is the operator associated to the repulsion energy that the nuclei exert on each other, $V_{eN} = \sum_\alpha\sum_i(Z_\alpha e^2/4\pi\varepsilon_0 r_{i\alpha})$ stands for the operator associated to the attraction energy between electrons and nuclei, and, last, $V_{ee} = \sum_i\sum_{j>i}(e^2/4\pi\varepsilon_0 r_{ij})$ corresponds to the operator associated to the repulsion interaction exerted by the electrons on each other.

As early as 1929, Dirac noticed that there was no way to find an analytic solution to the eigenvalue equation associated to such a Hamiltonian because of its complexity. Consequently, we are reduced to carry out approximations and to content ourselves with special approximate solutions.

— *Non relativistic approximation*

This is a good approximation for molecules consisting of light atoms. However, the electrons of the so-called inner shells move with very high velocities and thus are significantly affected by the relativistic mass increase with velocity. By interaction, the so-called external electrons are also perturbed. These relativistic effects may have a non negligible importance regarding the vibration effects of molecules containing heavy atoms, and will influence, for example, the spectroscopy of metal complexes. However, we shall ignore them in the rest of this book.

— *Born-Oppenheimer approximation (BOA, 1927)*

The electrostatic forces which act on nuclei and electrons are of the same order of magnitude. However, the mass of electrons being 1840 times weaker than that of the nuclei, electrons move much faster than the nuclei, offering the opportunity of decoupling their types of motion and to consider the nuclei as fixed in space. A physical picture, rather illustrative, is the one of the on-board camera: A camera attached to an electron would see the nuclei as motionless whereas, attached to a nucleus, it would record a spatial distribution of negative electricity.

Within the BOA, the Hamiltonian is obtained by neglecting T_N in Equation 3.3.12, giving

$$\mathrm{H}_{BO} \approx T_E + V_{eN} + V_{ee} + V_{NN} \tag{3.3.13}$$

As V_{NN} does not act on the spatial electronic coordinates, H_{BO} appears as a *purely electronic* operator, labeled H_e in the following. We seek the eigenvalues and eigenvectors by solving the equation

$$H_e|\psi_e\rangle = \mathrm{E}_e|\psi_e\rangle \tag{3.3.14}$$

for a given position of the nuclei defined by the spatial coordinate R (for the sake of simplicity, we consider the case of a diatomic molecule with an internuclear distance R). The function ψ_e depends explicitly on the spatial electronic coordinate (labeled **r**) and implicitly (parametrically) on R. Please observe that in writing Equation 3.3.14, the electronic energy E_e includes the energy of nuclear repulsion (V_{NN}). This energy is simply a constant number (since the internuclear distance R is fixed) and equal to $Z_1Z_2e^2/4\pi\varepsilon_oR$, where Z_1 and Z_2 are the atomic numbers of the two atoms.

Equation 3.3.14 can be solved with methods outlined in Section III.3.3.2 and provides the determination of $E_e(R)$. Next, we vary R in order to minimize the energy $E_e(R)$. The minimum of $E_e(R)$, obtained for $R = R_e$, provides an evaluation of the internuclear equilibrium distance R_e. The typical shape of the graph $E_e(R) = f(R)$ for a diatomic molecule is presented in Figure 3.3.6.

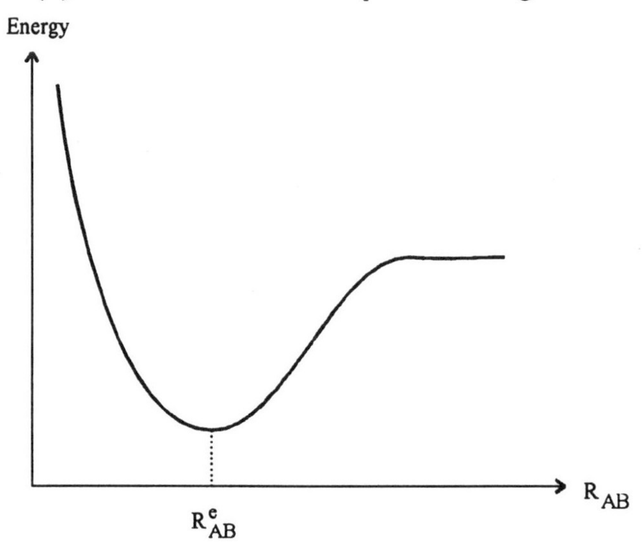

Figure 3.3.6: Sketch of the electronic energy curve of a diatomic molecule as a function of its internuclear distance R.

Having solved the electronic equation 3.3.14, we now want to find a solution of the whole equation

$$H|\psi\rangle = E\ |\psi\rangle \tag{3.3.15}$$

where H is the Hamiltonian given by Equation 3.3.12.

Since the motions of the electrons and of the nuclei are independent, we can express the total state vector ψ of the molecule as the product $\psi_e\psi_N$, where ψ_e and ψ_N are the state vectors relative to the electrons and to the nuclei, respectively. Vector $|\psi\rangle$ is written as

$$|\psi_{BO}\rangle = |\psi_e\psi_N\rangle \tag{3.3.16}$$

where ψ_e and ψ_N are the functions $\psi_e(\mathbf{r}, R)$ and $\psi_N(R)$.

We have

$$H|\psi\rangle = (H_e + T_N)|\psi_e\psi_N\rangle = H_e|\psi_e\psi_N\rangle + T_N|\psi_e\psi_N\rangle$$

and, according to Equation 3.3.14

$$H_e|\psi_e\psi_N\rangle = E_e|\psi_e\psi_N\rangle$$

Thus, we obtain from Equation 3.3.15

$$H|\psi\rangle = (E_e + T_N)|\psi_e\psi_N\rangle = E|\psi_e\psi_N\rangle$$

Since operator T_N acts only on $|\psi\rangle_N$, the equation becomes

$$(E_e + T_N)|\psi_N\rangle = E|\psi_N\rangle \tag{3.3.17}$$

This is the eigenvalue equation of a nuclear Hamiltonian where E_e acts as a potential term. Thus, according to BOA, the electronic energy acts as the potential term of the nuclear energy. Such a result is not surprising as the electrons adjust almost immediately their position after each movement of the nuclei (vibration or rotation movements), the E_e energy including the repulsion term V_{NN} can be viewed as the potential field acting on the nuclei. Last, solving Equation 3.3.17 provides $|\psi\rangle$ — and thus $|\psi_{BO}\rangle$ — and the total molecular energy E. However, if great precision is sought, it is no longer possible to uncouple the motions of the electrons and of the nuclei completely, and the coupling between these two motions, through the coupling of the angular momenta of electronic (orbital or spin) and of nuclear origin, must be taken into consideration. In this case, which we shall ignore in the following, the BOA is no longer valid.

Thus, the basic problem deals with finding a solution for Equation 3.3.14. The Born-Oppenheimer approximation provides a simplification in separating the study of the nuclei from that of the electrons. Concerning the latter, a Hamiltonian quite similar to the one corresponding to the many-electron atom is obtained, except that the electron-nucleus potential has been replaced by a summation over all the nuclei (term V_{eN}). Consequently, we can use the methods outlined in Section III.3.2 (for example, the self-consistent field treatment). We shall now investigate the various possible resolution methods.

III.3.3.2. Methods for solving the electronic eigenvalue equation

They rely on a minimization of the average energy (from a quantum mechanics point of view) given by postulate P4 (*cf.* Section III.1.2.2). The average energy is written as

$$E = \langle \psi_e | H | \psi_e \rangle / \langle \psi_e | \psi_e \rangle \tag{3.3.18}$$

(note that we obtain Equation 3.1.16 if ψ_e is normalized). This minimization is performed with the help of the variation theorem. First of all, therefore, we must express $|\psi_e\rangle$. Of course, and because of the antisymetrization postulate P9 (*cf.* Section III.1.2.6), $|\psi_e\rangle$ will be expressed in the form of a Slater determinant and, in the case of 2N electrons, written as

$$|\psi_e\rangle = \frac{1}{\sqrt{2N!}} \begin{vmatrix} |\psi_1(1)\alpha(1)\rangle & |\psi_1(1)\beta(1)\rangle & . & |\psi_N(N)\beta(1)\rangle \\ |\psi_1(2)\alpha(2)\rangle & |\psi_1(2)\beta(2)\rangle & . & |\psi_N(N)\beta(2)\rangle \\ . & . & . & . \\ |\psi_1(2N)\alpha(2N)\rangle & |\psi_1(2N)\beta(2N)\rangle & . & |\psi_N(2N)\beta(2N)\rangle \end{vmatrix}$$

This is a N×N determinant built from *spin-orbitals*. The ψ functions (which are in fact the ψ_e electronic functions) are called *molecular orbitals* (MO). The MO's are expressed as linear combinations of atomic orbitals (AO) whose coefficients are obtained from the energy minimization procedure.

— *Linear combination of atomic orbitals*

The linear combination of atomic orbitals (LCAO) suggests building molecular orbitals as linear combinations of atomic orbitals, where the jth state vector ($1 \le j \le N$) describing the ith electron ($1 \le i \le 2N$) can be written as

$$|\psi_j(i)\rangle = c_{sj} |\psi_s(i)\rangle \tag{3.3.20}$$

where the kets $|\psi_s(i)\rangle$ stand for atomic orbitals and c_{sj} are the coefficients to be determined. The number of coefficients to be determined can vary and it depends on the size of the chosen AO basis.

The advantage of this approach, which is outlined in Problem III.4.2, is that it can take into account the possible symmetry of the atoms in space by establishing a few relations between the coefficients c_{sj}. However, it also raises the fearsome problem of the most suitable choice for the atomic orbital basis. Intense research has been devoted to this question which will not be discussed here. Let us say however that to account for the electronic state of the molecule correctly, an extensive basis was needed. Fortunately, significant simplifications are often possible thanks to symmetry considerations.

— *Simplifications due to symmetry*

The full characterization of the electronic structure of a molecule requires the knowledge of the energy levels and of the state vectors which are attached to it. The state vectors are obtained from the resolution of the eigenvalue equation of the Hamiltonian, resolution which may sometimes prove to be quite cumbersome. However, we presented a few methods destined to restrict these difficulties. For example in the case of the atom, there is the perturbation approach which relies on establishing a hierarchy between the leading terms of the Hamiltonian, or also, there is the search for operators simpler than the Hamiltonian which commute with it so that they have a common set of eigenfunctions (see, for instance, how the electronic repulsions are taken into account in the many-electron atom problem through the Russel-Saunders coupling). All these simplifications, although they do not provide the full solution to the problem, offer a way to gather and sort its solutions, as in the problem of the many-electron atom whose state vectors were collected and labeled starting from the eigenvalues of operators $\boldsymbol{L}$ and $\boldsymbol{S}$ which commute with H.

Of course, such an approach also holds for the study of molecules. However, in this case, a new difficulty emerges, which was absent from the study of the isolated atom: How can we take account of the positions of the nuclei in order to find the electronic energy levels? The Born-Oppenheimer approximation tells us that the nuclei can be considered as motionless. Nevertheless, they will constrain the electronic state vectors and it is precisely in the determination of these constraints that symmetry arguments may prove to be quite powerful. Indeed, if the molecule possesses symmetry elements, these will necessarily appear in the electronic state vector whose square modulus stands for the volume density of the probability of presence of the electrons within the molecule.

Symmetry, as a tool to simplify the resolution of the eigenvalue equation, has not been used for the case of atoms or for free ions because of the spherical symmetry of the problem. However, in the case of molecules, it appears naturally, as a complement to the previously outlined methods. Two basic features characterize the molecular electronic structure:

(i) The Hamiltonian operator is *invariant* with respect to the set of symmetry operations R of the symmetry group of the molecule. As a result, H commutes with every operator O_{R} of the group

$$[H, O_{\mathrm{R}}] = 0 \tag{3.3.21}$$

(ii) Molecular orbitals $|\psi\rangle$ constitute *bases* for the irreducible representations of the group. This property results from the fact that the commutator $[H, O_R]$ vanishes. Indeed, if we multiply each member of the eigenvalue equation H on the left hand side by O_R we obtain

$$O_R H|\psi\rangle = \mathrm{E}O_R|\psi\rangle$$

and, according to Equation 3.3.21, we have

$$HO_R|\psi\rangle = \mathrm{E}O_R|\psi\rangle$$

equation showing that $O_R|\psi\rangle$ is an eigenvector of H, and thus it is proportional to $|\psi\rangle$. Moreover, since $|\psi\rangle$ is normalized and O_R is a unitary operator, $O_R|\psi\rangle$ is normalized. Consequently, we have

$$O_R|\psi\rangle = \pm|\psi\rangle$$

and thus $|\psi\rangle$ is a basis for an irreducible representation of order 1 of the symmetry group of the molecule. This fundamental result can be generalized to the case of degenerate energy levels. In this case, the degeneracy of E is equal to the order of the group representation of eigenstates of H which form a basis.

In the following, we shall consider only the operations acting exclusively on the position coordinates of the atoms (symmetry group of the molecule). It is natural to put together the energy levels of a molecule according to the irreducible representations whose state vectors associated to these levels form bases. Thus we will label an energy level as $^{2S+1}\Gamma$, where Γ stands for the Schoenflies notation of an irreducible representation of the symmetry group and $(2S + 1)$ is the spin multiplicity.

Of course, the energy levels of molecules change as soon as additional effects are considered. For instance, when the spin-orbit coupling is taken into consideration, the degeneracy of the spectroscopic terms is partly removed because the new state vectors of the extended Hamiltonian form a basis for another representation of the group.

—*Minimization of the average energy using Hückel's method (1930)*

This is a straightforward method based on an analytic type of minimization, which usually involves calculations which are not too tedious. It applies mainly to aromatic molecules and to their systems, consisting of 2N π electrons.

The valence electrons of a conjugated molecule are classified as σ type and π type electrons. σ type electrons are localized on the interatomic bond axis. On the contrary, π type electrons are, by essence, delocalized on the whole molecular lattice and play an important role in chemical reactions. This point leads us to distinguish the role of π type electrons from that of σ type electrons. More specifically, the energy of σ electrons is frequently assumed to be constant and energy minimization is only performed on π electrons. This distinction between π and σ electrons provides a way to split the electronic Hamiltonian into two Hamiltonians, each of them acting on one type of electrons

$$H_e = H_\sigma + H_\pi$$

Hückel suggests to expand the Hamiltonian concerning the π electrons as

$$H_\pi = \sum_{i=1}^{2N} H^{eff}(i)$$

where $H^{eff}(i)$ is the monoelectronic Hamiltonian of electron i, whose explicit form is not known. Choosing the molecular orbitals as linear combinations of atomic orbitals as

$$|\psi_i\rangle = c_{ri}|\chi_r\rangle$$

where $|\chi_r\rangle$ are atomic orbitals with π symmetry (for practical reasons, we shall use a minimal basis consisting of certain p orbitals of a plane conjugated molecule), the problem is solved as soon as the coefficients c_{ri} are known. Let us assume we are interested in a conjugated system formed by M carbon atoms. Every molecular orbital has M unknown coefficients c_{ri}. Substituting the expression of $|\psi\rangle$ in the relation for the average value <E> (Equation 3.3.18), we can express the minimum of <E> analytically starting from the derivative of <E> with respect to the M adjustable coefficients c_{ri}. We then end up with M linear equations of unknowns c_{ri} and <E>, written as

$$\sum_{s=1}^{M}\left[\left(H_{rs}^{eff} - S_{rs} < E >\right)c_{si}\right] = 0$$

with r = 1, 2, ..., M ; $H_{ij}^{eff} = \langle\chi_i|H|\chi_j\rangle$ and $S_{ij} = \langle\chi_i|\chi_j\rangle$.

This homogeneous system of linear equations has non zero solutions if

$$\left|H_{rs}^{eff} - S_{rs} < E >\right| = 0$$

In order to carry out the calculations, Hückel suggests the following simplifications:

$$
\begin{aligned}
H_{ii}^{eff} &= \langle \chi_i | H^{eff} | \chi_i \rangle = \alpha \quad (\alpha < 0) \\
H_{ij}^{eff} &= \langle \chi_i | H^{eff} | \chi_j \rangle = \beta \quad (\beta < 0) \ \text{for adjacent atoms} \\
H_{ij}^{eff} &= \langle \chi_i | H^{eff} | \chi_j \rangle = 0 \ \text{for non adjacent atoms} \\
S_{ij} &= \delta_{ij}
\end{aligned}
\qquad (3.3.22)
$$

Within the frame of these hypotheses, the system becomes much simpler and now depends only on two parameters: α (Coulomb integral) and β (resonance or bonding integral). Let us briefly outline the results when applied to ethylene and benzene.

(i) Ethylene

We want to find the electronic energy levels of the ethylene molecule, their number and their degeneracy. Using the scheme shown in Figure 3.3.7, the basis we used consists of the two AO atomic orbitals ($\varphi_1 = 2p_{1y}$ and $\varphi_2 = 2p_{2y}$), centered on each of the two carbon atoms, and the electronic state vectors are sought in the form of linear combinations $\psi = c_1\varphi_1 + c_2\varphi_2$.

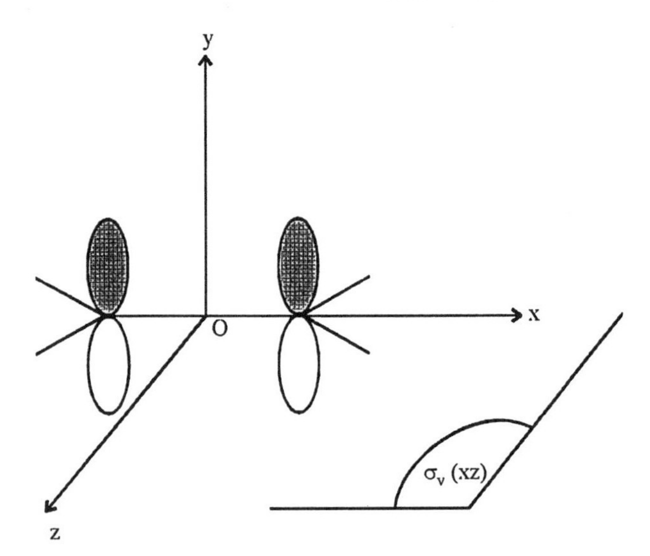

Figure 3.3.7: Geometrical representation of ethylene and of the two p-type orbitals centered on each of the carbon atoms.

Hückel's system of equations which provides coefficients c_1 and c_2 takes the form (in order to simplify the notation, we note E the average energy <E>)

$$c_1 (\alpha - E) + c_2 \beta = 0$$
$$c_1 \beta + c_2 (\alpha - E) = 0$$

Setting $k = (\alpha - E)/\beta$, we obtain

$$c_1 k + c_2 = 0$$
$$c_1 + c_2 k = 0$$

which is a homogeneous system which has non trivial solutions only if its determinant is zero

$$\begin{vmatrix} k & 1 \\ 1 & k \end{vmatrix} = 0$$

This particular case is quite straightforward, as the values of k can be found immediately, yielding for the values of the energy $E = \alpha \pm \beta$. The electronic state vectors are obtained by solving the system of equations for c_1 and c_2 after inserting the values of k. Finally, the energy levels $E_1 = \alpha + \beta$ and $E_2 = \alpha - \beta$ are, respectively, associated to state vectors $\psi_1 = \left(\sqrt{2}/2\right)\left(\varphi_1 + \varphi_2\right)$ and $\psi_2 = \left(\sqrt{2}/2\right)\left(\varphi_1 - \varphi_2\right)$.

We shall now re-examine the same problem using symmetry arguments. Figure 3.3.7 shows that the molecule belongs to the space group D_{2h} and that the functions (φ_1, φ_2) form the basis of a Γ representation of this group whose characters are given in Table 3.3.3.

Table 3.3.3 — **Characters of the representation Γ of group D_{2h}.**

D_{2h}	E	$C_2(z)$	$C_2(y)$	$C_2(x)$	i	$\sigma_v(xy)$	$\sigma_v(xz)$	$\sigma_v(yz)$
Γ	2	0	0	−2	0	2	−2	0

Representation Γ is reducible. Using Equation 3.2.4 (or by taking a look at the character table [*cf.* Appendix III.5.4]) we find the decomposition $\Gamma = B_{1g} \oplus B_{2u}$ which shows that all the electronic state vectors of ethylene fit into two groups of non degenerate state vectors (one dimensional representations). Thus, group theory provides a way to predict the *splitting* of the energy levels and the *symmetry* of the associated state vectors. For instance, energy level B_{1g} has an associated state vector which is symmetrical both with respect to C_2 (subscript 1) and with respect to the inversion (subscript g).

At this stage of the development, we know neither the *relative position* of the energy levels nor the expression of the associated state vectors. However, since the representations involved are one dimensional, character tables can provide useful information. Thus, when symmetry operation $\sigma_v(xz)$ with a character of one, acts on the vector ψ associated to representation B_{2u} , we must necessarily obtain $[\sigma_v(xz)]\psi = 1\times\psi$. Moreover, $[\sigma_v(xz)]\varphi_1 = \varphi_2$ and $[\sigma_v(xz)]\varphi_2 = \varphi_1$, and the equality $[\sigma_v(xz)]\psi = 1\times\psi$, applied to $\psi = c_1\varphi_1 + c_2\varphi_2$, implies that $c_1 = c_2$. Last, we obtain $k = -1$, confirming our first result (E_1, ψ_1). The same reasoning applied to representation B_{1g} would result in finding (E_2, ψ_2). The fundamental role of symmetry arguments is thus to simplify the secular determinant by replacing it by a determinant of lesser degree which is easier to manipulate. Of course, the example we considered here was quite special, in the sense that the representations for which we had to find bases were all of dimension 1. In this case, characters have a particular meaning since they directly tell the way the single basis function transforms. The matrices associated to geometric transformations are single numbers equal to the character of the representations.

For representations with dimensions greater than 1, simplifications can be obtained for the expression of the state vectors by using projection operators. By essence, the projection operator P_Γ applied to the elements of the basis generates a function which has the symmetry of the irreducible representation Γ. In the case of ethylene, the projection operator associated to representation B_{2u} takes the form

$$P_{B_{2u}} = 1\times E + (-1)\times C_2(z) + 1\times C_2(y) + (-1)C_2(x) + (-1)i$$
$$+1\times\sigma(xy) + (-1)\sigma(xz) + 1\times\sigma(yz)$$

Applying operator $P_{B_{2u}}$ to vector φ_1, we obtain

$$[P_{B_{2u}}]\,\varphi_1 = 1\times\varphi_1 + (-1)\times(-\varphi_2) + 1\times(\varphi_2) + (-1)(-\varphi_1) + (-1)(-\varphi_2)$$
$$+1\times(\varphi_1) + (-1)(-\varphi_1) + 1\times(\varphi_2)$$
$$[P_{B_{2u}}]\,\varphi_1 = 4(\varphi_1 + \varphi_2)$$

and we fall back on ψ_1 except for the normalization coefficient. Applying $P_{B_{2u}}$ to vector φ_2 would of course give a similar result, as B_{2u} is of dimension 1. In the same way, a vector proportional to ψ_2 would be obtained by applying operator $P_{B_{2u}}$ to either φ_1 or φ_2. The state vectors being known, the corresponding energies can be easily obtained and it is possible to sketch the energy diagram of ethylene (*cf.* Figure 3.3.8).

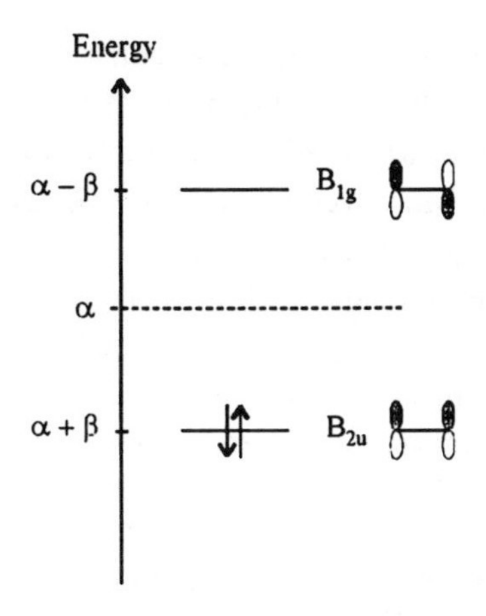

Figure 3.3.8: Schematic representation of the energy levels and of the molecular orbitals of ethylene in the ground state.

This — very fruitful — way to take advantage of symmetry properties to find the eigenvectors of the Hamiltonian, is used in Problem III.4.3 and in numerous problems of Chapter IV.

(ii) Benzene

Adopting the representation sketched in Figure 3.3.9, the basis we consider is formed by the six atomic orbitals $\varphi_i = 2p_{iy}$ centered on each of the six carbon atoms (labeled 1 to 6) and the electronic state vectors are sought in the form of linear combinations $\psi = c_i\varphi_i$.

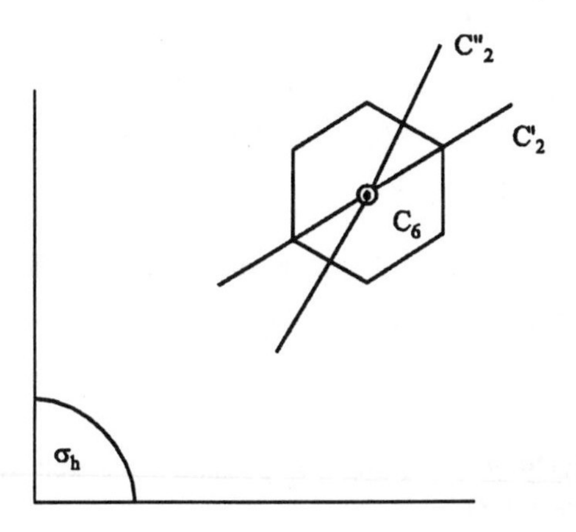

Figure 3.3.9: Schematic representation of the benzene molecule.

The secular determinant, and the associated eigenvalue equation, relative to benzene takes the form

$$\begin{vmatrix} \alpha - E & \beta & 0 & 0 & 0 & \beta \\ \beta & \alpha - E & \beta & 0 & 0 & 0 \\ 0 & \beta & \alpha - E & \beta & 0 & 0 \\ 0 & 0 & \beta & \alpha - E & \beta & 0 \\ 0 & 0 & 0 & \beta & \alpha - E & \beta \\ \beta & 0 & 0 & 0 & \beta & \alpha - E \end{vmatrix} = 0$$

and the average energy appears as a solution of an sixth degree equation whose analytic resolution is difficult. In this case also, the symmetry properties of the molecule turn out to be quite precious. The point group of benzene is D_{6h} and the six orbitals φ_i constitute the basis of a representation Γ of this group, whose characters are given in Table 3.3.4.

Table 3.3.4 — **Characters of the representation Γ of group D_{6h}.**

D_{6h}	E	$2C_6$	$2C_3$	C_2	$3C'_2$	$3C''_2$	i	$2S_3$	$2S_6$	σ_h	$3\sigma_d$	$3\sigma_v$
Γ	6	0	0	0	−2	0	0	0	0	−6	0	2

Representation Γ is reducible. Using Equation 3.2.4, Γ reduces according to $\Gamma = A_{2u} \oplus B_{2g} \oplus E_{1g} \oplus E_{2u}$ meaning that the electronic state vectors of benzene are divided in four groups of vectors, two of these being doubly degenerate (E_{1g}, E_{2u}). The use of the projection operators provides a way to express the bases of each of the irreducible representations. The basis of representation Γ is formed by six vectors and with the character of each representation the corresponding projection operator can be defined. Two cases may be encountered, depending on whether the representation we investigate is of dimension 1 or not. Indeed, for a 1 dimensional representation, applying the projection operator to any of the six basis functions of representation Γ will yield the basis function we are looking for. In the case of a representation of higher dimension, care will have to be taken to build the basis with orthogonal vectors. For example, building a basis for representation E_{1g} requires two functions. The first one, ψ_1, can be obtained by applying the projection operator to any of the six φ_i functions. The second one, which must be orthogonal to the first one, can be obtained by applying the projection operator to another of the functions φ_i. If the generated function ψ is neither equal to $\pm \psi_1$, nor orthogonal to ψ_1, then we must look for a linear combination of functions ψ_1 and ψ_2. Finally, the six electronic state vectors, orthogonal two by two, are

$$
\begin{aligned}
\psi(A_{2u}) &= (1/\sqrt{6})(\varphi_1 + \varphi_2 + \varphi_3 + \varphi_4 + \varphi_5 + \varphi_6) \\
\psi(B_{2g}) &= (1/\sqrt{6})(\varphi_1 - \varphi_2 + \varphi_3 - \varphi_4 + \varphi_5 - \varphi_6) \\
\psi_1(E_{1g}) &= (1/\sqrt{6})(2\varphi_1 + \varphi_2 - \varphi_3 - 2\varphi_4 - \varphi_5 + \varphi_6) \\
\psi_1(E_{2u}) &= (1/\sqrt{12})(2\varphi_1 - \varphi_2 - \varphi_3 + 2\varphi_4 - \varphi_5 - \varphi_6) \\
\psi_2(E_{1g}) &= (1/2)(\varphi_1 - \varphi_3 - \varphi_4 + \varphi_6) \\
\psi_2(E_{2u}) &= (1/2)(\varphi_1 - \varphi_3 + \varphi_4 - \varphi_6)
\end{aligned}
$$

The corresponding energy levels are calculated from the secular determinant expressed on this basis

$$
\begin{vmatrix}
\alpha + 2\beta - E & 0 & 0 & 0 & 0 & 0 \\
0 & \alpha - 2\beta - E & 0 & 0 & 0 & 0 \\
0 & 0 & \alpha + \beta - E & 0 & 0 & 0 \\
0 & 0 & 0 & \alpha + \beta - E & 0 & 0 \\
0 & 0 & 0 & 0 & \alpha - \beta - E & 0 \\
0 & 0 & 0 & 0 & 0 & \alpha - \beta - E
\end{vmatrix} = 0
$$

The simplification we obtained is obvious. In this basis, the determinant is equal to the product of the diagonal terms, and the values of the energies can be obtained immediately. We next derived the complete energy scheme of benzene as sketched in Figure 3.3.10.

— *Minimization of the average energy using the self-consistent field method of Hartree and Fock (HF – LCAO method, 1930)*

The Hückel method, though interesting because of its simplicity, remains of limited use. More generally, molecular electronic structures can be investigated using the self-consistent field method, developed by Hartree in 1927 for many-electron atoms (*cf.* Section III.3.2.1), and extended to molecules by Fock in 1930. As for atoms, this method results in an energy *shell* model and in energy values which are in very good agreement with the experimental ones obtained by *photoelectron spectroscopy.* This is a considerable advantage. This spectroscopy is described in Chapter VI, where this point will be discussed in detail.

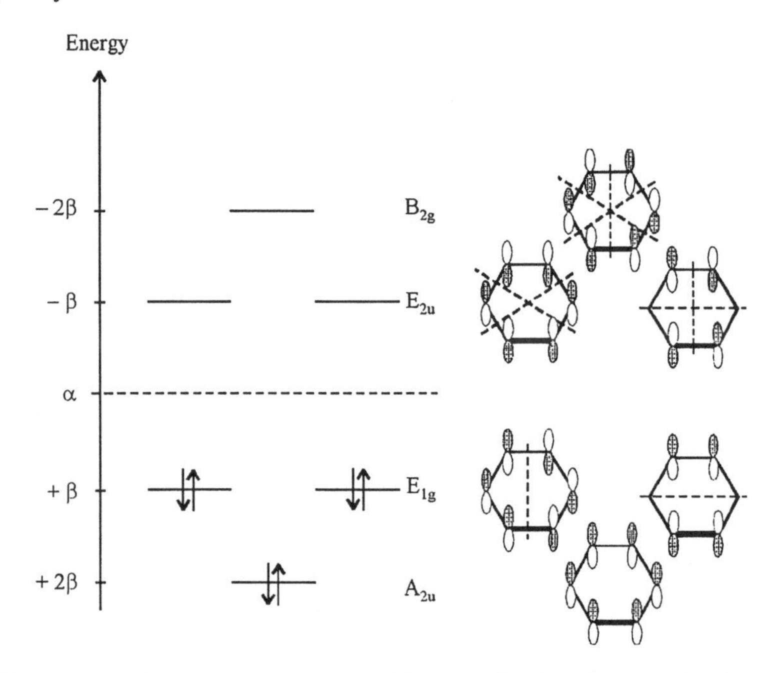

Figure 3.3.10: Schematic representation of the energy levels and of the molecular orbitals of benzene (arrows stand for the electrons in the ground state).

— *Beyond the self-consistent method: Electronic correlation and configuration interaction*

The methods described above does not consider the possible correlation between electrons with opposite spins (*cf.* Bibliographic reference 1.2.2). We shall illustrate this feature on a bielectronic microsystem, an example that will be revisited in Problem III.4.5.

The Hamiltonian describing a two electrons atomic or molecular system (for example helium) takes the form

$$H = H_1^o + H_2^o + H_{12}$$

where H_i^0 is an hydrogen like Hamiltonian and H_{12} is an operator describing the electronic repulsion, written as

$$H_{12} = (e^2/4\pi\varepsilon_0)\,(1/r_{12}) \qquad r_{12} = |\mathbf{r}_1 - \mathbf{r}_2|$$

Let us assume that the electrons of the system are described by two distinct state vectors, φ_a and φ_b for example. It is therefore in an excited state. The complete state vector must be expressed as the product of the total spin function by the space function, result which is typical of a two electron system, since the Slater determinant cannot in general be written as the product " space × spin ". Postulate P9 requires the total state vector to be antisymmetrical for the exchange of two electrons. Consequently, the space and spin functions must be symmetrical (respectively antisymmetrical) and antisymmetrical (respectively symmetrical). The four possible state vectors describing the electrons, antisymmetrical with respect to exchange, are thus expressed as

$$\psi_1 = (1/\sqrt{2})[\varphi_a(\mathbf{r}_1)\varphi_b(\mathbf{r}_2) - \varphi_a(\mathbf{r}_2)\varphi_b(\mathbf{r}_1)] \times \alpha(1)\alpha(2)$$

$$\psi_2 = (1/\sqrt{2})[\varphi_a(\mathbf{r}_1)\varphi_b(\mathbf{r}_2) - \varphi_a(\mathbf{r}_2)\varphi_b(\mathbf{r}_1)] \times 1/\sqrt{2}\,[\alpha(1)\beta(2) + \alpha(2)\beta(1)]$$

$$\psi_3 = (1/\sqrt{2})[\varphi_a(\mathbf{r}_1)\varphi_b(\mathbf{r}_2) - \varphi_a(\mathbf{r}_2)\varphi_b(\mathbf{r}_1)] \times \beta(1)\beta(2)$$

$$\psi_4 = (1/\sqrt{2})[\varphi_a(\mathbf{r}_1)\varphi_b(\mathbf{r}_2) + \varphi_a(\mathbf{r}_2)\varphi_b(\mathbf{r}_1)] \times 1/\sqrt{2}\,[\alpha(1)\beta(2) - \alpha(2)\beta(1)]$$

The total spin of the system is characterized by two quantum numbers S and M_S. The values of M_S, for all the state vectors, are obtained from the sum of each of the possible values of the numbers m_{s1} and m_{s2} associated to the two electrons. Therefore we have $M_S = 1, 0, -1$ and 0 for vectors ψ_1, ψ_2, ψ_3 and ψ_4. Due to the symmetry of their spin part, vectors ψ_1, ψ_2 and ψ_3 are of course the three components of the triplet (S = 1), and ψ_4 corresponds to the singlet state (S = 0).

The average energy is obtained from the Hamiltonian defined above, through

$$\langle \psi | H | \psi \rangle = E_1^0 + E_2^0 + H_{12}$$

with

$$H_{12} = (1/2) \times \langle \varphi_a(\mathbf{r}_1)\varphi_b(\mathbf{r}_2) \pm \varphi_a(\mathbf{r}_2)\varphi_b(\mathbf{r}_1) | e^2/4\pi\varepsilon_0 r_{12} | \varphi_a(\mathbf{r}_1)\varphi_b(\mathbf{r}_2) \pm \varphi_a(\mathbf{r}_2)\varphi_b(\mathbf{r}_1) \rangle$$

H_{12} can be written as

$$H_{12} = J \pm K$$

with

$$J = \langle \varphi_a(\mathbf{r}_1)\varphi_b(\mathbf{r}_2) | e^2/4\pi\varepsilon_0 r_{12} | \varphi_a(\mathbf{r}_1)\varphi_b(\mathbf{r}_2) \rangle = \langle \varphi_a(\mathbf{r}_2)\varphi_b(\mathbf{r}_1) | e^2/4\pi\varepsilon_0 r_{12} | \varphi_a(\mathbf{r}_2)\varphi_b(\mathbf{r}_1) \rangle$$

$$K = \langle \varphi_a(\mathbf{r}_1)\varphi_b(\mathbf{r}_2) | e^2/4\pi\varepsilon_0 r_{12} | \varphi_a(\mathbf{r}_2)\varphi_b(\mathbf{r}_1) \rangle = \langle \varphi_a(\mathbf{r}_2)\varphi_b(\mathbf{r}_1) | e^2/4\pi\varepsilon_0 r_{12} | \varphi_a(\mathbf{r}_1)\varphi_b(\mathbf{r}_2) \rangle$$

where J (Coulomb integral) and K (exchange integral) are two positive quantities. Last, by considering electrostatic interactions, the singlet state proves to have an energy greater than that of the triplet state. These energy states are sketched on Figure 3.3.11.

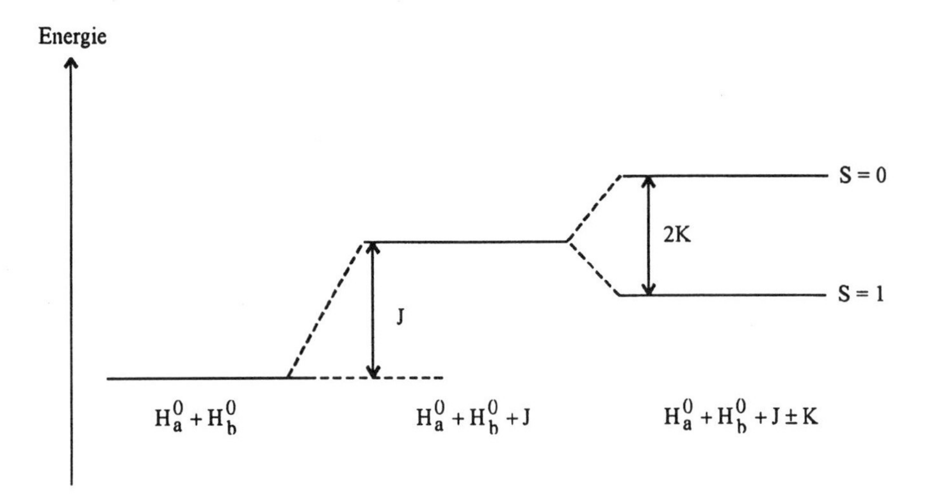

Figure 3.3.11: Energy levels of an excited bielectronic system.

More precisely, we notice that the single Coulomb integral would account for electrostatic repulsion if the antisymmetrisation principle of the wave function did not hold (in this case, one might consider the state vector $\varphi_a[\mathbf{r}_1]\varphi_b[\mathbf{r}_2]$). In fact, K originates in the antisymmetrization principle and describes the spin correlation which exists between the two electrons. This correlation is clearly revealed on this example. The space vector of the triplet state is antisymmetrical and consequently the probability to find electron 1 at $\mathbf{r}_1$ and electron 2 at $\mathbf{r}_2 = \mathbf{r}_1$ is zero. Such a case does not occur for the space function of the singlet state, which is symmetrical (by the way, we notice that this approach does not consider the correlation between electrons with opposed spins).

Thus, a correlation, due to spin, indeed exists between electrons and it is independent of the purely electrostatic interaction. This type of interaction is very difficult to take into consideration in calculations of the HF-LCAO type and it is the foundation of the so-called *configuration interactions* method, which we shall not develop in this book.

III.3.4. Vibrational and rotational structure of electronic levels in atoms and molecules

An additional complication, typical for molecules and absent in the case of atoms, originates in the fact that the electronic energy depends on the position of nuclei (rotations and vibration movements). This case has already been discussed within the frame of the Born-Oppeinheimer approximation. At a fixed temperature, the nuclei vibrate and so modify their relative distances, and therefore also the energy of the molecule. More generally, the energy of a molecule has several contributions: On the one hand, a kinetic energy of translation related to temperature and whose effect will appear macroscopically through pressure in the case of a fluid phase, and on the other hand, the characteristic energy E of the nucleons and electrons which make up the molecule. The energy E itself depends on an electronic term E_e, related to the Coulomb interactions undergone by the electrons, on a vibrational part E_v originating in the vibrations of the nuclei about their equilibrium position, and on a rotational contribution E_r arising from the rotational movements of the molecule about its center of mass. Assuming, as a first approximation, that all these movements are independent, the total energy takes the form

$$E = E_r + E_v + E_e$$

and, within this approximation, the associated state vector is the product $|\psi\rangle = |\varphi_r \varphi_v \varphi_e\rangle$ of the state vectors relative to each type of energy. All these kinds of energy are quantified and depend on the geometry of the molecule.

III.3.4.1. Vibrational structure

The simplest way to describe the vibrations of a diatomic molecule AB is to imagine that the two nuclei of mass M_A and M_B are connected together by a spring whose force constant is k. This system is characterized by its own vibration frequency ν given by

$$\nu = (1/2)\sqrt{k/\mu}$$

where μ is the reduced mass, defined as $\mu = M_A M_B/(M_A + M_B)$.

The Hamiltonian of such a system can be built from the correspondence principle P7 and from the classical expression of the potential energy of the harmonic oscillator $V(x) = (1/2)kx^2$, where x is the displacement of a fictitious particle of mass μ with respect to its equilibrium position. The resolution of the eigenvalue equation shows that the energy is quantified and that it depends on a quantum number v (the vibration quantum number) according to the expression

$$E_v = (v + 1/2)(h/2\pi)\sqrt{k/\mu} \quad \text{with } v = 0, 1, 2, \ldots$$

The potential energy curve for a real diatomic molecule (Figure 3.3.6), whose interatomic equilibrium distance is $R = R_e$, though it is quite similar to that of the harmonic oscillator for small values of the energy (parabolic), clearly differs from it for higher values of the energy. Thus, for values of R smaller than R_e, the energy increases sharply, expressing the difficulty of the nuclei to get indefinitely closer, whereas for large values of R, the energy tends asymptotically towards a limit value D_e, characteristic of the bond rupture. These variations of V are well reproduced by substituting the Morse function to the harmonic potential

$$V = D_e \{1 - \exp[-a(R - R_e)]\}^2$$

This anharmonicity results in a narrowing of the vibrational energy levels E_v until they finally form a continuum as v tends to infinity. The vibrational energy is written as

$$E_v = (v + 1/2)\, h\nu - (v + 1/2)^2 x_e\, h\nu$$

where x_e is the *anharmonicity coefficient*. This difference is sketched on Figure 3.3.12.

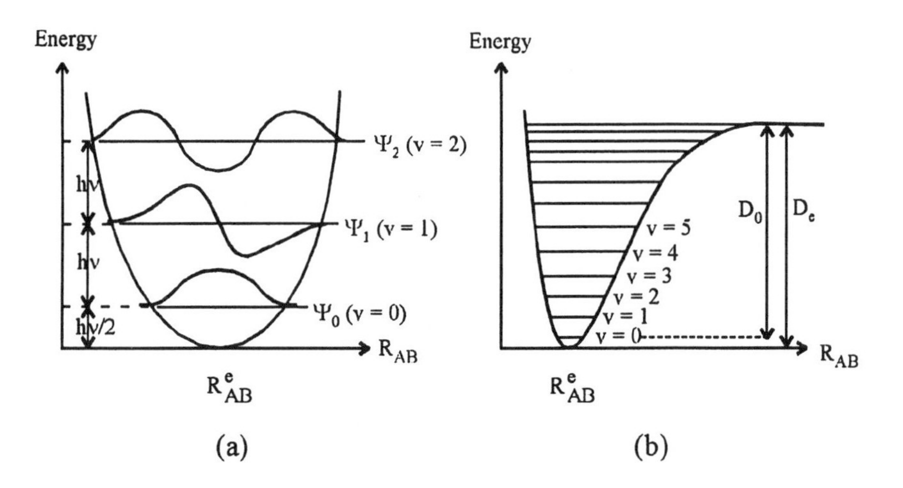

Figure 3.3.12: Overview of the potential energy curve of a diatomic molecule and of the vibrational energy levels. (a) Case harmonic approximation (ψ_0, ψ_1, ψ_2 standing for the vibrational wave functions for v = 0, 1 and 2, respectively) ; (b) Case of the anharmonic approximation for which the vibrational levels are no longer equidistant.

The eigenvectors associated to the harmonic oscillator can be expressed as the product of an exponential term by a Hermite polynomial. Some of them are given in Table 3.3.5.

For polyatomic molecules with N nuclei, counting and analyzing the vibration modes is more difficult. Since each of the N atoms has three Cartesian coordinates, without considering translation and rotation about the center of mass, there are $3N - 6$ possible vibration modes ($3N - 5$ for linear molecules).

Table 3.3.5 — **Analytical vibrational eigenfunctions of a diatomic oscillator (q stands for the deviation with respect to the equilibrium position, $\alpha = 2\pi / \sqrt{\mu k}$).**

v	ψ_v
0	$(\alpha/\pi)^{1/4} \exp(-\alpha q^2/2)$
1	$(4\alpha^3/\pi)^{1/4}\, q \exp(-\alpha q^2/2)$
2	$(\alpha/4\pi)^{1/4} (2\alpha q^2 - 1) \exp(-\alpha q^2/2)$
3	$(\alpha^3/9\pi)^{1/4} (2\alpha q^3 - 3q) \exp(-\alpha q^2/2)$

III.3.4.2. Rotational structure

Because the vibration velocity is much greater than the rotation velocity, we can assume, as a first approximation, that, averaged over a complete " vibration ", the molecules rotate while keeping their interatomic distances fixed. This is the same as to consider the molecules as a rigid rotators, characterized by a bond length equal to the equilibrium distance between the two nuclei.

The simplest model to describe rotation is that of a particle which is free to move over a sphere of constant radius r. In the case of a diatomic molecule, we can study the movement of a fictitious particle of reduced mass μ. The quantum treatment of this problem shows that the angular momentum σ is quantified and given by $\sigma = J(J+1)(h/2\pi)$, where J is the quantum number of rotation (similar to the secondary quantum number l quantifying the angular momentum of the electron). A diatomic molecule has a vanishing moment of inertia about the molecular axis and two equal and non zero moments of inertia I about the axis perpendicular to the axis of rotation. Its rotation energy takes the form

$$E_r = (\sigma^2/I) = (h^2/8\pi^2 I)\, J(J+1)$$

It is customary to define the *rotational constant* $B = h^2/8\pi^2 I$. The rigid rotator model leads to a series of quantified energy levels, separated from each other by 2B, 4B, 6B, ... Generally, scientists use the cm^{-1} as an energy unit, so that the rotational constant writes

$$\overline{B}(cm^{-1}) = B/hc = h/8\pi^2 Ic$$

The hypothesis of the rigid rotator may turn out to not to be appropriate when the rotation velocity is such that the molecule can be distorted because of the centrifugal force, thus causing an increase of its moment of inertia. This effect leads us to introduce a corrective term in the rotation energy, which then takes the form

$$E_r = \left(h^2 / 8\pi^2 I\right) J(J+1) - DJ^2(J+1)^2$$

where D is the *centrifugal distortion constant*, whose value is related to the force constant k of the oscillator. For example, for HCl, $\overline{B} = 10.35\ cm^{-1}$ and $D = 0.0004\ cm^{-1}$.

The state vectors associated to the energy levels are obtained by solving the corresponding eigenvalue equation. No radial equation exists in the case of the rigid rotator, as the distance r is fixed, and the eigenfunctions are the spherical harmonics which we already studied.

For polyatomic molecules with N nuclei, three different moments of inertia must be taken into consideration, and the quantification conditions of the energy levels becomes more difficult.

III.3.4.3. Vibration-rotation interaction

Before starting the study of the energy levels of molecules, we assumed the rotation and vibration movements to be independent. This assumption leads to a value of the vibration-rotation energy $E_{V\text{-}R}$ given by

$$E_{V\text{-}R} = h\nu\,(v + 1/2) + BJ(J+1)$$

However, the anharmonicity of the mechanical potential of the oscillator results in an increase of the average interatomic distance when the vibration quantum number increases, leading thus to a decrease of the rotational $\overline{B}$ (decrease exclusively due to the vibration, and independent of the centrifugal distortion). If $\overline{B}_e$ stands for the rotational constant corresponding to the minimum of the potential, $\overline{B}_e$ is expressed by

$$\overline{B} = \overline{B}_v = \overline{B}_e - \alpha(v + 1/2)$$

α being a positive number. The vibration-rotation energy becomes

$$E_{V\text{-}R}/hc = \bar{\nu}(v+1/2) - \bar{\nu}x_e(v+1/2)^2 + \bar{B}_e J(J+1) - \alpha(v+1/2)J(J+1)$$

The last term of the above expression depends on the product of the vibration and rotation quantum numbers and constitutes the *vibration-rotation interaction term*. Of course, if in the energy we include the energy due to the vibration-rotation coupling, the associated state vectors can no longer be written as $|\psi\rangle = |\psi_r\psi_v\rangle$. Moreover, the centrifugal distortion effect can also be included in the energy by adding the term $-(D/hc)J^2(J+1)^2$ to the expression of the rotation energy.

III.3.5. Conclusions and consequences about electronic levels

The complete structure of the energy levels of a molecule is sketched on Figure 3.3.13.

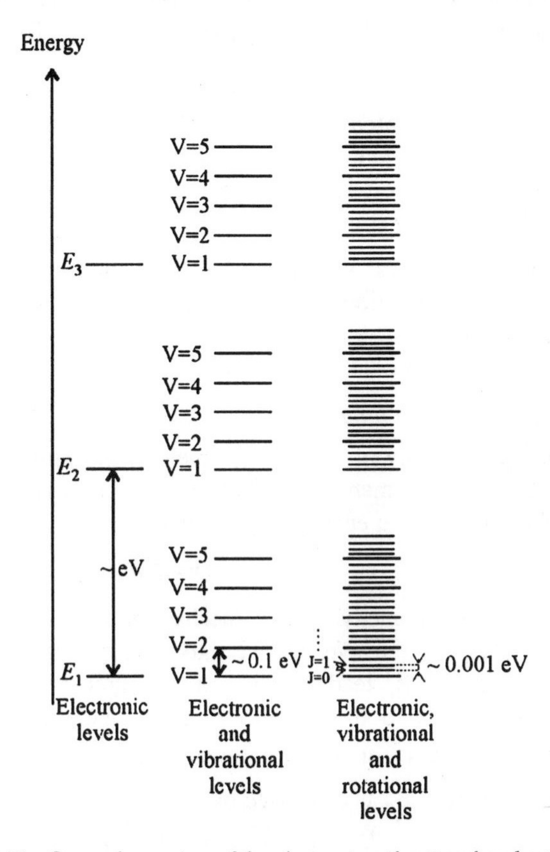

Figure 3.3.13: General overview of the electronic, vibrational and rotational levels of a molecule.

According to the values of the energy involved, it is clear that splitting the energy in purely electronic, vibrational and rotational parts is only an approximation since the values of the energy levels are not rigorously independent. Consequently, writing the wavefunction as a product related to exclusively electronic, vibrational and rotational contributions is also an approximation. We already discussed the physical reasons for the coupling between vibration and rotation. Similarly, coupling can exist between the electronic and the vibrational energies: This is the *vibronic coupling*. Taking into account all the types of couplings between the various kinds of energy, though it may turn out to be very difficult, is sometimes necessary for the interpretation of specific spectroscopic data.

III.4. Exercises and problems

III.4.1. Questions about symmetry: True or false ?

1. Diatomic homonuclear molecules belong to the $C_{\infty v}$ symmetry group.
2. A non polar molecule always has an inversion center.
3. A molecule with an improper axis of symmetry is not chiral.

1: F 2: F 3: T

III.4.2. General aspects of LCAO theory (from Agreg. C; C; 1982)

We want to find the energy levels of the electrons of a molecule, using the LCAO theory in its simplified form. To do so, we will make two simplifying assumptions.

— The first one deals with the structure of the wavefunction ψ (1, 2, ...) describing the behavior of all the electrons. We assume that $\psi(1, 2, ...) = \varphi_1(1)\,\varphi_2(2) ...$, the molecular wavefunctions φ_i are monoelectronic and describe the behavior of a single electron. According to the LCAO method, every molecular orbital φ_i may be written as $\varphi_i = \sum_r c_{ir}\,\chi_r$, the χ_r being atomic orbitals.

— The second one deals with the Hamiltonian H (1, 2, ...) of the system, which can be written as $H(1,2,...) = \sum_i h(i)$, every partial Hamiltonian h (i) acting on a single electron.

1. General questions:

a. Give the expression of the energy E of all electrons together and relate it to the energies e_i of the molecular orbitals φ_i.

b. Give the expression of e_i considering the atomic orbitals χ. For the sake of simplicity, we set

$$h_{rs} = \langle \chi_r | h | \chi_s \rangle \text{ and } S_{rs} = \langle \chi_r | \chi_s \rangle$$

c. Applying the variational principle, write the system of equations which give the c_{ir} coefficients.

d. Deduce the secular equation giving the values of e_i. We shall simplify the use of the secular equation by making additional hypotheses.

e. The overlap integral S_{rs} takes on very simple values when assuming that the atomic orbitals χ are orthogonal. Specify these values and indicate whether the orbitals considered are identical or different.

f. We set $h_{rr} = \alpha_r$, an integral we name the Coulomb integral. Explain this name. On the other hand, $h_{rs} = \beta_{rs}$ is different of zero only if the considered atomic orbitals (χ_r et χ_s) concern chemically bonded atoms. Justify this approximation too. β_{rs} is named bonding or exchange integral. Explain this name.

2. Example: Study of formol

We want to find the π electron energy levels. For the C and O atoms, we set

$$\alpha_C = \alpha + 0.2\beta \ ; \quad \alpha_O = \alpha + 0.7\beta \ ; \quad \beta_{CO} = 1.1\beta$$

(α and β are Coulomb and exchange integrals).

a. Write the secular equation and solve it.

b. Write the molecular orbitals φ_i.

c. Give the expression of the wavefunction in the ground state.

1.a. Assuming the wavevector $|\psi\rangle$ to be normalized, then $E = \langle \psi | H | \psi \rangle$ (postulate P4). Moreover, remembering that the operators h are monoelectronic, the energy takes the form of a sum $E = \sum_i e_i$ with $e_i = \langle \varphi_i | h(i) | \varphi_i \rangle$.

b. Thus $e_i = \left\langle \sum_r c_{ir}\, \chi_r \,|\, h_i \,|\, \sum_r c_{ir}\, \chi_r \right\rangle$ and, taking into account the expressions of h_{rs} and of S_{rs}, we have $e_i = \dfrac{\langle \varphi_i | h_i | \varphi_i \rangle}{\langle \varphi_i | \varphi_i \rangle} = \dfrac{\sum_r \sum_s c_{ir}\, c_{is}\, h_{rs}}{\sum_r \sum_s c_{ir}\, c_{is}\, S_{rs}}$.

c. Minimizing the quantum average of the energy with respect to c_{is}, we obtain a system of N linear equations (here N is the number of AO's taking part in the spin-orbital φ_i) with (N + 1) unknowns, which writes $\sum_r c_{ir}\,(h_{rs} - e_i S_{rs}) = 0$.

d. This system has a non trivial solution if, and only if

$$|h_{rs} - eS_{rs}| = 0$$

e. This approximation consists in writing $S_{rs} = \delta_{rs}$.

f. The reader should refer to Section III.3.3.2.

2.a. The secular equation can be written as $\begin{vmatrix} \alpha_C - E & \beta_{CO} \\ \beta_{CO} & \alpha_O - E \end{vmatrix} = 0$. E stands for the quantum average energy <E>. The solutions of this equation are

$$E_1 = \alpha + 1.57\,\beta \quad \text{and} \quad E_2 = \alpha - 0.67\,\beta.$$

b. The MO's can be written as $\varphi_1 = 0.62\ \chi_C + 0.78\ \chi_O$ and $\varphi_2 = 0.85\chi_C - 0.53\ \chi_O$.

c. The wavefunction for the ground state reads $\psi\,(1,2) = \varphi_1(1)\,\varphi_1(2)$.

III.4.3. Study of the orbitals of 1-3 butadiene

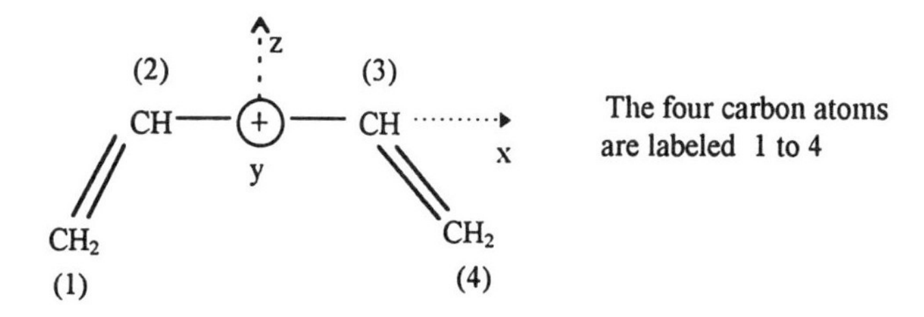

Quantum chemistry calculations provide the following results:

MO energies

E_4

E_3

E_2

E_1

OM

$\varphi_4 = 0.37\,(p_1 - p_4) - 0.60\,(p_2 - p_3)$

$\varphi_3 = 0.60\,(p_1 + p_4) - 0.37\,(p_2 + p_3)$

$\varphi_2 = 0.60\,(p_1 - p_4) + 0.37\,(p_2 - p_3)$

$\varphi_1 = 0.37\,(p_1 + p_4) + 0.60\,(p_2 + p_3)$

1. In the Hückel approximation, calculate the values of the energies E_1, E_2, E_3 and E_4 and justify the expressions suggested for the MO's φ_1, φ_2, φ_3 et φ_4 (p stands for the p_y orbitals of the carbon atoms).

2. Show that symmetry considerations provide similar results.

1. In the Hückel approximation, we neglect the wavefunction overlap and set S = 0. Taking $x = (\alpha - E) / \beta$ (and, for the sake of simplicity, $< E > = E$), and choosing the 4 dimensional basis formed by the four AO's p_y of the four carbon atoms, we obtain, by minimization of the average energy (from a quantum point of view), a system of four equations with five unknowns which has non trivial solutions only if E is solution of the secular equation

$$\begin{vmatrix} x & 1 & 0 & 0 \\ 1 & x & 1 & 0 \\ 0 & 1 & x & 1 \\ 0 & 0 & 1 & x \end{vmatrix} = 0$$

This equation has double roots x = ± 0.62 and x = ± 1.62. Thus, the four values for the energy are $E_1 = \alpha + 1.62\beta$; $E_2 = \alpha + 0.62\beta$; $E_3 = \alpha - 0.62\beta$; $E_4 = \alpha - 1.62\beta$.

Inserting these four values successively in the above system, we find four series of coefficient corresponding to the following four MO's, classified by increasing value of energy

$\varphi_1 \propto (p_1 + p_4) + 1.62\,(p_2 + p_3)$ $\varphi_2 \propto (p_1 - p_4) + 0.62\,(p_2 - p_3)$

$\varphi_3 \propto (p_1 + p_4) - 0.62\,(p_2 + p_3)$ $\varphi_4 \propto (p_1 - p_4) - 1.62\,(p_2 - p_3)$

2. In order to indicate the contribution of the group theory, we easily find that 1-3 cis-butadiene belongs to the C_{2v} point group. From the character table of this group (*cf.* Section III.2.4), and using the procedure previously discussed (*cf.* Section III.2.3), we find the reducible representation 4, 0, – 4, 0. It is reduced in $2A_2 + 2B_2$.

a. MO of A_2 symmetry:

We use a practical and very convenient procedure, directly inspired from the basic concepts of group theory, detailed on this example.

For the irreducible representation A_2 of dimension 1, the trace (character) of operator σ'_v (yOz plane, normal to the molecular plane) is – 1.

$\sigma'_v |p_1\rangle = |p_4\rangle$ $c_1 = -c_4 = c$ and thus $c\,(p_1 - p_4)$

We have:

$\sigma'_v |p_2\rangle = |p_3\rangle$ $c_2 = -c_3 = c'$ and thus $c'\,(p_2 - p_3)$

Last, the generic form of the MO belonging to A_2 is $c\,(p_1 - p_4) + c'\,(p_2 - p_3)$.

Reporting these results in the four equations with four unknowns, we obtain

$$xc + c' = 0 \quad \text{and} \quad c + x\,c' - c' = 0 \quad \text{i.e.} \quad c + (x - 1)\,c' = 0$$

The compatibility between the first and the third equation implies

$$\begin{vmatrix} x & 1 \\ 1 & x-1 \end{vmatrix} = 0$$ whose roots are $x = -0.62$ and $x = 1.62$.

The two corresponding energies are

$$E_2 = \alpha + 0.62\,\beta \quad E_4 = \alpha - 1.62\,\beta$$

and the associated MO read

$$\varphi_2 \propto (p_1 - p_4) + 0.62\,(p_2 - p_3) \quad \text{and} \quad \varphi_4 \propto (p_1 - p_4) - 1.62\,(p_2 - p_3)$$

b. MO of B_2 symmetry:

For this representation, the character of σ'_V is + 1. We have $c_1 = c_4 = c$ and $c_2 = c_3 = c'$. We finally get the equations $xc + c' = 0$ and $c + (x + 1) c' = 0$.

The secular equation takes the form

$$\begin{vmatrix} x & 1 \\ 1 & x+1 \end{vmatrix} = 0 \quad \text{whose roots are} \quad x = -1.62 \quad \text{and} \quad x = 0.62.$$

The values for the energy are $E_1 = \alpha + 1.62\,\beta$ and $E_3 = \alpha - 0.62\,\beta$.

Thus, the corresponding MO's read

$$\varphi_1 \propto (p_1 + p_4) + 1.62\,(p_2 + p_3) \quad \text{and} \quad \varphi_3 \propto (p_1 + p_4) - 0.62\,(p_2 - p_3).$$

III.4.4. Energy of singlet and triplet states of a two-electron system (from Agreg. C; C; 1982)

We shall consider a two electrons system, whose wavefunction is ψ (1, 2).

1. Spectral multiplicity

a. Which property does the ψ (1, 2) function exhibit with respect to the exchange of the two electrons? Write the ψ (1, 2) function for the ground state correctly using the spin functions labeled α and β (not to be confused with the Coulomb and exchange integrals). What is the spectral state of the ground state?

b. Investigate, with the help of simple sketches, the various excited states. Show that they can be classified in states with different spectral multiplicities.

c. Using the antisymmetry property of function ψ (1, 2) in the excited state, show that four functions can be written in the excited state and classify them according to whether the function is symmetrical or antisymmetrical with respect to the exchange of two electrons. Relate these four wavefunctions to the spin multiplicity.

d. Establish that the triplet states have the same energy.

2. Calculation of E_0: We want to calculate the energy of the ground state. We call the correct wavefunction of the ground state ψ_0.

a. Show that it is not possible to consider the Hamiltonian $H(1, 2)$ as being the sum of monoelectronic Hamiltonians and that that we have

$$H(1, 2) = h(1) + h(2) + h(1, 2).$$

Justify the meaning of these three terms qualitatively.

b. Calculate the expression of E_0 (it is the sum of three terms).

We set $I_i = \langle \varphi_i | h | \varphi_i \rangle$ and $J_{ij} = \langle \varphi_i^2 | h(1, 2) | \varphi_j^2 \rangle$

3. Calculation of E_S: We want to calculate the energy of the singlet excited state.

We call the corresponding wavefunction ψ_S. Furthermore, we write

$$K_{ij} = \langle \varphi_i \varphi_j | h(1, 2) | \varphi_i \varphi_j \rangle$$

Show that we have $E_S = I_1 + I_2 + J_{12} + K_{12}$.

4. Calculation of E_T: We want to evaluate the energy of the excited state. Calculate E_T as a function of I_1, I_2, J_{12} and K_{12}.

5. Energy diagram:

a. Show that $K_{12} > 0$.

b. Position the various values for E_0, E_S and E_T on an energy axis.

1.a. The state vector must be antisymmetrical with respect to the exchange of two electrons (postulate P8 of quantum mechanics, *cf.* Section III.1.3).

Taking into account the spin (states α and β), we write

$$\varphi(1,2) = \left(1/\sqrt{2}\right)\varphi_1(1)\,\varphi_1(2)\left[\alpha(1)\,\beta(2) - \alpha(2)\,\beta(1)\right]$$

Antisymmetry arises from the spin. We have S = 0. 2S + 1 = 1. The state is a singlet.

b. In fact, we have two MO's, φ_1 and φ_2. We can obtain two singlets and two triplets.

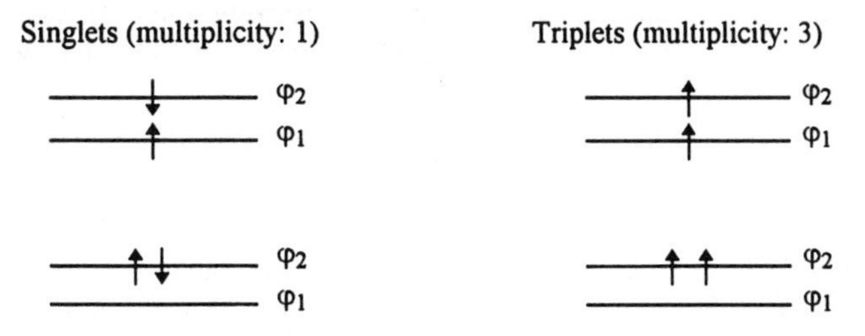

c. We can define four functions:

Singlet state:

$$\frac{1}{2}\left[\varphi_1(1)\,\varphi_2(2) + \varphi_1(2)\,\varphi_2(1)\right]\left[\alpha(1)\,\beta(2) - \alpha(2)\,\beta(1)\right]$$

Triplet states:

$$\begin{cases} \dfrac{1}{\sqrt{2}}\left[\varphi_1(1)\,\varphi_2(2) - \varphi_1(2)\,\varphi_2(1)\right]\alpha(1)\alpha(2) \\ \dfrac{1}{\sqrt{2}}\left[\varphi_1(1)\,\varphi_2(2) - \varphi_1(2)\,\varphi_2(1)\right]\beta(1)\beta(2) \\ \dfrac{1}{2}\left[\varphi_1(1)\,\varphi_2(2) - \varphi_1(2)\,\varphi_2(1)\right]\left[\alpha(1)\,\alpha(2) + \beta(1)\,\beta(2)\right] \end{cases}$$

d. The Hamiltonian operator acts only on space coordinates. However, the three states have the same orbital and thus correspond to the same energy.

2.a. The Hamiltonians h_1 and h_2 represent the negative attraction energies of the two electrons by the nuclei. The positive term $h_{12} = e^2/4\pi\varepsilon_0 r_{12}$ stands for the Coulomb interelectronic repulsion energy.

b. The energy can be written as $\langle \psi(1,2) | H(1,2) | \psi(1,2) \rangle$. Thus $\langle \varphi_1(1)\varphi_1(2) | h_1 + h_2 + h_{12} | \varphi_1(1)\varphi_1(2) \rangle$. A simple calculation gives

$$E_0 = 2I_1 + J_{11}$$

3. We have

$$E_S = \frac{1}{2}\langle \varphi_1(1)\varphi_2(2) + \varphi_1(2)\varphi_2(1) | h_1 + h_2 + h_{12} | \varphi_1(1)\varphi_2(2) + \varphi_1(2)\varphi_2(1) \rangle$$

then $$E_S = I_1 + I_2 + J_{12} + K_{12}$$

4. We obtain the same equation as above providing we substitute two – signs for the two + signs in the bra and in the ket.

$$E_T = I_1 + I_2 + J_{12} - K_{12}$$

5.a. K_{12} is a positive repulsion term.

b. We have $E_0 < E_T < E_S$.

III.5. Appendices

III.5.1. Main properties of linear transformations and matrices

Linear transformations

Let (α, β, γ) and (i, j, k) be two bases. Let x be the components of a vector and c the directional cosines of the transformations. We can write

$$x_i = c_{i\alpha} x_\alpha \qquad x_\alpha = c'_{\alpha i} x_i$$

The directional cosines are such that

$$c_{i\alpha} = \text{minor } c'_{\alpha i} / |(c'_{\alpha i})| \qquad c'_{\alpha i} = \text{minor } c_{i\alpha} / |(c_{i\alpha})|$$

— Minor: Cofactor affected with the sign of $(-1)^{(l+c)}$ (l and c are the line and column indices with respect to the directional cosines matrix).

— Cofactor: Determinant of the matrix obtained by suppression of the line and of the column of the chosen element.

Fundamental properties:

$$|(c_{i\alpha})||(c'_{\alpha i})| = 1 \qquad c_{i\alpha} c'_{\alpha j} = \delta_{ij}$$

Linear orthogonal transformations

Fundamental properties:

— $|(c_{i\alpha})| = \pm 1$ (+ : dextrorotatory ; – : levorotatory transformations)

— $|(c_{i\alpha})| = |(c'_{\alpha i})|$

— $c_{i\alpha} = c'_{\alpha i}$; $c_{i\alpha} c_{j\alpha} = \delta_{ij}$

Matrix properties

— Adjoint of a matrix (M^{adj}): Transposition with respect to the principal diagonal and substitution of each element by its minor.

— Inverse matrix (M^{-1}) (if $|(M)| \neq 0$):

$$\left(M^{-1}\right) = (1)/(M) = \left(M^{adj}\right)/|(M)|$$

— Hermitian matrix ($M^{\dagger}$):

$$(M^{\dagger}) = (M)$$

The operation † defines a transposition followed by a conjugation.

— Unitary matrix: Is such that

$$(M^{\dagger}) = (M^{-1})$$

III.5.2. Representation of an operator by a matrix

Let $|u_n\rangle$ be the discrete, orthonormed and finite-sized (order n) basis of eigenvectors of O, and let a_n be the corresponding eigenvalues:

$$O|u_n\rangle = a_n|u_n\rangle$$

Let A be an operator acting on $|\psi\rangle = c_m|u_m\rangle$:

$$A|\psi\rangle = |\psi'\rangle$$

We have

$$|\psi'\rangle = c'_m|u_m\rangle = c_m A|u_m\rangle$$

Multiplying the left hand side by the bra $\langle u_{m'}|$ leads to

$$c'_m\langle u_{m'}|u_m\rangle = c'_m\delta_{m'm} = c'_{m'} = c_m\langle u_{m'}|A|u_m\rangle = c_m A_{m'm}$$

(m': column index; m: line index)

We finally obtain the two following equivalent representations:

1.
$$\begin{Bmatrix} c'_1 \\ \vdots \\ c'_m \\ \vdots \\ c_n \end{Bmatrix} = \begin{pmatrix} A_{11} \cdots\cdots A_{1n} \\ \vdots \qquad \vdots \\ A_{m1} \cdots A_{mm'} \cdots A_{mn} \\ \vdots \qquad \vdots \\ A_{n1} \cdots\cdots A_{nn} \end{pmatrix} \begin{Bmatrix} c_1 \\ \vdots \\ c_m \\ \vdots \\ c_n \end{Bmatrix}$$

column-vector (dimension n)	matrix m: line; m': column (dimension n × n)	column-vector (dimension n)

2.

$\|\psi'\rangle$	=	A	$\|\psi\rangle$
ket		operator	ket

III.5.3. Table of the postulates of quantum theory

DESCRIPTION	P_1	State of the system described at point **r** and time t by $\|\psi\rangle$ $\psi \in$ F (vectorial space; functions everywhere defined, continuous and whose square can be summed over all space).
	P_2	Observable O ⟶ Operator O O acts on $\|\psi\rangle$ Eigenvectors $\|u_n\rangle$ of $O \rightarrow$ basis of F
MEASUREMENT	P_3	Measurement of O ⟶ one eigenvalue a_n of O
	P_4	Probability $P(a_n)$ to find a_n when a measurement of O is performed $P(a_n) = \|\langle u_n \| \psi\rangle\|^2$ Consequence: $\langle a\rangle = \langle\psi\|O\|\psi\rangle$
	P_5	Measurement of O $\|\psi\rangle$ ⟶ $\|u_n\rangle$ Before Measured value a_n After
EVOLUTION	P_6	$(ih/2\pi)\|\dot{\psi}\rangle = H\|\psi\rangle$ H: Hamiltonian
CORRESPONDENCE	P_7	In the Schrödinger representation: O obtained by replacing in adequately symmetrized O: $\mathbf{r} \rightarrow r$; $\mathbf{p} \rightarrow p = -(ih/2\pi)\, \partial/\partial\mathbf{r}$; $t \rightarrow t$
SPIN	P_8	Electron characterized by an intrinsic kinetic momentum of spin S $S_z\|\alpha\rangle = (1/2)(h/2\pi)\|\alpha\rangle$; $S_z\|\beta\rangle = (-1/2)(h/2\pi)\|\beta\rangle$ $\|\alpha\rangle$ and $\|\beta\rangle$ → Two-dimensional space Fs Complete descriptional space: F × Fs
ANTI SYMMETRIZATION	P_9	For a system consisting of N electrons: $\|\psi(1, \ldots, N)\rangle$ ⟶ Slater determinant

III.5.4. Tables of characters of the symmetry groups used in this book

The tables of characters of the irreducible representations of the symmetry groups used in this book are presented below (all the tables can be found in the Bibliographic reference 3.6.2). Some operations, when they belong to the same class, are indicated jointly (for example, in the C_{3v} group: $2C_3$ instead of C_3 and C_3^2).

The last two columns indicate the symmetry of the translation functions of first and second degree, together with the rotation functions about the axes Ox, Oy and Oz. These columns can be used to obtain the symmetry of the atomic orbitals.

C_{2v}	E	C_2	$\sigma_v(xz)$	$\sigma_v(yz)$		
A_1	1	1	1	1	z	$x^2 ; y^2 ; z^2$
A_2	1	1	–1	–1	R_z	xy
B_1	1	–1	1	–1	x ; R_y	xz
B_2	1	–1	–1	1	y ; R_x	yz

C_{3v}	E	$2C_3$	$3\sigma_v$		
A_1	1	1	1	z	$x^2 + y^2 ; z^2$
A_2	1	1	–1	R_z	
E	2	–1	0	(x, y) ; (R_x, R_y)	($x^2 - y^2$, xy) ; (xz, yz)

C_{4v}	E	$2C_4$	C_2	$2\sigma_v$	$2\sigma_d$		
A_1	1	1	1	1	1	z	$x^2 + y^2 ; z^2$
A_2	1	1	1	–1	–1	R_z	
B_1	1	–1	1	1	–1		$x^2 - y^2$
B_2	1	–1	1	–1	1		xy
E	2	0	–2	0	0	(x, y) ; (R_x, R_y)	(xz, yz)

C_{2h}	E	C_2	i	σ_h		
A_g	1	1	1	1	R_z	$x^2 ; y^2 ; z^2$; xy
B_g	1	–1	1	–1	R_x; R_y	xz ; yz
A_u	1	1	–1	–1	z	
B_u	1	–1	–1	1	x ; y	

D_{2h}	E	$C_2(z)$	$C_2(y)$	$C_2(x)$	i	$\sigma(xy)$	$\sigma(xz)$	$\sigma(yz)$		
A_g	1	1	1	1	1	1	1	1		$x^2 ; y^2 ; z^2$
B_{1g}	1	1	–1	–1	1	1	–1	–1	R_z	xy
B_{2g}	1	–1	1	–1	1	–1	1	–1	R_y	xz
B_{3g}	1	–1	–1	1	1	–1	–1	1	R_x	yz
A_u	1	1	1	1	–1	–1	–1	–1		
B_{1u}	1	1	–1	–1	–1	–1	1	1	z	
B_{2u}	1	–1	1	–1	–1	1	–1	1	y	
B_{3u}	1	–1	–1	1	–1	1	1	–1	x	

D_{3h}	E	$2C_3$	$3C_2$	σ_h	$2S_3$	$3\sigma_v$		
A_1'	1	1	1	1	1	1		$x^2 + y^2$; z^2
A_2'	1	1	−1	1	1	−1	R_z	
E'	2	−1	0	2	−1	0	(x, y)	$(x^2 - y^2, xy)$
A_1''	1	1	1	−1	−1	−1		
A_2''	1	1	−1	−1	−1	1	z	
E''	2	−1	0	−2	1	0	(R_x, R_y)	(xz, yz)

D_{4h}	E	$2C_4$	C_2	$2C'_2$	$2C''_2$	i	$2S_4$	σ_h	$2\sigma_v$	$2\sigma_d$		
A_{1g}	1	1	1	1	1	1	1	1	1	1		$x^2 + y^2$; z^2
A_{2g}	1	1	1	−1	−1	1	1	1	−1	−1	R_z	
B_{1g}	1	−1	1	1	−1	1	−1	1	1	−1		$x^2 - y^2$
B_{2g}	1	−1	1	−1	1	1	−1	1	−1	1		xy
E_g	2	0	−2	0	0	2	0	−2	0	0	(R_x, R_y)	(xz, yz)
A_{1u}	1	1	1	1	1	−1	−1	−1	−1	−1		
A_{2u}	1	1	1	−1	−1	−1	−1	−1	1	1	z	
B_{1u}	1	−1	1	1	−1	−1	1	−1	−1	1		
B_{2u}	1	−1	1	−1	1	−1	1	−1	1	−1		
E_u	2	0	−2	0	0	−2	0	2	0	0	(x, y)	

D_{6h}	E	$2C_6$	$2C_3$	C_2	$3C'_2$	$3C''_2$	i	$2S_3$	$2S_6$	σ_h	$3\sigma_d$	$3\sigma_v$		
A_{1g}	1	1	1	1	1	1	1	1	1	1	1	1		$x^2 + y^2$; z^2
A_{2g}	1	1	1	1	−1	−1	1	1	1	1	−1	−1	R_z	
B_{1g}	1	−1	1	−1	1	−1	1	−1	1	−1	1	−1		
B_{2g}	1	−1	1	−1	−1	1	1	−1	1	−1	−1	1		
E_{1g}	2	1	−1	−2	0	0	2	1	−1	−2	0	0	(R_x, R_y)	(xz, yz)
E_{2g}	2	−1	−1	2	0	0	2	−1	−1	2	0	0		$(x^2 - y^2, xy)$
A_{1u}	1	1	1	1	1	1	−1	−1	−1	−1	−1	−1		
A_{2u}	1	1	1	1	−1	−1	−1	−1	−1	−1	1	1	z	
B_{1u}	1	−1	1	−1	1	−1	−1	1	−1	1	−1	1		
B_{2u}	1	−1	1	−1	−1	1	−1	1	−1	1	1	−1		
E_{1u}	2	1	−1	−2	0	0	−2	−1	1	2	0	0	(x, y)	
E_{2u}	2	−1	−1	2	0	0	−2	1	1	−2	0	0		

T_d	E	$8C_3$	$3C_2$	$6S_4$	$6\sigma_d$		
A_1	1	1	1	1	1		$x^2+y^2+z^2$
A_2	1	1	1	−1	−1		
E	2	−1	2	0	0		$(2z^2-x^2-y^2, x^2-y^2)$
T_1	3	0	−1	1	−1	(R_x, R_y, R_z)	
T_2	3	0	−1	−1	1	(x, y, z)	(xy, xz, yz)

O_h	E	$8C_3$	$3C_2$	$6C_4$	$6C'_2$	i	$8S_6$	$3\sigma_h$	$6S_{64}$	$6\sigma_d$		
A_{1g}	1	1	1	1	1	1	1	1	1	1		$x^2+y^2+z^2$
A_{2g}	1	1	1	−1	−1	1	1	1	−1	−1		
E_g	2	−1	2	0	0	2	−1	2	0	0		$(2z^2-x^2-y^2, x^2-y^2)$
T_{1g}	3	0	−1	1	−1	3	0	−1	1	−1	(R_x, R_y, R_z)	
T_{2g}	3	0	−1	−1	1	3	0	−1	−1	1		(xy, xz, yz)
A_{1u}	1	1	1	1	1	−1	−1	−1	−1	−1		
A_{2u}	1	1	1	−1	−1	−1	−1	−1	1	1		
E_u	2	−1	2	0	0	−2	1	−2	0	0		
T_{1u}	3	0	−1	1	−1	−3	0	1	−1	1	(x, y, z)	
T_{2u}	3	0	−1	−1	1	−3	0	1	1	−1		

III.6. Bibliography

3.6.1. The two fundamental books concerning the teaching of quantum mechanics are:

COHEN-TANNOUDJI C., DIU B. AND LALOE F. *Quantum Mechanics*, John Wiley & Sons, New York, 1979.

MESSIAH A. *Quantum Mechanics*, North Holland Publishing Company, John Wiley & Sons, New York, 1965.

3.6.2. Applications of group theory in quantum mechanics are presented in the following reference book:

COTTON F.A. *Chemical Applications of Group Theory*, John Wiley & Sons, New York, 1962.

Didactic papers are those by:

FALTYNEK R.A. — Group Theory in Advanced Inorganic Chemistry, *J. Chem. Educ.*, 72, 20 (1995).

ORCHIN M. et JAFFE H.H. — Symmetry, point groups and character tables, *J. Chem. Educ.*, 47, 246, 372 et 510 (1970).

SANNIGRAHI A.B. — Projection Operators and Their Simple Applications in Group Theory, *J. Chem. Educ.*, 52, 307 (1975).

WHITE E.J. — An Introduction to Group Theory for Chemists, *J. Chem. Educ.*, 44, 128 (1967).

3.6.3. Atomic and molecular structure is detailed in the books indicated in the Bibliography I.2 and in:

ELLIS R. L. et JAFFE H. H. — The Symmetries and Multiplicities of Electronic States in Polyatomic Molecules, *J. Chem. Educ.*, 48, 92 (1971).

KAHN O. — *Structure electronique des elements de transition*, Presses Universitaires de France (1977).

LELLAND HOLLENBERG J. — Energy States of Molecules, *J. Chem. Educ.*, 47, 2 (1970).

RIVAIL J.L. — *Elements de chimie quantique à l'usage des chimistes*, InterEditions et CNRS Editions, 2e ed. (1994).

PART TWO

OPTICAL SPECTROSCOPIES OF ELECTRONIC ABSROPTION

Chapter IV

UV-visible Absorption Spectroscopy

IV.1. Generalities

The radiation described in Chapter II is chosen in the near-ultraviolet and in the visible range of the spectrum. The energy of the wave (in microscopic classical theory) or of the photon (in Einstein's macroscopic theory and in the semi-classical and quantum theories) lies in the domain corresponding to wavelengths between 200 and 800 nm, which is equivalent to one or several electron-volts (*cf.* Table 4.1.1).

Table 4.1.1 — **Limits of near-UV and visible spectra.**

	Near-UV starting from	visible (start)	visible (end)
wavelength (nm)	200	400	800
frequency (Hz)	1.5×10^{15}	7.5×10^{14}	3.75×10^{14}
wave number (cm^{-1})	5×10^{6}	2.5×10^{6}	1.25×10^{6}
energy (eV)	6.21	3.10	1.55

However, we have learned in Chapter III that these energies correspond to the energies of the external electrons of atoms and molecules: The UV-visible absorption spectroscopy will therefore naturally give us information concerning the properties of the so-called "external" or "optical" electrons of atoms and molecules. These properties have been described in Section III.3. We will describe light-matter interaction in Section IV.3, where a simple classical presentation is followed by a semi-classical presentation and completed by many exercises and problems presented in Section IV.4. But first, we will take a few pages to present an example of a recent realization of a UV-visible spectrometer: New apparatus developed during the last few years have reached a high degree of refinement, both in the quality of the optic and electronic components and the extend of their automation. They are used very extensively, not only by the fundamental and applied scientific communities, but also in secondary schools and in universities. This presentation, which can at times be a little technical, is found in Section IV.2. Finally, Section IV.6 offers a short bibliography, limited to recent books about spectroscopy in general, or more specifically about optical spectroscopy.

IV.2. Experimental techniques

In UV-visible absorption spectroscopy, two main complementary techniques are used:

– So-called UV-visible spectroscopy, in which a polychromatic light first undergoes wavelength dispersion (Problem IV.4.1 presents a detailed evaluation of the usual scattering techniques), and then it is separated in bands which should be as monochromatic as possible. These bands then cross the sample one after the other and the transmission is recorded for each band. The global result of these recordings forms the absorption spectrum of the sample.

– Fourier transform spectroscopy, whose principle is presented in detail in Problem IV.4.2 but which we shall not encounter anymore in the rest of this chapter.

Let us present successively the UV-visible spectrometer and the practical preparation of a sample.

IV.2.1. The UV-visible spectrometer

Figure 4.2.1 shows the principle of a recent commercial apparatus. We willfully selected a model of average performances so that can be considered to represent the whole commercially available range.

IV.2.1.1. Radiation source

Two sources are used. The first, used for recording in the near-UV range (in our case from 190 nm to 400 nm), is a deuterium lamp. Other models, requiring high intensities, use arc or xenon lamps. The second source, which can be used up

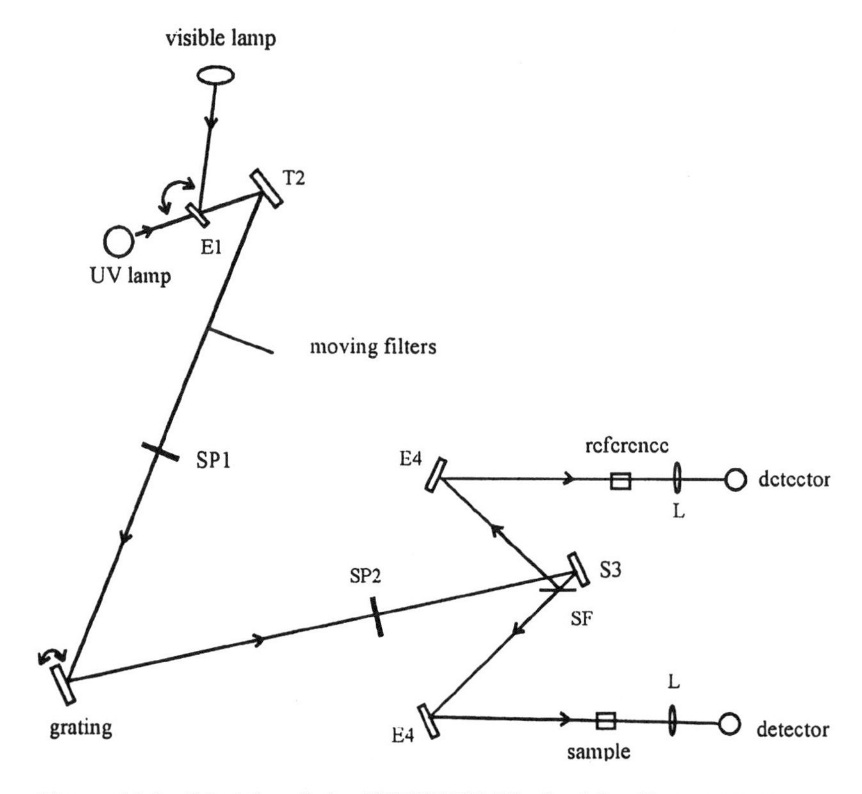

Figure 4.2.1: Principle of the PERKIN-ELMER Lambda 40 spectrometer (reproduced thanks to the authorization of Perkin-Elmer SA). E1: mirror for lamp change; T2: toroïdal mirror or grating playing the part of a first monochromator; SP1: entrance slit; SP2: exit slit; S3: spherical mirror; SF: beam splitter; E4: elliptical mirror; L: lens.

to 1100 nm, is a tungsten-halogen arc lamp. The change of lamp is automatic and lamps are pre-aligned by the manufacturer.

IV.2.1.2. Dispersing elements

In our case, the dispersing element is a holographic concave grating with 1 053 grooves per mm. Prisms are no longer used at present.

IV.2.1.3. Detectors

In the near-UV and the visible ranges, the detector almost always consists of an electron photomultiplier or of a photodiode. In the near-IR range, the apparatus use lead sulfide detectors, cooled with the Peltier effect. In order to reduce low frequency noise, which is very annoying in experiments, it is often better to introduce a high frequency modulation in the detection device.

IV.2.1.4. Computer assisted monitoring

The apparatus has a very sophisticated computer assisted monitoring system, as well for the collection of information as for its treatment. A microcomputer drives all the operations, including a measurement of the noise signal, which is treated and taken into account with the utmost care. In our example, the UV Winlab software offers many different treatment options. It is also possible to add specialized products, which can be used for example for biochemical analysis of DNA, of enzymes and proteins, or for the analysis of water in environment chemistry, or again for the study of printer inks. Many accessories can be adapted to the apparatus, such as an experimental device which can measure 218 samples successively, or an optical fiber system permitting the collection of information from relatively inaccessible samples.

IV.2.1.5. Specifications

The apparatus described here offers a choice of four frequency band-widths: 0.5; 1; 2 and 4 nm in the 190-1 100 nm spectral domain. Stray light is reduced to 0.005 % by the use of a grating placed before the monochromator. Although this noise still remains about three orders of magnitude larger that the shot noise, of quantum origin, that constitutes the theoretical limit, this important reduction of stray light guarantees a very appropriate linearity over about *four units* in absorbance.

IV.2.2. The sample and its conditioning

Solutions are carefully dusted in Pyrex glass (visible and near IR) and fused quartz (for the near UV) vessels. These vessels are currently of spectroscopic quality (by the quality of the material used, by the polishing and flatness of the faces of up to $\lambda/4$, etc.). The general solution to increase the sensitivity of the detection is to lengthen the optical path in the sample using a many-pass cell. The light beam is "folded" with the help of two (Herriott cell) or three (White cell) mirrors. Absorption lengths of several hundreds of meters can thus be obtained.

V.3. Absorption of the optical wave by molecules

IV.3.1. Brief classical presentation

IV.3.1.1. Microscopic description

An illustration of the microscopic description is given in Problems IV.4.3 to 5.

We already pointed out that at optical frequencies, the interaction between light and matter mainly concerns the less bounded electrons of atoms and molecules. Let us represent the electron by the simple image of a linear harmonic oscillator. Light can then be described classically as the component — for instance the x-component — of an electric field, expressed in its complex form as

$$E_X(t) = \mathrm{Re}\left\{\underline{E}_X(\omega)\exp\left[i(\omega t + \varphi)\right]\right\} \tag{4.3.1}$$

ω is the angular frequency of the radiation and Re denotes the real part of the expression. Figure 4.3.1 illustrates this model.

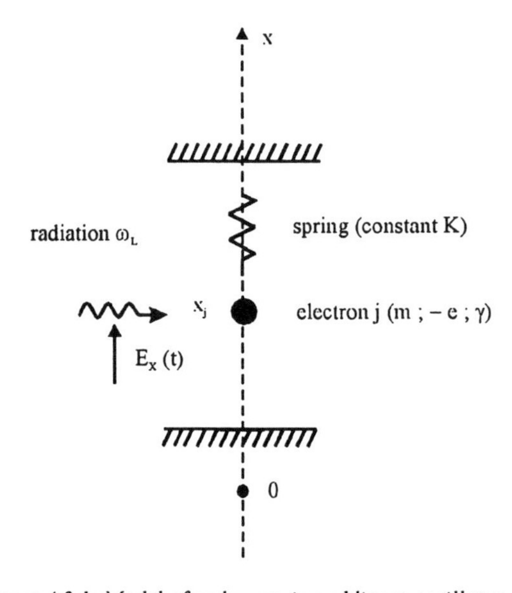

Figure 4.3.1: Model of an harmonic and linear oscillator.

m and – e are, respectively, the mass and the electric charge of the electron, and γ is the damping coefficient, damping being proportional to the speed. K is the spring constant, and the spring is assumed to obey Hooke's law. We shall study the motion of mass m successively in the absence, and in the presence of radiation.

In the absence of radiation

The motion of the jth electron is traditionally described by the equation

$$m\ddot{x}_j(t) = -Kx_j(t) - m\gamma\dot{x}_j(t) \tag{4.3.2}$$

The solution of this equation contains two constants, describing the amplitude and the phase of the oscillation, and it can be written as

$$x_j(t) = x_{j0} \exp[(-\gamma / 2 + i\omega'_0)(t - t_0) + i\varphi_{j0}] \quad (4.3.3)$$

with $\omega'^2_0 = \omega^2_0 - (\gamma / 2)^2$ and $\omega^2_0 = K / m$.

The resonant angular frequency, defined in terms of the strength of the link and the oscillating mass, is of the order of 10^{15} rd.s^{-1}. The motion of the electron is a damped oscillating motion. The energy — the sum of a kinetic energy and a potential energy — relaxes in time with a constant $\gamma^{-1} = \tau$, according to the relation

$$U_j(t) = (1/2)m\left|\dot{x}_j(t)\right|^2 + (1/2)K\left|x_j(t)\right|^2 = U_{j0} \exp[-\gamma(t - t_0)] \quad (4.3.4)$$

These losses, which cause a frequency broadening of the line (*cf.* Problem IV.4.6), are traditionally classified in two main groups:

— Radiative type losses: They correspond to "antenna-type" radiations, due to the fact that the electron radiates an electromagnetic wave in all directions;

— Nonradiative type losses: They take place when the electron is not isolated in vacuum and they are due to a dissipative coupling between the electron and its surrounding (inelastic collisions for instance).

The damping constant γ therefore includes both the radiative and the nonradiative losses and can be written as $\gamma = \gamma_{rad} + \gamma_{nrad}$.

In the presence of radiation

A force equal to – eEx0(t) must be added to Equation 4.2 describing the interaction between electron and radiation so that it now writes

$$m\ddot{x}_j(t) = -Kx_j(t) - m\gamma\dot{x}_j(t) - eE_x(t) \quad (4.3.5)$$

We neglect the transitory state. Let us look for a particular solution of the equation with second member. Such a solution can be written as

$$x_j(t) = [-eE_x(t)/m]\{1/[(\omega^2_0 - \omega^2) + i\gamma\omega]\} \quad (4.3.6)$$

The electron undergoes an oscillating motion which is forced by the optical field of the wave, and it gains an additional mean power given by (*cf.* Problem IV.4.5)

$$P = \tau(e\omega E_0)^2 / 2m[\tau^2(\omega^2_0 - \omega^2)^2 + \omega^2] \quad (4.3.7)$$

This energy transfer from radiation to matter constitutes the very first description of the phenomenon of electronic absorption. It is, of course, greatest at resonance.

Now we need to consider the more realistic case of an interaction between the radiation and a material system consisting of a large number of electrons.

— A first and fundamental problem concerns the phase of the forced oscillation. No physical consideration imposes the existence of a relation between the phases of the different oscillators. The elastic and inelastic collisions in fluid surroundings, as well as the lattice vibrations in solids and the dipole coupling of oscillators in dense media, create a partial incoherence of the oscillators. A generally adopted solution consists in keeping complete coherence mathematically, but correcting the ensuing physical non relevance of such reasoning by arbitrarily increasing the radiative part of the damping constant. We thus write

$$\gamma_{rad} = \gamma_{rad1} + \gamma_{rad2} = \tau_1^{-1} + \tau_2^{-1}$$

And so, in addition to the first time constant τ_1 describing the natural width of the electronic oscillation, a second time constant appears, called τ_2, which is due to the partial incoherence of oscillators. Together, these two effects lead to the *homogeneous width*. Table 4.3.2 gives some orders of magnitude of these time constants.

Table 4.3.2 — **Some values of time constants τ_1 and τ_2.**

Material	τ_1(s)	τ_2(s)
dyes in liquids	10^{-9} to 10^{-7}	10^{-12} to 10^{-11}
gas under reduced pressure	10^{-7} to 10^{-5}	10^{-9} to 10^{-8}
doped solids	10^{-4} to 10^{-3}	10^{-12} to 10^{-11}

The importance of the partial incoherence clearly appears. For example, for an Nd^{3+} ion in a solid crystalline matrix of yttrium and aluminum garnet (YAG), a material very often used in laser technology, the absorption spectral width is almost totally due to the incoherence of oscillators. For this material

$$\tau_1^{-1} = 2\,500 \text{ Hz}; \ \tau_2^{-1} = 120 \text{ GHz}.$$

— A second problem arises from the fact that there is no reason either for the angular frequency resonances ω_0 to be the same for all the electrons. In fact, for the observer in the laboratory frame, there exists a distribution of the resonance frequencies in dilute and dense surroundings, due for instance to the Doppler effect, leading to a Maxwell distribution of the particle velocities. This effect is especially important in gases and liquids. In solids, the large variety of local environments, due to imperfections of the crystal lattice, leads to a similar result. So an inhomogeneous broadening appears, which conceals the intrinsic electronic property we are interested in.

After these two modifications of the damping constant to adjust it to a more realistic description of the materials submitted to the radiation, we can now proceed to a description of the interaction of the radiation with a material sample containing N electrons per unit volume.

Equation 4.3.6 gives the displacement x(t) of each radiation driven electron. The ensemble of electrons gives rise to a polarization density, which is equal to the sum of all the individual electric dipole moments, written as

$$P_x(t) = \sum_{j=1}^{N} p_{xj}(t) = -Nex_j(t) \tag{4.3.8}$$

Then, by using Equation 4.3.6, we obtain

$$P_x(t) = (e^2 NE_x(t)/m)\left\{1/[(\omega_0^2 - \omega^2) + i\gamma\omega]\right\} \tag{4.3.9}$$

where we used the relation implied by Equation 4.3.1:

$$P_x(t) = \mathrm{Re}[\underline{P}_x(\omega)\exp(i\omega t)] \tag{4.3.10}$$

So now we can define a susceptibility (*cf.* constitutive Maxwell relations, Section II.2.1.1) as follows

$$\underline{\chi}(\omega) = \chi'(\omega) + i\chi''(\omega) = \underline{P}_x(\omega)/\varepsilon_0 \underline{E}_x(\omega) \tag{4.3.11}$$

and, in a same way, a complex refractive index written as

$$\underline{n} = n' + in'' \approx 1 + (\chi'/2) + i(\chi''/2) \tag{4.3.12}$$

Please note that for simplicity, we choose to describe a material for which the modulus of the susceptibility is small compared to 1.

The imaginary part the refractive index is going to play a fundamental role in this description of the absorption. Indeed, starting from Equation 4.3.1, and remembering that the phase term can be written as

$$\exp(-i2\pi n'' x/\lambda) = \exp(\pi\chi'' x/\lambda)$$

then using Equations 4.3.9 and 4.3.11, we can derive a propagation term for the wave field. We find:

$$\exp-(\gamma\omega\pi e^2 N/m\lambda)\left\{1/\left[(\omega_0^2 - \omega^2)^2 + \gamma^2\omega^2\right]\right\}x \tag{4.3.13}$$

In this equation, we see that the amplitude of the field — and therefore its intensity which is proportional to the square of the amplitude — undergoes an exponential attenuation with the x coordinate. This is a classical result of the *Beer-Lambert* law: The optical density of a sample increases linearly with its thickness and is proportional to N, the density of the absorbers. The proportionality constant, called *absorption coefficient*, has a resonant behavior. Figure 4.3.2 represents the variations of the resonant part of the susceptibility $1/(1+\Delta\omega^2)$ with $\Delta\omega$, where $\Delta\omega = 2(\omega-\omega_0)/\gamma$.

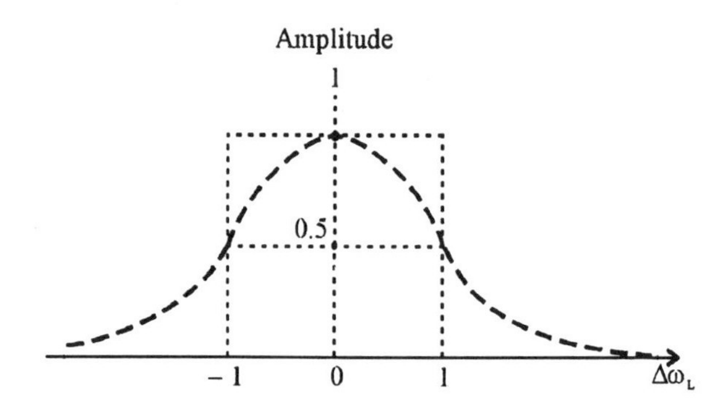

Figure 4.3.2: Variations of the resonant part of the absorption coefficient as a function of the relative distance to the resonant frequency of the microscopic system.

However, there is something very important we should bear in mind. In this model, the elastic force of the spring is proportional to its elongation. This leads to a linear model where the electron displacement is proportional to the electric field while the susceptibility remains constant. Of course, this is a crude approximation. Forces are never proportional to displacements, nor displacements to fields, and in general, the susceptibility is a complex function of the electric field carried by the light wave. In this more complex case, of which evidence is now often found in laser physics, the Beer-Lambert law no longer applies.

Nevertheless, this very simple classical microscopic model is a good way to understand the absorption phenomenon. An energy exchange occurs between the wave and the electron, which causes a progressive attenuation of the intensity of the wave during its propagation in the sample. It is even gives some insight beyond the absorption phenomenon: We can understand that the electron can relax part of its energy in a radiative or in a nonradiative way. The "antenna" type radiation corresponds to spontaneous emission, of which we shall see more in the next paragraph. Unfortunately, the microscopic classic description cannot account for stimulated emission. This point constitutes its fundamental limitation.

IV.3.1.2. Phenomenological description

The phenomenological description was introduced by Einstein in 1917. In this description, absorption appears as one of the three possible forms of interaction between radiation and matter, the other two being spontaneous emission and stimulated emission. Only two notions from the theory of quantum mechanics, developed in the beginning of the century, are kept: That of the *discrete distribution* of energy levels in atoms and molecules and that of a corpuscular description for the microscopic system. The radiation consists of particles called *photons*. Each of the three exchanges is accompanied by (i) the creation or the annihilation of a photon and (ii) the change of energy level in the microscopic system. Figure 4.3.3 describes the three exchange processes between radiation and matter. We call $E_1 = h\nu_1$ and $E_2 = h\nu_2$ the energies of the two discrete levels 1 and 2. The energy of the photon is equal to $h\nu$, which gives an exclusively mathematical definition of ν since we do not use a wave-like description of light in this theory. Frequency ν is assumed to be of the order of the frequency difference $\nu_{21} = (\nu_2 - \nu_1)$.

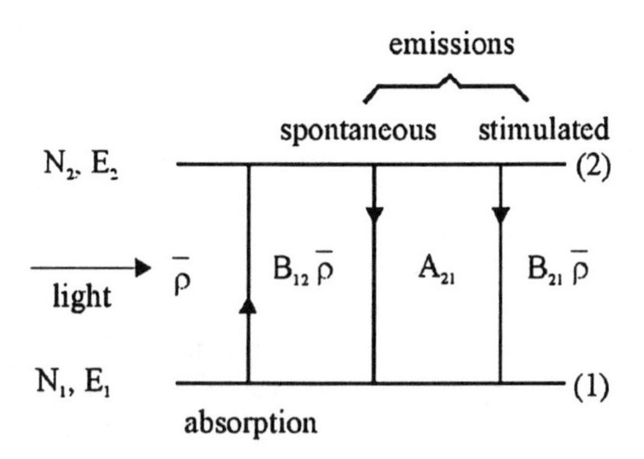

Figure 4.3.3: Matter-radiation exchanges in the phenomenological theory.

With this assumption, we can limit the infinite distribution of energy levels of the microscopic systems to the two levels labeled 1 and 2. We do this on the ground of the principle that *systems can interact only if their energies are almost identical*. This is why, in this rest of this book dealing with absorption, we shall always represent microscopic systems by *two-level systems*, chosen according to the energy of the applied radiation. Table 4.3.2 indicates the changes occurring during the three processes.

Figure 4.3.4 specifies the model more quantitatively.

Table 4.3.2 — **Characteristics of the three Einstein processes.**

Process	Photon	Microscopic system
Absorption	Destroyed	1-2 level change with gain of energy
Spontaneous emission	Created by interaction with the electromagnetic radiation of the vacuum; random phase and direction	2-1 level change with loss of energy
Stimulated emission	Second photon created by the interaction of the first one with the sample. It has the same frequency, the same direction, and the same phase as the first photon.	2-1 level change with loss of energy

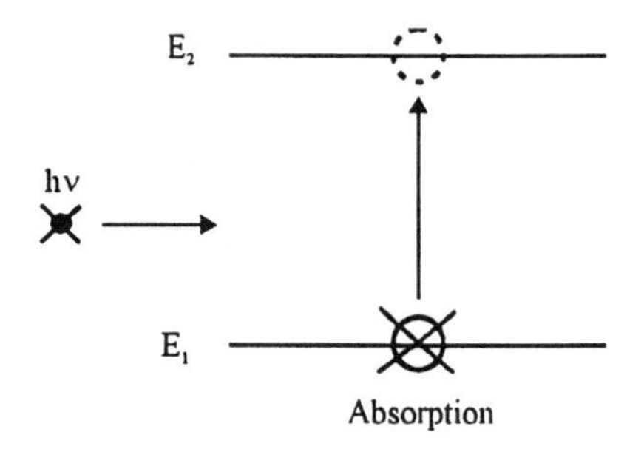

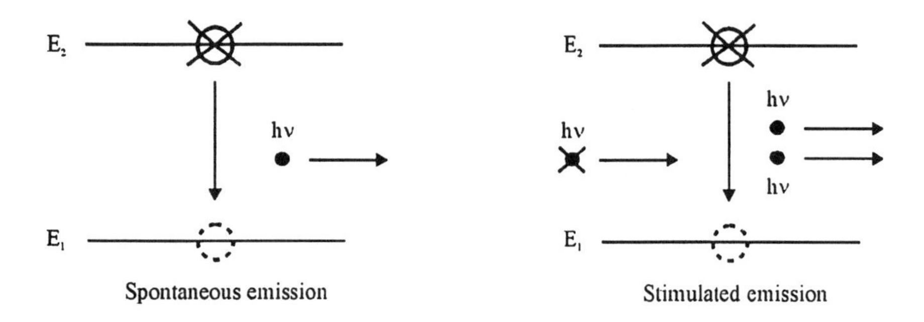

Figure 4.3.4: Illustration of the radiation-matter interaction processes in the phenomenological theory.

In all the following, except when indicated otherwise, we shall assume that states 1 and 2 have the same degeneracy (*cf.* Section III.3.1). The radiation is characterized by a time-average energy density, expressed in $J.m^{-3}$. The three terms B_{12}, A_{21} and B_{21} represent the transition probabilities per unit time (they are expressed in s^{-1}). If only absorption is taken into account, the rate equations express the rate of change of the populations with time (the population is the number of microscopic systems in each state per unit volume). These rate equations can be written as

$$\dot{N}_1 = -\dot{N}_2 = -N_1 B_{12} \bar{\rho} \tag{4.3.14}$$

In classical physics the two constant B_{12} and B_{21} can be shown to be equal (*cf.* Bibliographic reference 4.5.1).

The fundamental problem of this phenomenological theory is that we cannot specify the real physical content of the Einstein constants so that we are unable to express them analytically. Table 4.3.3 shows the advantages and the drawbacks of these two classical descriptions.

Table 4.3.3 — **Advantages and drawbacks of the two classical descriptions.**

	Advantages	Drawbacks
Classical microscopic theory (CMT)	Simple description of the interaction using the susceptibility and the refractive index	Incomplete description which ignores stimulated emission
Einstein's phenomenological theory (EPT)	Simple and efficient	No information concerning the physical content of the coefficients

This table shows us that the two previous descriptions are not satisfactory. One solution, which has long been used, consists in combining the two descriptions to associate their advantages while avoiding their drawbacks.

Multiplying Equation 4.3.14 by hν, setting the average power of the radiation equal to $-\dot{N}_1 h\nu$ (Equation 4.3.7), and writing $\bar{\rho} = (1/2)\varepsilon|\underline{E}(\omega)|^2$ gives us

$$B_{12} = (2\pi\tau e^2\omega / \varepsilon m h)\left\{1/\left[\tau^2(\omega_0^2 - \omega)^2 + \omega^2\right]\right\} \tag{4.3.15}$$

Thus, we gave meaning to the phenomenological constant with the help of the classical microscopic theory. Among others, we can see that this constant has a resonant character and that close to resonance, it is proportional to the time relaxation constant, which is inversely proportional to the width of the resonance.

But, of course, the two previous descriptions are not really satisfactory. Indeed, as said earlier, they do not describe the radiation-matter interaction in terms of atomic or molecular parameters. However, although the energy of the incoming radiation favors a direct interaction with the electrons, it must be well understood that the properties of these electrons are closely tied to those of the whole microscopic system, which remains the true interaction partner. This observation has led physicists of the beginning of this century to introduce the arbitrary notion of *oscillator force* in order to focus on a few electronic sites and so as to be able to take into account in a correct way the very large diversity of experimentally observed absorption cases.

Currently, only a quantum treatment can lead to a satisfactory theoretical description of the phenomenon of atomic and molecular absorption.

IV.3. 2. Semi-classical description

IV.3.2.1. General features: Useful approximations

Chapter III introduced us to non relativistic quantum mechanics. In that chapter, we spelled out its principles and specified their consequences concerning the description of the electronic properties of atoms and molecules. We indicated that radiation and matter cannot be separated from each other and should therefore be treated identically. Radiation and matter are both quantified and must be considered as partners in an single treatment which takes into account all the properties of the photon-material system. In Bibliographic reference 4.6.3, there are some remarkable examples of the complete quantum treatment of several cases of radiation-matter interaction. However, such a treatment uses a relatively complex formalism. We chose to remain in the so-called semi-classical approximation in which matter alone is quantified (semi-classical approximation SCA), and the radiation is treated classically (*cf.* Section II.2). Moreover, we shall assume that the described phenomena are non relativistic (so we can use the non-relativistic approximation, NRA). This second approximation is valid because at 10^{18} Hz, three orders of magnitude beyond the highest frequency limit indicated in Table 4.1.1, the speed of the electron is still not greater than c/10. Finally, and before starting the detailed presentation of the absorption phenomenon, we must indicate two additional approximations used here:

— *Two-levels system approximation* (TLSA): The classical presentation, as seen earlier, showed us that absorption was a resonant phenomenon, of importance only when the frequency of the incoming radiation is close to the characteristic frequency of the system. However, quantum mechanics teaches us that the energy of a microscopic system comes in discrete levels, each level n having energy

$E_n = h\nu_n$. As in the phenomenological theory, the absorption of the radiation will drive the system in a state of higher energy, noted for example $E_m = h\nu_m$. If ν represents the frequency of the wave, the phenomenon will be important only if $\nu \approx \nu_m - \nu_n$. So therefore, the frequency ν of the wave selects only two levels in the infinity of possible energy levels of the system, namely those levels n and m — if they exist — whose frequency difference has a value close to ν. In this way, quantum treatment will be limited to the treatment of a two-level system. However, exchanges with other, non specified levels may also have to be considered in some situations. In these cases though, we shall step out of the quantum model. We shall strictly reserve the quantum model to describe the exchanges between the two indicated levels, as described in detail below.

— *Electric dipole approximation* (EDA): This approximation is treated in detail in Problem IV.4.7, where the reader will find a complete demonstration. We have previously indicated the fundamental rôle played by the Hamiltonian operator in the description of microscopic systems (*cf.* Section III.1.2). Principle P7 tells us how to construct this operator starting from the expression of the energy of the system in interaction with the radiation. Before applying the correspondence principle, the energy should be calculated classically. In the case of an atom with atomic number Z, with the origin of the coordinate frame assumed to be on the nucleus, the Hamiltonian function (function of the spatial coordinate $\mathbf{r}_j$ of the jth electron, of its time derivative $\dot{\mathbf{r}}_j$ and of time t) is written as

$$H(\mathbf{r}_j, \dot{\mathbf{r}}_j, t) = \sum_j \left[\left(m_e \dot{\mathbf{r}}_j^2 \right) - \left(e\Phi(\mathbf{r}_j)/2 \right) \right] + Ze\Phi(O)/2 - \mathbf{pE}(0,t) \qquad (4.3.16)$$

In this expression, $\mathbf{p}$ represents the electric dipole moment of all the electrons and takes the form

$$\mathbf{p} = -e\sum_j \mathbf{r}_j \qquad (4.3.17)$$

The interpretation of the first three terms of Equation 4.3.16 is quite simple. Without difficulty, we identify successively the kinetic energy of electrons, the potential energy of the electrons imbedded in the static electric field created by the electric charge distribution of the system [we must not forget that here we are in the *stationary state* hypothesis (*cf.* III.1.3.2) and that therefore a static distribution of electricity exists all around the nucleus], and, last, the potential energy of the nucleus placed at the coordinate origin. In this reference frame, we do not need to consider the kinetic energy of the nucleus. The last term represents the radiation-atom interaction in the electric dipole approximation (EDA).

In Bibliographic reference 3.6.1 the reader will find the demonstration of the relation leading to this form of the interaction energy. It can be written in the form $e\mathbf{r}_j \mathbf{A}(\mathbf{r}_j, t)$, where **A** is the potential vector driving the electric field **E** carried by the optical wave (*cf.* Section II.2.1.1). This field must not be confused with the just mentioned static field which is due to the distribution of static electric charges in the atom and which is independent of the applied radiation. When we express the potential vector by a power expansion as a function of **r** around the origin of the coordinates, we see that only the first term **A**(0,t) is important. The second term $\mathbf{r}_j\left[\nabla\mathbf{A}(\mathbf{r}_j, t)\right]_{\mathbf{r}_j=0}$ is very small when compared to the first. Indeed, the potential vector has a characteristic dimension of a few hundreds nanometers (radiation wavelength) and its spatial gradient is very small in the vicinity of the nucleus defined by the spatial extension of coordinate **r** (a few tenths of nanometers). Three orders of magnitude lie between these two characteristic lengths, and the second term of the power expansion can be neglected with respect to the first. Of course, the same is true for the next terms. We can therefore understand why the last term of the interaction in the right member of Equation 4.3.16 is localized at the origin O of the coordinates. Going from expression $e\mathbf{r}_j \mathbf{A}(0, t)$ to the one written in Equation 4.3.16 is not trivial. To do this correctly, one must remember that the Lagrange function, often used in mechanics, and from which the previously defined Hamilton function can be derived, preserves its main property — to satisfy the Lagrange equations — when certain types of functions are added to it. In particular, one can add to the Lagrangian the function $-ed[\sum_j \mathbf{r}_j\mathbf{A}(0,t)]/dt$, where the symbol d/dt indicates the complete derivation with respect to time. In our case, the interaction term can be expressed in this form

$$-e\sum_j \mathbf{r}_j \dot{\mathbf{A}}(0,t) = e\sum_j \mathbf{r}_j\mathbf{E}(0,t) = -\mathbf{pE}(0,t)$$

since the field derives from the potential vector by the relation $\mathbf{E} = -\dot{\mathbf{A}}$ (*cf.* Section II.2.1).

Let us note that in the above power expansion we neglected (i) higher order terms (i.e. higher than zero) corresponding to multipolar interactions, which are usually small when compared to the dipole interaction term and (ii) terms describing the interaction between electrons and the magnetic field induced by the wave, which are usually negligible (*cf.* Problem IV.4.3). We will see later that these interactions, though they are about two orders of magnitude smaller than the electric dipole interaction, sometimes play a certain part in optical absorption spectroscopy.

Let us now calculate the absorption coefficient.

IV.3.2.2. Calculation of the absorption in the semi-classical theory

The microscopic classic treatment shows us how to proceed. We need to calculate the absorption coefficient, equal to twice the value of the argument of the exponential of Equation 4.3.13 divided by Nx. This is the *Beer-Lambert coefficient* (*cf.* Problem IV.4.8). This coefficient is proportional to the imaginary part of the refractive index, and to that of the susceptibility, so therefore also to that of the polarization induced by the electric field of the optical wave.

So it is the polarization that we must express by using SCT. By polarization, we mean the average polarization per unit volume containing N microscopic systems. The word average deserves to be commented on: (i) it is a *quantum* kind of average, in the sense that a measurement of the physical property of polarization can usually only result in an average value (*cf.* Section III.1.2), and (ii) it is also *statistical*, because the N systems are probably not in the same quantum state. However, Section III.1.1 taught us that we can in fact use two possible descriptions: One based on the *state vector formalism* and another using the *density operator*. Which should we choose? The polarization is very simply defined in the state vector description. However, the representation using the density operator presents a considerable advantage that we shall explain presently.

Let us consider a collection of $N = N_1 + N_2$ microscopic systems in which N_1 systems are in state $|\psi_1\rangle$ and N_2 systems in state $|\psi_2\rangle$. The average probability $P_N(a_n)$, reduced to one system, to find the eigenvalue a_n of operator A corresponding to eigenstate $|u_n\rangle$ when we measure the unspecified physical property A, is given by

$$\begin{aligned} P_N(a_n) &= (1/N)\left[N_1 P^{(1)}(a_n) + N_2 P^{(2)}(a_n)\right] \\ &= p_1\left|c_n^{(1)}\right|^2 + p_2\left|c_n^{(2)}\right|^2 \end{aligned} \qquad (4.3.18)$$

The coefficients c are the expansion coefficients of the two state vectors on base $|u_n\rangle$; $p_1 = N_1/N$ and $p_2 = N_2/N$ are the statistical weights of the two states. We can look for a possible representation of the mixed statistical state by constructing an average state vector which would be written as

$$|\psi\rangle = \lambda_1|\psi_1\rangle + \lambda_2|\psi_2\rangle = \sum_n (\lambda_1 c_n^{(1)} + \lambda_2 c_n^{(2)})|u_n\rangle \qquad (4.3.19)$$

Don't forget that a representation of the N systems by such an average state means that *each* of the systems must be correctly represented by this state vector. This system then has the probability $|\lambda_1|^2$ to be in state $|\psi_1\rangle$, and $|\lambda_2|^2$ to be in state $|\psi_2\rangle$. We can ask if, by choosing $|\lambda_1|^2 = p_1$ and $|\lambda_2|^2 = p_2$ the wave vector would describe the mixed state correctly. To check this, let us calculate the probability $P_N(a_n)$ of state $|\psi\rangle$ written as

$$\begin{aligned} P_N(a_n) &= |\langle u_n|\psi\rangle|^2 = |\lambda_1\langle u_n|\psi_1\rangle + \lambda_2\langle u_n|\psi_2\rangle|^2 \\ &= |\lambda_1|^2|c_n^{(1)}|^2 + |\lambda_2|^2|c_n^{(2)}|^2 + \lambda_1\lambda_2^* c_n^{(1)} c_n^{(2)*} + \text{c.c} \end{aligned} \qquad (4.3.20)$$

We clearly see that the rectangular terms of the expansion imply that we cannot represent the mixed state by an average vector, in spite of the fact that the mixed state truly describes the real physical situation. This is a fundamental limitation of the state vector representation, so we prefer to use the representation by the *population rate operator* (an extension of the density operator), after showing that it does not have this particular drawback.

In Section III.1, we defined the density operator ρ, and we showed that its evolution was governed by the Liouville equation. We also showed how to express the average value of an observable using the trace of the product of the density operator by the operator describing the observable. Let us take the expression of the probability defined by Equation 4.3.20, and let us rewrite it by factoring the square of its modulus

$$P_N(a_n) = \langle\psi|u_n\rangle\langle u_n|\psi\rangle$$

It turns out to be the quantum average, calculated in state $|\psi\rangle$, of an operator called *projector*, and defined by

$$P_n = |u_n\rangle\langle u_n| \qquad (4.3.21)$$

So we obtain

$$P_N(a_n) = \text{Tr}(\rho P_n) \qquad (4.3.21\text{bis})$$

This equation is very interesting. Whereas Equation 4.3.20 showed that the probability is expressed as a quadratic function of coefficients of the state vector expansion on the chosen base, this relation shows the linearity of the same probability with respect to the density operator. Then, if we generalize our binary distribution to a multi-state distribution, characterized by the statistical probability p_k to find the system in state $|\psi_k\rangle$ (and of course with $\sum_k p_k = 1$), we obtain

$$P_N(a_n) = \sum_k p_k \mathrm{Tr}(\rho_k P_n) = \mathrm{Tr}\left[(\sum_k p_k \rho_k) P_n\right]$$

This is identical to Equation 4.3.21 if the average density operator, called ρ to keep things simple, is defined by

$$\rho = \sum_k p_k \rho_k \tag{4.3.22}$$

Having solved the problem of the statistical distribution of the state vectors representing the microscopic systems, we can now generalize the notion of average density operator in order to take into account two very important physical phenomena that we will need to introduce into our SCA treatment: They are the phenomena of *relaxation* and of *population increase*. But let us first look at TLSA in absence of radiation. Let $|a\rangle$ and $|b\rangle$ be two levels, with $|a\rangle$ the level of lowest energy, and let us calculate the components of the time derivatives of the density operator on this binary orthonormal basis. If for each state we multiply the Liouville equation on the left by the *bra*, and on the right by the corresponding *ket* (Equation 3.1.23), we obtain

$$\begin{aligned} \dot{\rho}_{aa} &= 0 \\ \dot{\rho}_{bb} &= 0 \\ \dot{\rho}_{ba} &= -i\omega_0 \rho_{ba} \end{aligned} \tag{4.3.23}$$

Remembering the definition of the density operator, the reader will easily see that $\rho_{aa} = |\langle a|\psi\rangle|^2$ and $\rho_{bb} = |\langle b|\psi\rangle|^2$ represent the probabilities of finding the system in states $|a\rangle$ and $|b\rangle$, respectively, when it is in the average state $|\psi\rangle$. These values are called the *populations* of the two states. The third term, $\rho_{ba} = \langle b|\psi\rangle\langle\psi|a\rangle$ represents the quantum links which exist between the two states and is called the *coherence term*. As could have been predicted for this model of an isolated system, we see that the populations remain constant whereas the coherence oscillates at the resonant angular frequency $\omega_0 = \omega_b - \omega_a$ of the system. Still, this description remains very unrealistic even though the spectacular progress of recent years in the manipulation of atoms by lasers could well mean that in the near future we shall be able to isolate atoms by optical means. As a matter of fact, the quantum model we just discussed needs some adjustments. Figure 4.3.5 indicates the extra elements we added to our first model.

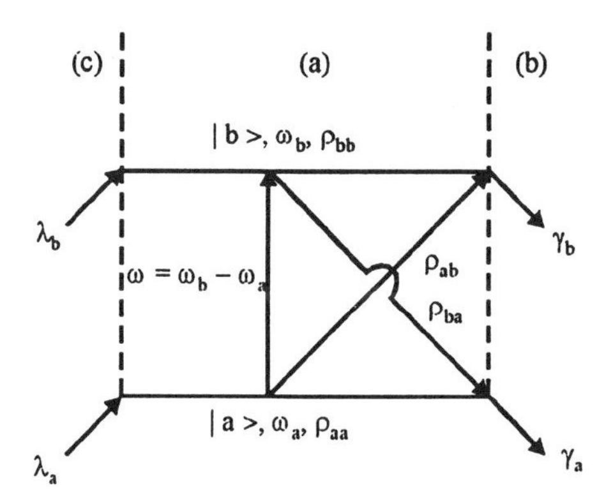

Figure 4.3.5: Phenomenological modelling of relaxation and population.
(a) Isolated system; (b) relaxation; (c) population.

We have introduced:

— The phenomenon of relaxation (constants γ_a, γ_b, γ_{ba}) describing, among others, the elastic and inelastic collisions undergone by the microscopic particles, their coupling with the thermal radiation and the electromagnetic field of the vacuum (the latter being responsible for spontaneous emission), etc.

— The phenomenon of population increase (through the rate constants λ_a and λ_b), describing a possible preparation of levels a and b. Indeed more and more often, absorption spectroscopy experiments are performed under constraint, i.e. the system is correlatively submitted to some external radiation that modifies the populations of the levels concerned by the interaction. We will take into account the rate of change of the population by replacing the density operator by the population rate operator, which we will also call ρ since in the rest of this book, it will be used systematically instead of the operator density (*cf.* Appendix IV.5.1).

Thus, Equation 4.3.23 becomes

$$\dot{\rho}_{aa} = 0 - \gamma_a \rho_{aa} + \lambda_a$$

$$\dot{\rho}_{bb} = 0 - \gamma_b \rho_{bb} + \lambda_b \qquad (4.3.24)$$

$$\dot{\rho}_{ba} = -i\omega_0 \rho_{ba} - \gamma_{ba} \rho_{ba}$$

and the reader can check that the Liouville equation can be replaced by

$$\dot{\rho} = (2\text{p} / \text{ih})[H_0, \rho] - (1/2)(\Gamma\rho + \rho\Gamma) + \chi \tag{4.3.25}$$

where the relaxation and population matrixes are, respectively,

$$\Gamma = \begin{pmatrix} \gamma_a & 0 \\ 0 & \gamma_b \end{pmatrix} \quad \text{relaxation matrice}$$
$$\chi = \begin{pmatrix} \lambda_a & 0 \\ 0 & \lambda_b \end{pmatrix} \quad \text{population matrice} \tag{4.3.26}$$

ρ now represents the *population rate operator (cf.* Appendix IV.5.1).

Please note that this equation yields $\gamma_{ba} = (\gamma_a + \gamma_b)/2$. The relaxation constant of the coherence can be easily calculated from the relaxation constants of the populations. But in fact, some times there is an additional, intrinsic contribution to this term, in particular in the case of elastic collisions (which preserve populations while modifying the coherence), so we prefer to keep the more general notation γ_{ba}.

Let us now calculate the average value of the polarization. We can write it as

$$<|\mathbf{P}|> = \text{N} <|\mathbf{p}|> = -\text{e}\,\text{N} <|\mathbf{r}|> = -\text{eN}\,\text{Tr}(\rho \boldsymbol{r}) \tag{4.3.27}$$

In the above defined binary base, operator ρ is non diagonal: $\boldsymbol{r}$ is an even spatial function and the integrals $<\text{a}|\boldsymbol{r}|\text{a}>$ et $<\text{b}|\boldsymbol{r}|\text{b}>$ are even functions, whatever the parities of vectors a et b. The integration over all space therefore vanishes (we have assumed that the Hamiltonian describing the isolated system commutes with the parity operator, i.e. that the parities of the vectors a and b are well defined). The trace is easy to calculate. By defining p as $\text{p} = <\text{a}|\boldsymbol{p}|\text{b}> = <\text{b}|\boldsymbol{p}|\text{a}>$, i.e. by assuming that p is a real parameter (a condition which can always be satisfied), we obtain

$$<|\text{P}|> = \text{Np}(\rho_{ba} + \rho_{ab}) = 2\text{Np}\,\text{Re}(\rho_{ba}) \tag{4.3.28}$$

Therefore, we need to calculate the coherence, then to take its real part. Equation 4.3.24 needs to be modified since the Hamiltonian must take into account the wave-matter interaction, expressed by using the EDA approximation. The complete Hamiltonian writes

$$H = H_0 - \boldsymbol{p}\mathbf{E} \tag{4.3.29}$$

and as an exercise the reader can check that Equation 4.3.24 takes the form

$$\dot{\rho}_{aa} = \lambda_a - \gamma_a \rho_{aa} + (2\pi i / h)\left[pE(0,t)\rho_{ba} - c.c\right]$$
$$\dot{\rho}_{bb} = \lambda_b - \gamma_b \rho_{bb} - (2\pi i / h)\left[pE(0,t)\rho_{ba} - c.c\right] \tag{4.3.30}$$
$$\dot{\rho}_{ba} = -(i\omega_0 + \gamma_{ba})\rho_{ba} + (2\pi i / h)pE(0,t)(\rho_{aa} - \rho_{bb})$$

Let us write the electric field of the wave as

$$E(0,t) = (E_0/2)[\exp(i\omega t) + \exp(-i\omega t)] \tag{4.3.31}$$

Let us then integrate the third equation of 4.3.20. We obtain

$$\rho_{ba}(t)\exp\left[(i\omega_0 + \gamma_{ba})t\right] \approx (i\pi / h)$$
$$\times \int_{-\infty}^{t} dt' pE_0 \exp(-i\omega t')\exp\left[(i\omega_0 + \gamma_{ba})t'\left[\rho_{aa}(t') - \rho_{bb}(t')\right]\right]$$

We only kept the second term of the right hand member of Equation 4.3.31. Indeed, when integrated over t′, this term gives a divergent contribution in $1/i(\omega_0 - \omega)$, because the integration is performed near resonance. The first term would give a nonresonant contribution in $1/i2\omega_0$. In this new approximation, known under the name of *the rotating wave approximation* (RWA), we can take into account only one of the two circular waves described by Equation 4.3.31. Moreover, the diagonal terms of derivatives with respect to time of the population rate operator (Equation 4.3.30) do not have a pure quantum contribution, as opposed to the non diagonal term. As a result, we can, in a first approximation, assume the population difference to be constant. This is called the *rate equations approximation* (REA). We obtain

$$\begin{aligned} \rho_{ba}(t) \approx (i\pi / h)pE_0\left[\rho_{aa}(t) - \rho_{bb}(t)\right] \\ \times \exp(-i\omega t) / \left[i(\omega_0 - \omega) + \gamma_{ba}\right] \end{aligned} \tag{4.3.32}$$

We can report this result in the first two equations of 4.3.30, and calculate the populations and their difference D. We find

$$\begin{aligned} D &= \left[\rho_{aa}(t) - \rho_{bb}(t)\right] \\ &= D_0 / \left\{1 + S / \left[1 + (\omega_0 - \omega)^2 / \gamma_{ba}{}^2\right]\right\} \end{aligned} \tag{4.3.33}$$

Where S is a saturation parameter defined as

$$S = (1/2)(2\pi pE_0 / h)^2(\gamma_a + \gamma_b) / \gamma_a \gamma_b \gamma_{ba} \tag{4.3.34}$$

Equation (IV.3.33) expresses a fundamental result. If, before the application of the wave, only the ground state was populated, we find $D_0 = \rho_{aa}$. S being proportional to E^2, the intensity of the wave, the optical field decreases the population of ground state a and the systems are lifted into state b, which we shall call the *excited state*. The closer one is to resonance, the more this description is true. If S is infinitely large, D reaches the value $D_0/2$. Thus, the best we can obtain is equal populations for levels a and b. As a fundamental consequence, we see that it is impossible to obtain populations inversion in a two-levels system and therefore laser effect is excluded in this case.

Let us look again at the definition of the saturation parameter because it encloses a second very important result. It contains two kinds of parameters: (i) those connected to the optical wave (E_0) and (ii) those describing the properties of matter (all other parameters). The population decrease of the ground state induced by the absorption therefore depends not only on the intensity of the applied radiation, but also on the properties of the sample and, in particular, on parameter p called the *transition moment*. If the value of the transition moment is very large, a wave of weak intensity will be sufficient to decrease the population of the ground level to the benefit of the excited state.

By reporting Equations 4.3.32 and 4.3.33 into Equation 4.3.28, we can express the average polarization induced by the field as

$$\begin{aligned} <|P|> &= (2\pi p^2 E_0 D_0 / h\gamma_{ba}) \\ &\times \left[(\omega_0 - \omega)/\gamma_{ba} + i\right] / \left\{1 + S + \left[(\omega_0 - \omega)/\gamma_{ba}\right]^2\right\} \end{aligned} \tag{4.3.35}$$

From this expression, we can extract successively the susceptibility (Equation 4.3.11), the refractive index (Equation 4.3.12), and the absorption coefficient (Equation 4.3.13) relative to the intensity, which is written as

$$\begin{aligned} \alpha &= -(4\pi^2 N p^2 D_0 / \lambda \varepsilon_0 h \gamma_{ba}) \\ &\times 1 / \left\{1 + S + \left[(\omega_0 - \omega)/\gamma_{ba}\right]^2\right\} \end{aligned} \tag{4.3.36}$$

The transmitted light intensity I_t of the wave varies with coordinate r as

$$I_t = I_0 \exp(\alpha r) \tag{4.3.37}$$

Equation 4.3.36 is very important for the rest of this chapter because it contains all the parameters describing the phenomenon of absorption. We shall study each of them in detail, indicating the relevant parameter of the equation in parentheses for every case.

IV.3.2.3. The absorption phenomenon and its physical content

— Absorption condition (D_0)

Absorption occurs if $I_t < I_0$, that is, if α is a strictly negative quantity. All the parameters included in Equation 4.3.36 are positive, excepted D_0 whose sign depends on the population difference between the two levels. For absorption to occur therefore, we need to have $\rho_{aa}(0) > \rho_{bb}(0)$, i.e. the population of the ground state must be greater than that of the excited state. We have already seen that this is practically always the case in two-levels systems. However, we should not forget that we used a model. Reality is rather different because in fact a microscopic system has an infinite number of excited levels. If, on the contrary, we have $\rho_{aa}(0) < \rho_{bb}(0)$, the excited state will be more populated than the fundamental state, D_0 will be negative and the positive value of the absorption coefficient expresses a gain of intensity of the radiation, because in this case stimulated emission is dominant.

— Importance of the intensity of the wave (S)

The absorption coefficient is directly linked to the amplitude of the electric field E, and therefore to the intensity of the applied wave. If E increases, S also increases and α decreases. The absorption phenomenon disappears progressively when the intensity of the wave increases: This is the *phenomenon of saturation*. But here we should bear in mind an important point. In fact, it is the product pE that appears in S. If the transition moment p is large (a system very sensitive to radiation), even a weak optical electric field may lead to saturation. Let us assume we are at resonance and let us examine two limit cases:

$S \ll 1$ (optical wave of very weak intensity, or (and) a sample not very sensitive to the radiation, i.e. with a very small transition moment p):

In this case, α does not depend on E. The absorption is weak and does not modify the population of the ground state very much ($\rho_{aa} \gg \rho_{bb}$). In these conditions, the concerned populations remain constant and the Beer-Lambert law applies. In fact, this law should only be considered as a limit law, and to avoid false interpretations, conditions should be tested carefully before applying it.

$S \gg 1$ (optical wave of very large intensity and (or) sample very sensitive to the radiation, i.e. with a large transition moment):

In this case, the optical wave is so strong that it has practically erased the absorption phenomenon ($\alpha = 0$). D is now vanishingly small (Equation 4.3.33). In fact, the populations of the two levels are equal and the absorption is exactly compensated by the stimulated emission. In a first approximation, this mechanism yields an approximate description of *induced and self-induced transparencies*: A wave with the same frequency (or the same wave) can cross the thus prepared sample without any apparent attenuation of its energy. These phenomena have been widely studied in recent years in laser physics. They have found many applications, especially in optoelectronics and in laser technology where the saturation phenomenon is used to built the passive modulators needed to obtain the so-called Q-switched and mode-locked regimes (*cf.* Problem VII.5.2).

— *Role of the microscopic system density (N)*

The absorption coefficient is proportional to the number density of the microscopic systems (number of microscopic systems per unit volume).

— *Role of the wavelength of the radiation ($\omega_0 - \omega$)*

The absorption coefficient is a *resonant* parameter (for all finite linewidths). Figure 4.3.6 illustrates this kind of frequency dependence.

The same figure also shows the phenomenon of saturation. The effect of the wave on the sample lessens progressively as the frequency of the optical wave moves away from resonance.

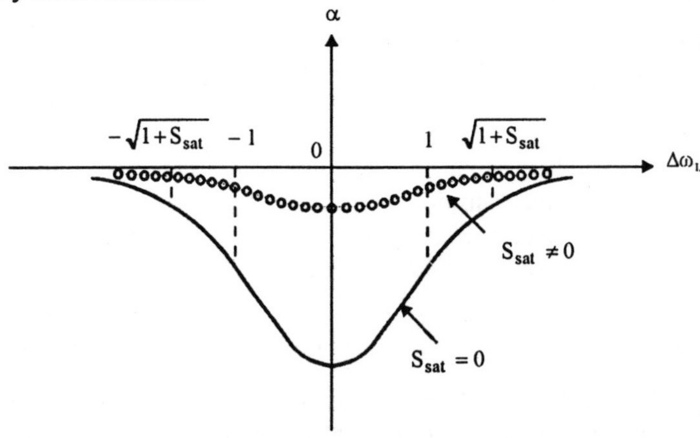

Figure 4.3.6: Variations of the absorption coefficient with frequency and the effect of saturation.

— *Role of the relaxation parameter (linewidth γ_{ba})*

What do we observe, theoretically, when recording an absorption spectrum? The absorption coefficient is divided by two when $\omega_0 - \omega = \sqrt{1+S}\,\gamma_{ba}$. Recorded curves, which can be linear (transmission) or logarithmic (absorbance) are characterized by a linewidth proportional to $\omega_0 - \omega = \sqrt{1+S}\,\gamma_{ba}$. *The linewidth always depends on the intensity of the applied wave.* Only when S = 0 can the linewidth be used to determine the dipole moment and the coherence relaxation.

So here emerges a new cause of homogeneous broadening, which was not predicted by classical physics. In fact, other causes for broadening exist, mostly nonradiative ones, which were not mentioned in the previous classical approach. Together, they result in the absorption *linewidth*. For simplicity, we shall give a brief summary and classification of the contributions to the linewidth, referring the reader to the bibliography for more information.

(i) natural linewidth. It is the linewidth corresponding to a single completely isolated microscopic system. The relaxation constant is $\gamma_{ba,rad1}$. It is linked to the *natural lifetime* of the transition (previously noted τ_1) in the following way: $(1/\tau_1) = \gamma_{ba,rad1}$.

(ii) homogeneous broadening. It is mainly due to:

— The radiation intensity (already mentioned),

— Phase loss phenomena, described in the microscopic approach, which are taken into account by the phenomenological constants τ_2 and $\gamma_{ba,rad2}$, related to each other by $(1/\tau_2) = \gamma_{ba,rad2}$.

— Homogeneous nonradiative phenomena, due, for instance, to collisions, mainly occurring in gases. The number of these collisions is closely linked to the pressure. In a first approximation, the time separating two collisions is inversely proportional to the pressure, and the broadening is therefore proportional to the pressure. This effect plays an especially important part in ionic gas lasers (*cf.* Problem IV.4.6). This phenomenon is taken into account by the constant γ_{nrad}.

Altogether, homogeneous broadening can be described by the global constant:

$$\gamma_{ba} = \gamma_{ba,rad1} + \gamma_{ba,rad2} + \gamma_{ba,nrad} \tag{4.3.38}$$

The natural line has a *Lorentzian* shape described by Equation 4.3.36.

(iii) inhomogeneous broadening. It corresponds to a kind of broadening which is not the same for all the microscopic systems and it is mainly due to:

— In the solids, the diversity of particle environments induced by the imperfections of the lattice,

— In liquids and gases, the Doppler effect which we mentioned previously when presenting the microscopic classical treatment. In this case, Equation 4.3.36 is no longer valid. Because of the Doppler effect, the observer detects a continuous distribution of resonance frequencies in the reference frame of the laboratory. Or, to express it in another way, each micro-system as it moves in its own frame of reference, sees a wave of angular frequency ω' such that

$$\omega' = \omega - \mathbf{kv} \tag{4.3.39}$$

where **k** and **v** are, respectively, the wave vector and the velocity of the system. According to the relative directions of the wave vector and of the velocity, ω' can be larger or smaller than ω. Defining v as the component of **v** along **k**, Maxwell-Boltzmann statistics give the distribution of the speeds around the value v = 0 in the following Gaussian form:

$$f(v) = (1/\sqrt{\pi}v_0)\exp\left[-(v/v_0)^2\right] \tag{4.3.40}$$

v_0 being a characteristic velocity defined by $v_0^2 = 2kT/m = 2RT/M$ (k and R are, respectively, the Boltzmann constant and the perfect gas constant, m and M the molecular and molar masses, and T the temperature).

If in Equation 4.3.36, we affect the absorption for each frequency ω' with the number of microsystems which can interact at this frequency which is f(v), where v is the speed corresponding to ω' in Equation 4.3.39 we obtain a curve with a Gaussian profile. This Gaussian shape, which completely occults the Lorentzian shape of the homogeneously broadened line, is observed in practically all experimental situations involving atomic and molecular gases. As a simple demonstration of this result, let us take the case of linear motion along the Oz axis. To find the average polarization, we sum Equation 4.3.35 over the space of velocities:

$$\begin{aligned} <\mathbf{P}> &= \left(2\pi P^2 E_0 D_0 / h\gamma_{ba}\sqrt{\pi}v_0\right) \\ &\times \int_{-\infty}^{+\infty} \exp(-v_z/v_0)^2 \left\{\left[(\omega_0 - \omega + kv_z)/\gamma_{ba}\right] + i\right\} \\ &\times 1/\left\{1 + S + \left[(\omega_0 - \omega + kv_z)/\gamma_{ba}\right]^2\right\} \end{aligned} \tag{4.3.35bis}$$

To obtain the Gaussian absorption, we integrate the previous equation at the Doppler limit, i.e. when $kv_0 >> \gamma_{ba}\sqrt{1+S}$, and express the imaginary part of the refractive index:

$$\begin{aligned} \alpha &= -\left(2\pi\sqrt{\pi}D_0P^2/\varepsilon_0 h v_0\right) \\ &\times \left(1/\sqrt{1+S}\right)\exp\left\{-\left[(\omega_0 - \omega)/kv_0\right]^2\right\} \end{aligned} \tag{4.3.36bis}$$

This Doppler broadening is studied in detail in Problems IV.4.4 and IV.4.6, and Chapter VII presents a few so called " Doppler Free " absorption spectroscopy techniques.

— Significance of the transition moment (p); selection rules

We have seen that, with the EDA, the transition moment is nonvanishing only if the two states concerned by the absorption have different parities. This constitutes the first selection rule. However, we may wonder whether this condition is sufficient. The answer is no. A transition between two states of different parities can still be forbidden because the transition moment has a zero value. Therefore, the rule needs to be refined. This certainly constitutes the most difficult problem of this chapter. The difficulty resides essentially in the fact that we must combine relatively easy cases (the hydrogen atom, for instance) with extremely complex cases (a heavy complex ion, for instance, for which we do not even know how to write the state vector correctly). We are going to try to classify, in order of increasing difficulty, the three cases constituted by the one-electron atom, the many-electron atom and the molecule. We must keep in mind that at present we are interested only in one-photon transitions. Multi-photon transitions, which will be described briefly in Chapter VII, obey to more complex selection rules but these can be derived from the same concepts.

(i) One-electron atom: This means the hydrogen atom or atoms in the so-called hydrogen-atom approximation *(cf.* Section III.3.1). We have already seen that their states are described by vectors of type

$$\Phi = R_{n,l}(r)Y_{l,m}(\theta,\varphi) \tag{4.3.41}$$

where R is a function derived from Laguerre polynomials and Y are spherical harmonics. r, θ and φ are the spherical coordinates. For a $|\Phi\rangle \to |\Phi'\rangle$ transition to be allowed , we must have

$$\langle \Phi'|\boldsymbol{r}|\Phi\rangle \neq 0 \tag{4.3.42}$$

that is, the three following equations must be satisfied:

$$\begin{aligned}
&\int_0^{+\infty} r^3 R_{n',l'} R_{n,l} dr \neq 0 \\
&\int_0^{\pi} P_{l',m'} \begin{bmatrix} \sin\theta \\ \sin\theta \\ \cos\theta \end{bmatrix} P_{l,m} \sin\theta d\theta \neq 0 \\
&\int_0^{2\pi} \exp(-im'\varphi) \begin{bmatrix} \cos\varphi \\ \sin\varphi \\ 1 \end{bmatrix} \exp(im\varphi) d\varphi \neq 0
\end{aligned} \tag{4.3.43}$$

Column vectors represent the successive projections on the three 0x, 0y and 0z axes. Let us look at each one in detail.

z component: A nonzero value of the third integral implies $m' = m$, or $\Delta m = 0$. In this case, the integration factor of the second integral is $P_{l',m} \cos\theta P_{l,m}$. The properties of Legendre polynomials are such that $(2l+1)\cos\theta = (l-m+1)P_{l+1,m} + (l+m)P_{l-1,m}$. Here, two integration factors appear: $P_{l',m}P_{l+1,m}$ and $P_{l',m}P_{l-1,m}$. The two integrals can have non-zero values only if $l' = l \pm 1$ or $\Delta l = \pm 1$. The first integral is non-zero for all integral values of Δn.

x and y components: The third integrals given by Equation 4.3.43 have two integration factors that can be written as $\exp(-im'\varphi)\cos\varphi \exp(im\varphi)$ and $\exp(-im'\varphi)\sin\varphi \exp(im\varphi)$. These factors can also be expressed as $\exp\{-i\varphi[m'-(m+1)]\}$ and $\exp\{-i\varphi[m'-(m-1)]\}$. The two integrals can be nonvanishing only if $m' = m \pm 1$ or $\Delta m = \pm 1$. In this case, the second integral has the integration factor $P_{l',m\pm1} \sin\theta P_{l,m}$. Another property of Legendre polynomials is that $(2l+1)\sin\theta P_{l,m} = (P_{l+1,m+1} - P_{l-1,m+1})$. Using this yields integration factors $P_{l',m+1} P_{l+1,m+1}$; $P_{l',m+1} P_{l-1,m+1}$; $P_{l'+1,m} P_{l,m}$, and $P_{l'-1,m} P_{l,m}$. Nonzero values of the integrals imply that $l' = l \pm 1$ or $\Delta l = \pm 1$. As in the previous case, the first integral always has a nonzero value.

In short, transitions are allowed only between states of different parities and whose quantum number differences satisfy the conditions

$$\Delta n \text{ is an integer; } \Delta l = \pm 1 \text{; } \Delta m = 0 \text{ or } \pm 1 \quad (4.3.44)$$

Comments: 1) Although no restriction exists concerning the value of jump Δn of the principal quantum number, the amplitude of the first integral decreases rapidly when Δn increases, and the corresponding transitions become more and more difficult to observe.

2) Usually, we don't take into account the spin of the electron in the case of the one-electron atom, though there is no doubt about its existence. As a matter of fact, a fine structure of doublets due to the coupling of the orbital electronic kinetic moment to the spin can be observed in the spectrum of the hydrogen atom. But its effect is small and comparable to other relativistic effects not reported here. In the conjecture of a weak coupling energy (small compared to the total energy), we take account of the spin by using a method based on successive approximations. Figure 4.3.7 shows an illustration of the absorption spectrum of the hydrogen atom in its ground state.

The inclusion of the fine structure imposes an additional selection rule, $\Delta j = 0$

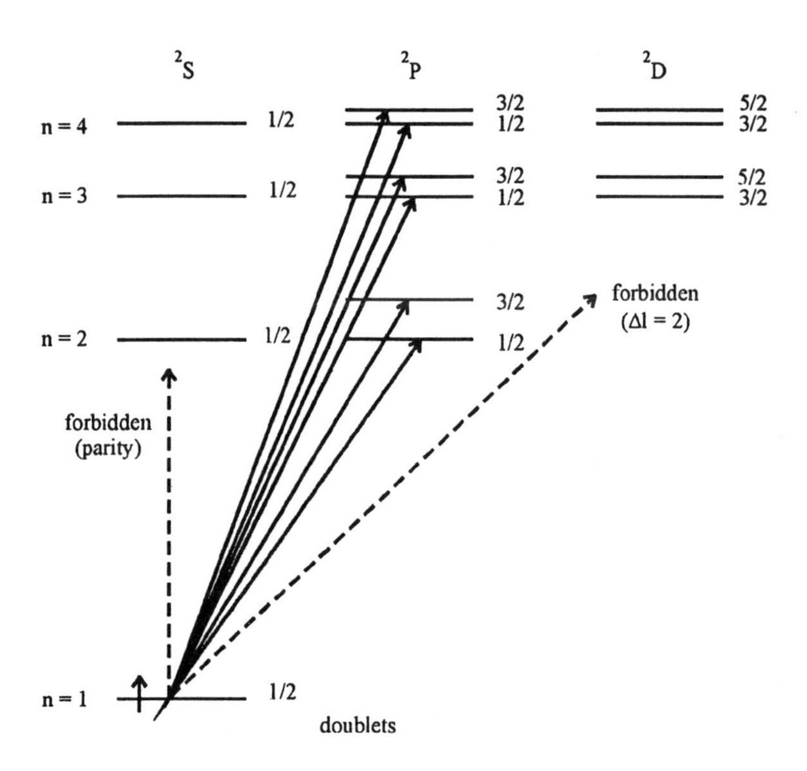

Figure 4.3.7: Illustration of the absorption spectrum of the hydrogen atom in its ground state.

or ±1, which can be easily derived from the first rules (**j** = **l** + **s**). Some very important points should be mentioned:

— Each energy level is twice degenerate (s = 1/2; 2s+1 = 2; this is indicated by the number 2 at the top left of the state symbol, denoted by a capital letter).

— The p and d states are split in two (l = 1; s = 1/2; j = 3/2 or 1/2 for state p; indicated by the fraction placed at the lower right hand side of the level),

— The spectrum consists of *doublets* (which must not be confused with the value 2 of the degeneracy of the states),

— The rule $\Delta l = \pm 1$ only allows s-p transitions,

— The s-s transitions have a zero transition moment in the EDA (identical parities of states), and are therefore not visible.

Only the s-p doublets persist.

(ii) Many-electrons atoms: Their correct description is more complicated. We need to include the spin. In the case of a two-electron atom (helium atom), the spin state vectors $|\alpha\rangle$ and $|\beta\rangle$ lead to the so-called parallel spin and antiparallel spin states, so that a choice is possible (which was not the case for hydrogen, where only one spin state was to be taken into consideration). So, since the electronic spin must necessarily be taken into account, we need to know more about its behavior. Should we consider the various electron spins individually or collectively before coupling them to the orbital angular moments? Moreover, we must first think about whether the orbital angular moments should be considered individually or collectively. Or in other words, can we use the Russell-Saunders coupling (*cf.* Section III .3.2.2) without restriction? Finally, the electronic state representation will be certainly be more difficult to describe than in the previous case and we shall have to use the Slater representation (*cf.* Section III.3.2).

The RS coupling: We can generally use it, except in the case of the so-called heavy atoms. In the latter case, it is preferable to couple **l** and **s** for every electron, then to add the different **j** vectors. This is called *j-j coupling*. Of course, selection rules depend on the type of coupling used in the description.

The question of spin: To illustrate our subject, let us look at the case of the helium atom. The atom has two electrons. Let us assume the hypothesis of RS coupling. The value of the spin S is 0 or 1. States therefore have a 1 or 3 fold multiplicity. They are *singlet* or *triplet* states (whereas the states of the hydrogen atom were doublets). We are now going to describe them using the Slater determinants. In this case, they are as follows:

For the singlet states

$$\psi_S = (1/\sqrt{2})\begin{vmatrix} \varphi(1)\alpha(1) & \varphi(2)\alpha(2) \\ \varphi(1)\beta(1) & \varphi(2)\beta(2) \end{vmatrix}$$

For the triplet states

$$\psi_T = (1/\sqrt{2})\begin{vmatrix} \varphi_1(1)\alpha(1) & \varphi_1(2)\alpha(2) \\ \varphi_2(1)\alpha(1 & \varphi_2(2)\alpha(2) \end{vmatrix}$$

For singlet states, the global antisymmetry comes from the spin, whereas for triplets it comes from the orbital part of the wave function.

Let us write the transition moment. Omitting the – e coefficient, we have

$$p = \langle \psi' | \boldsymbol{r}_1 + \boldsymbol{r}_2 | \psi \rangle$$

The electric dipole moment operator only acts on space variables, not on spin variables. On the other hand, operator $\boldsymbol{r}_1$ only transforms the orbitals relative to electron 1, and $\boldsymbol{r}_2$ only those relative to electron 2. A simple calculation shows that, because the spin α and spin β functions are orthogonal, the braket of the spin part is equal to unity for the S-S and T-T transitions, while it vanishes for the S-T transitions.

This very simple result can be generalized and leads to the first selection rule $\Delta S = 0$.

But, as we already mentioned, the RS coupling is no longer valid in the case of heavy atoms. In this case, the S-S, T-T dichotomy, characteristic of the RS coupling, no longer holds, and *intersystem* transitions corresponding to $\Delta S = \pm 1$ can be observed.

Now let us calculate the transition moment. As for the helium case, we want to calculate the S-S transition moment. We find

$$\langle \varphi'(1)|r_1|\varphi(1)\rangle\langle \varphi'(2)|\varphi(2)\rangle + \langle \varphi'(2)|r_2|\varphi(2)\rangle\langle \varphi'(1)|\varphi(1)\rangle$$

Let us look at the first term of the sum. We identify, in the left part of the equation, the classical expression of the transition moment $\varphi \rightarrow \varphi'$ concerning electron 1. It is different from zero only if $\langle \varphi'(2)|\varphi(2)\rangle$ has a nonzero value, i.e. if $\varphi'(2)$ is identical to φ (2) (orthogonal functions φ). *This result shows that the first term of the sum describes a one-electron hydrogen-like transition of electron 1, electron 2 remaining in the initial state j.* The same is true for the second term of the sum, which shows a one-electron hydrogen-like transition of electron 2, electron 1 remaining in its initial state j. Thus, we will describe the electronic transitions of many-electron atoms in terms of the one-electron spin-orbitals used in the Slater determinants (whose orbital part will very often be an orbital of the hydrogen atom). The result will be generalized to the case of molecular transitions.

We can now understand why the selection rules concerning many-electron atoms are in fact mixed rules, combining requirements of one- and many-electron cases.

For instance, the criterium $\Delta l = \pm 1$ must be satisfied for each electron undergoing a transition before establishing the selection rules resulting from the Slater determinant and which concern the global electronic state vector. Here, we just give the resulting selection rules without going into the calculations. Transitions are allowed exclusively between states of different parities satisfying:

$$\begin{aligned} &\Delta l = \pm 1 \\ &\Delta L = 0 \text{ or } \pm 1 \ (\text{except the transition } L = 0 \rightarrow L = 0) \\ &\Delta S = 0 \\ &\Delta J = 0 \text{ or } \pm 1 \ (\text{except the transition } J = 0 \rightarrow J = 0) \end{aligned} \qquad (4.3.45)$$

Figure 4.3.8 is an incomplete and schematic illustration of the helium spectrum. In the case of atoms with higher atomic numbers, some S-T transitions can appear.

(iii) Molecules: For molecules, we use the same kind of approach as we used for atoms. The electronic state vector is represented by a Slater determinant build with spin-orbitals consisting of linear combinations of atomic orbitals. As before, transitions will be one-electron type transitions. By absorbing a photon, an electron will undergo a spin-orbital jump, while the other electrons remain in their initial states. The energy of these molecular orbitals will be calculated using the HF-LCAO method.

Of these selection rules, only those concerning the spin and the parity should be used with care. Some S-T transitions may appear on experimental spectra, for instance

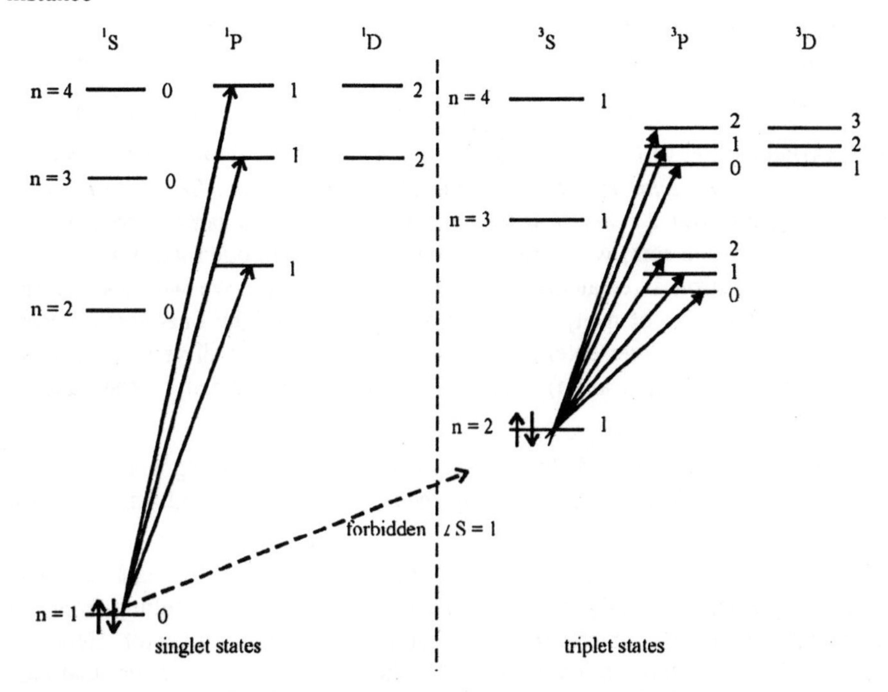

Figure 4.3.8: Simplified electronic absorption spectrum of the helium atom.

in benzene where a weak absorbance exists in the 300 nm region. Here, the parity rule is called the *Laporte law*. It is not always respected. A famous example occurs in complex ions. In these ions, vibrations can modify the symmetry of the electronic state so that some Laporte forbidden transitions can become allowed. Problem IV.4.9 deals with the important subject of the spectroscopy of transition metals and of their complexes. Rules using quantum numbers are only used for the description of very simple molecules, such as the di-hydrogen H_2. In this case, a more complex symbolism is used (*cf.* Bibliographic reference 4.5.1). Fortunately, *group theory* (*cf.* Section III.2) will be of great help in our search for allowed transitions in molecules.

Contribution of the group theory: Let ψ and ψ' be the MO's representing the initial and the final state of the transition. We know that these two MO's either form an irreducible representations of the symmetry group of the molecule by themselves, or they belong to one. We will designate them by the symbols associated to these representations. Moreover, the x, y and z components of the position vector **r** have similar properties. The direct product of the three representations associated to the transition moment $\langle \psi' | \boldsymbol{r} | \psi \rangle$ forms a reducible representation, which we know how to decompose on the set of irreducible representations of the group. Moreover, the transition moment is a characteristic entity of the studied molecule. It must be invariant under the action of every symmetry operator of the group, and its reducible representation (which we just constructed) must then belong partially or totally to the totally symmetrical representation of the group. If it does not, the transition is forbidden. Problems IV.4.10-12 provide several examples of such calculations. One should bear in mind, though, that a contribution of the totally symmetrical representation constitutes a necessary condition, but that it is not sufficient to prove the existence of the transition, which can be forbidden for other reason, as mentioned earlier (spin or Laporte rule). Let us now see how we can write the MO's.

Writing the MO and its consequences; vibrational and rotational structures: In a polyatomic molecule, the state vector depends on the spatial nuclear coordinates $\mathbf{R}_i$ and on the electronic coordinates $\mathbf{r}_\alpha$. With the Born-Oppenheimer approximation (BOA), it can written as a product

$$\psi = \psi_e(\mathbf{r}_\alpha, \mathbf{R}_i)\psi_N(\mathbf{R}_i) \tag{4.3.46}$$

In the electronic part of the vector, the dependence on $\mathbf{R}_i$ is parametric and the transition moment takes the form

$$p = -e\left\langle \psi'_e(\mathbf{r}_\alpha, \mathbf{R}_i) \middle| \sum_\alpha r_\alpha \middle| \psi_e(\mathbf{r}_\alpha, \mathbf{R}_i) \right\rangle \left\langle \psi'_N(\mathbf{R}_i) \middle| \psi_N(\mathbf{R}_i) \right\rangle \Big| \tag{4.3.47}$$

This equation shows that the many-electron transition moment that we already calculated is now multiplied by the *overlap factor* of the nuclear parts of the state vector. Since the modulus of this factor depends on the nuclear motion, we understand why the electronic absorption spectrum includes bands which are modulated by the vibrational and the rotational motions of molecules. These two motions are quite obvious in gases. When molecules are studied in dilute solutions, only the vibrational structure is recorded at usual resolutions. As an illustration, let us take the case of a diatomic molecule. This case has been very extensively studied (*cf.* Bibliographic reference 4.5.1). Figure 4.3.9 shows the dependence of the electronic energy on the distance between nuclei for each of the two concerned states.

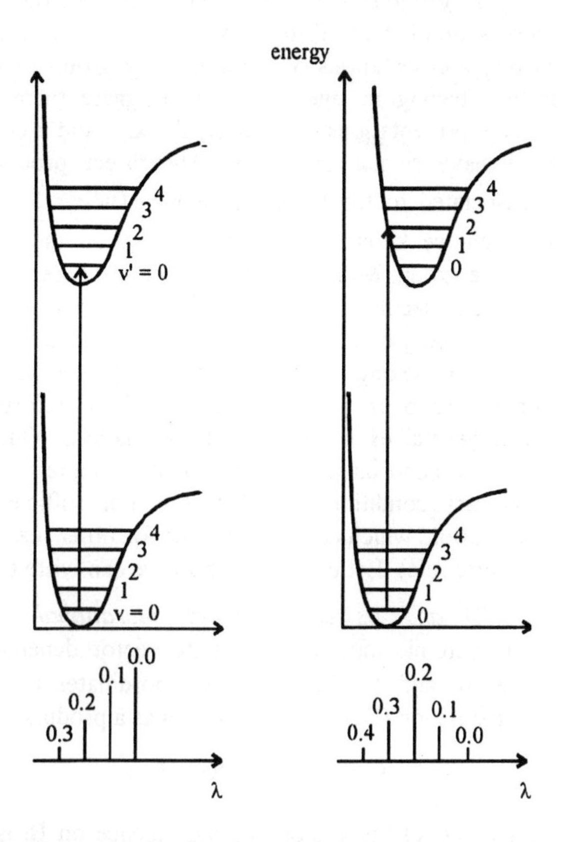

Figure 4.3.9: Change of the electronic energy of a diatomic molecule with distance between nuclei (for the ground and the excited states); right: the equilibrium distances are the same for both states; left: the excited state has a larger equilibrium distance than the ground state. The shapes of the spectra are shown underneath.

We have also represented the associated vibrational structure, characterized by the numbers v (ground state) and v′(excited state). Since the photon impact is of very short duration as compared to the vibrational period, the distance between the nuclei does not change during the interaction and the transition is called a *vertical transition.* This constitutes the *Frank-Condon principle*. In the first case illustrated on the figure, equilibrium distances are identical for the two states and the overlap is greatest for the $v = 0 \to v' = 0$ transition. The modulation is greatest on the right hand part of the band (long wavelengths). In the second case, the equilibrium distance in the excited state is larger and maximum overlap occurs for the $v = 0 \to v' = 3$ transition.

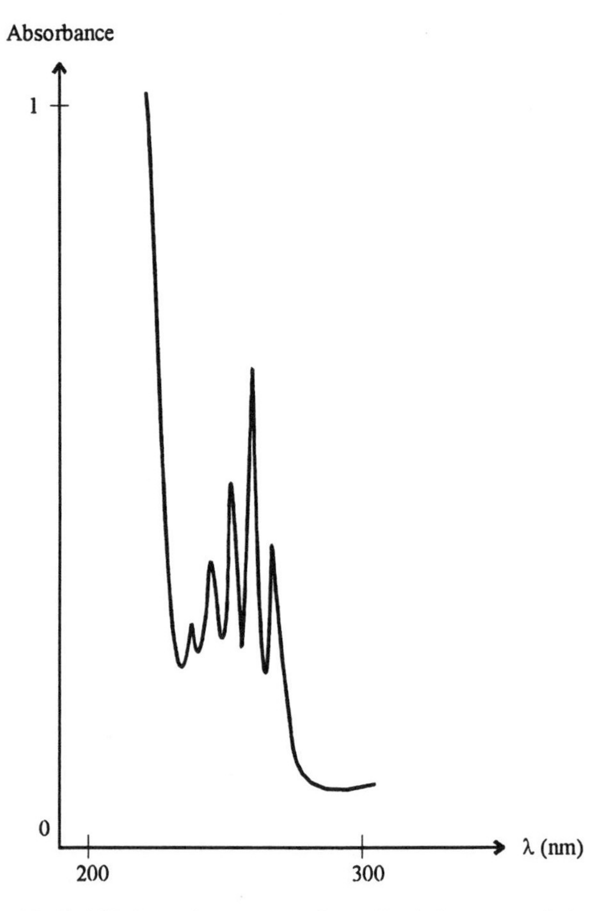

Figure 4.3.10: UV absorption spectrum (limited) of a benzene solution; solvent: cyclohexane; molarity: 3×10^{-3} mol.l^{-1}; optical path: 1 cm.

Figure 4.3.10 shows a recording of transition $^1A_{1g} \rightarrow {}^1B_{2u}$, attributed to the *π–electron* system of a benzene solution.

The mean absorbance is about 0.5 and corresponds to an absorption coefficient of about 150 $l.mol^{-1}.cm^{-1}$. The vibrational structure is very important and we clearly distinguish six equidistant bands (weak anharmonicity). The frequency of the nuclear vibration of the excited state is about 2.8×10^{13} Hz. We can see that this band concerns a transition between a gerade singlet and an ungerade singlet and is allowed both by the Laporte and by the spin selection rules. But when we calculate the transition moment using the group theory with the D_{6h} group of the benzene molecule, we see that the two direct products $B_{2u}\oplus A_{2u}\oplus A_{1g}$ and $E_{2u}\oplus E_{1u}\oplus A_{1g}$ corresponding, respectively, to the z and to the (x, y) components of the transition moment, are equal to B_{1g} and to E_{2g}. The completely symmetrical representation A_{1g} is not involved so that the transition is forbidden, which explains the weakness of its absorption coefficient.

Figure 4.3.11 illustrates the influence of ortho, meta and para di-substitution of two H atoms by two methyl groups CH_3 on the same $\pi \rightarrow \pi^*$ benzene band.

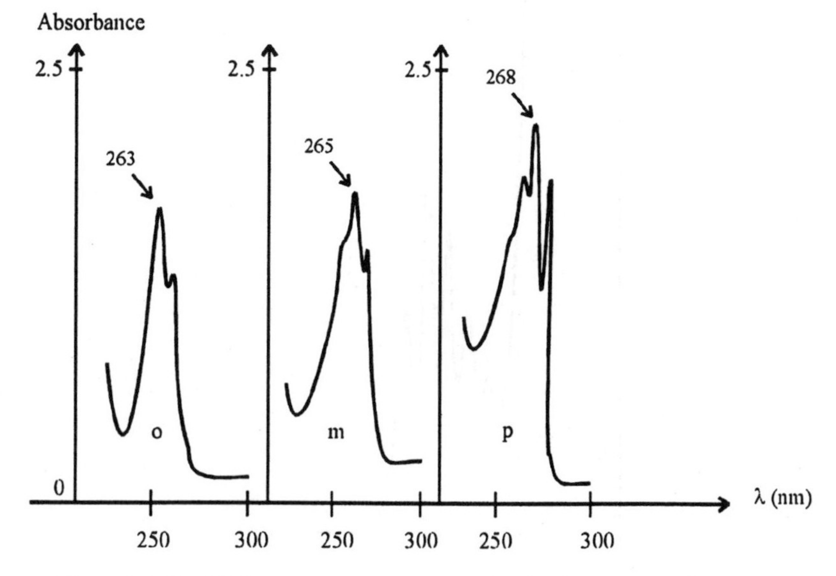

Figure 4.3.11: UV absorption spectrum (limited) of ortho-, meta- and para-xylenes; solvent: cyclohexane; molarity: $4,8\times10^{-3}$ mol.l^{-1}; optical path: 1 cm.

o-, m-, and p-xylenes solutions are prepared with the same solvent (cyclohexane) and at the molarity of 4.8×10^{-3} mol.l^{-1}. We notice marked modifications of the vibrational structure, (i) a so-called *bathochromic* displacement of the absorption maximum towards the longer wavelengths and (ii) a so-called *hyperchromic* increase of the absorption coefficient when the di-substitution changes from ortho to para.

Figure 4.3.12 illustrates the influence of the para di-substitution of two carbon atoms of the aromatic cycle of benzene by two nitrogen atoms (the molecule is called pyrazin) and it shows how essential the choice of the solvent is.

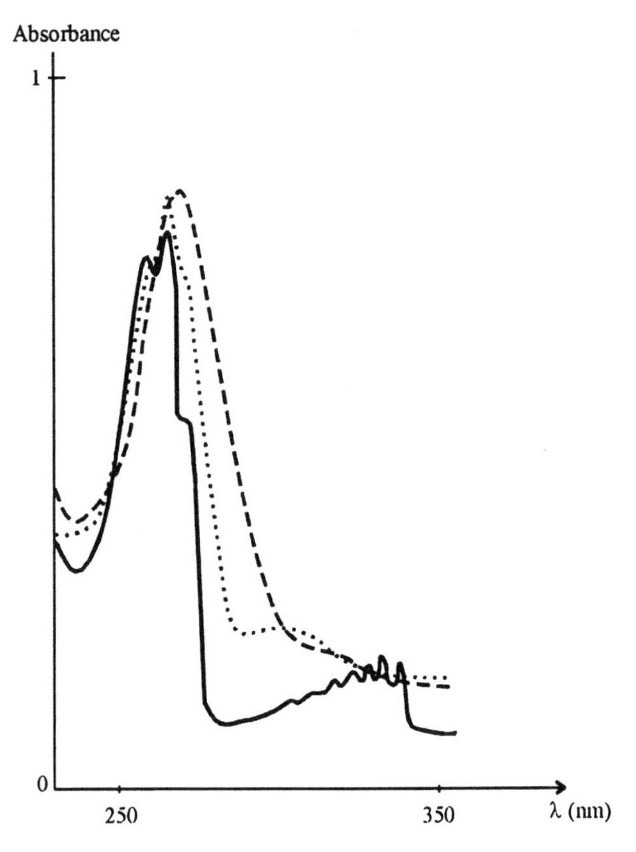

Figure 4.3.12: UV absorption spectra (limited) of a pyrazin solution; optical path: 1 cm. Continuous line: 1.34×10^{-4}mol.l^{-1} solution in n-hexane. Dotted line: 8.5×10^{-5} mol.l^{-1} solution in water (pH = 7). Dashed line: 8.5×10^{-5} mol.l^{-1} solution in water (pH = 2).

The $\pi \rightarrow \pi^*$ band does not change after the di-substitution. We notice batho- and hyperchromic effects when the polarity of the solvent increases (n-hexane < water < acidulous water), as well as a progressive disappearance of the vibrational structure. But this time, a new band appears, called n $\rightarrow \pi^*$ band, which is due to the free electronic doublet of the nitrogen atom. It exhibits a strong vibrational structure in a non-polar solvent. This band seems to be strongly rubbed when water is used as solvent, and it disappears totally in acids. The wavelength of its maximum decreases when going from n-hexane to water (*hypsochromic effect*), while its intensity decreases with solvent polarity (*hypochromic effect*).

Chemists explain the disappearance of this band by invoking a preferential formation of a so-called *semipolar bond* between the nitrogen atom and the H_3O^+ ion in polar solvents whose concentration increases in acid solutions.

Problems IV.4.13 to 18 familiarize the reader with the spectroscopy of the main organic absorbing chromophores, while Problem IV.4.19 proposes an extension to the visible part of the spectrum of these studies concerning the near-UV domain. Last, Problem IV.4.20 summarizes all these questions.

In conclusion, a few very strict selection rules have been established for the hydrogen atom. As soon as the number of electrons increases, quantum mechanics becomes confronted to some difficulties that it attempts to overcome, more or less successfully, by using approximations. Theoretical UV-visible spectroscopy then becomes less rigorous and totally forbidden transitions no longer exist, in agreement with experiments which reveal absorption coefficients ranging from 0.1 to 100 000, spread over practically six orders of magnitude !

But theoretical treatment cannot be done away with, since it allows us to understand and to anticipate the order of magnitude of the observed transitions.

IV.4. Exercises and problems

IV.4.1. Study of the various real dispersing devices (from Agreg. P; A; 1992)

A. Scattering in daily life

1. Give three observations taken from every day life of the decomposition of white light. For every case, explain the physical origin of this decomposition in one sentence.

2. According to Descartes and Newton, the rainbow is due to the reflection of the light of the sun on the spherical water drops.

a. Calculate the deviation D of the emergent beam after a reflection on a drop of radius R, as a function of the incidence angle α. Express the minimum value of D for drops with refractive index n = 1.33, and for a wavelength of 600 nm chosen in the middle of the visible spectrum. Infer from this some characteristics of the observed bow (position, order of succession of the colors).

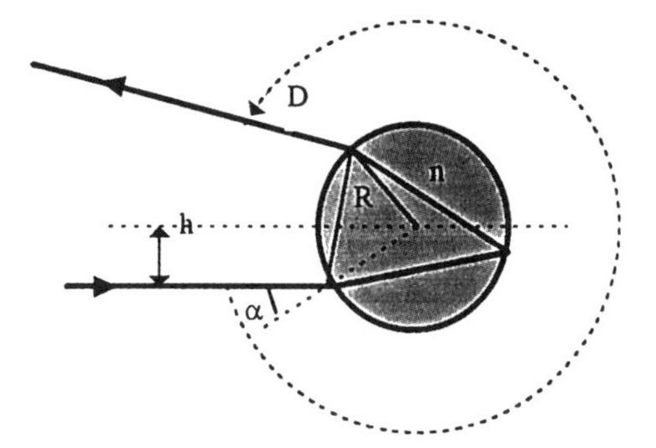

b. It frequently happens that two rainbows occur side by side. How can you explain the second one? Draw the path of the beam corresponding to the second bow inside a drop. What is the position of the second bow relative to the first; What is the order of succession of its colors?

c. Could a third rainbow exist? Argue the answer.

B. Scattering in the laboratory

1. Give the diagram of an experiment in which the spectrum of a white light source can be obtained with the help of a prism. Indicate the function of the various elements in the drawing and their possible orientation.

2. Same question for scattering induced by a grating.

3. We want to disperse light in the horizontal direction with a grating and, at the same time, in the vertical direction with a prism. This is the so-called crossed prism and grating experiment. Describe this experiment, draw the picture(s) obtained on the observation screen. What is (are) the didactical value (s) of such an experiment? How does one choose the prism? the grating?

4. Prism.

a. Define the conditions which give rise to the minimum deviation of a prism of angle A, of refractive index n, illuminated under a variable incident angle i by a radiation of wavelength λ. What advantage is there to use the minimum of deviation?

b. In a prism spectrometer, the prism is illuminated with parallel light under a fixed angle of incidence i. Give the expression of $dD/d\lambda$, the variation of the deviation with the wavelength for this prism, as a function of $dn/d\lambda$, the dispersion of the glass the prism is made of.

c. We choose to operate at the deviation minimum for a fixed wavelength. To obtain a parallel incident beam, a slit of width e and a lens L_1 of focal length f are placed between the source and the prism, parallel to the edge of the prism. The spectrum is collected on a photographic plate in the focal plane of a lens L_2 with the same focal length f. The basis of the illuminated part of the prism has a length b, the prism is illuminated up to its edge. The width of the emergent beam (emerging at angle i') in the principal plane of the prism is called a. Express $dD/d\lambda$ as a function of a, b and $dn/d\lambda$.

d. What image would you get on the photographic plate if the spectrum contained only two wavelengths λ and λ' ? For a fixed value of the width of the source slit, the resolution is defined by $R = \lambda/\Delta\lambda$, where $\Delta\lambda$ is such that the distance between the centers of the slit images is equal to the width of the images, and $\lambda' = \lambda + \Delta\lambda$. Give the expression of R in this case. What is the main phenomenon which limits the resolution LR of the prism? Give an evaluation of LR.

Numerical application: f = 20 cm, a = 3 cm, e = 20 μm, b = 5 cm, $dn/d\lambda = -105\ m^{-1}$.

What is the value of the resolution of the set up for $\lambda = 600$ nm? Calculate LR.

5. Grating

a. Give two processes used to manufacture optical gratings.

b. Recall the constructive interference condition (at infinity) between two vibrations diffracted by two successive equivalent sites of a grating used in reflection (we note "a" the distance between two successive grooves of the grating). Same question for a grating used in transmission. Specify which sign convention you choose for the beam angles.

c. When a grating with a distance a between successive grooves is illuminated at incidence i and with a light of wavelength λ, how many different orders can be observed?

Numerical application: l = 600 nm. The grating is assumed to be illuminated under normal incidence. First do the calculations for a grating with 570 grooves per millimeter, then for one with 50 grooves per millimeter.

d. Calculate and represent graphically the intensity of the interference produced by a whole grating (distance between two grooves: a ; width: L = Na), as a function of the phase difference between waves diffracted by two successive grooves of the grating. Deduce the resolution limit LR of the grating. If τ is the time-path difference for the outer parallel beams reaching the grating, what does $1/\tau$ represent with respect to a principal maximum? What is physical limit of the grating? Give an order of magnitude for τ and give the order of magnitude of the width L needed to analyze the emission bands of a discharge lamp.

e. Explain the blaze phenomenon. What should the grating profile look like? Now we would like to observe the whole visible spectrum (the maximum intensity lying at about 600 nm) in one single order using the blaze phenomenon, and with extinctions taking place at 350 nm and at 850 nm in the same order. The grating has 570 grooves per millimeter. What would the profile of this grating look like? In what order of interference, approximately, would the intensity maximum be observed?

f. In a grating spectrometer, the source we want to study is placed in front of the entrance slit. With the help of a collimating system (usually a mirror) the grating is illuminated with a parallel light beam. A second mirror focuses the diffracted beam onto the exit slit, where a detector collects the transmitted light. Usually, the entrance and exit slits have the same width. Draw a diagram illustrating the principle of this device.

g. The constructor indicates a resolution as x nm per slit millimeter in the first order. Using the ideas of question B.4.d, calculate the resolution (in nm) of a grating spectrometer with a focal length of 1 m, equipped with a grating with 570 grooves per millimeter, when entrance and exit slits are opened to 50 μm. What is the influence of the incidence angle i? Assume normal incidence. What resolution can be obtained in the mth order?

h. When is it better to open up the slits? To close them? For a certain slit opening, the system reaches its limiting resolution (calculated in question B.5.d). Using the method of question g, calculate the slit opening which gives rise to the resolution limit for the same grating of 570 grooves per millimeter, of width L = Na = 10 cm, and for λ = 600 nm. Draw the curve representing the resolution of this grating as a function of the slit width, for this wavelength and for the first diffraction order.

C. Filters

Another way to separate a continuous spectrum in its components is to use one or several filters.

1. On what principle are colored glasses manufactured? What do their transmission profiles look like, expressed in wavelengths? Where does their thickness come in, and are there other parameters?

2. Same questions for an interference filter. Recall the constructive interference condition. Give typical orders of magnitude for the transmission and for the bandpass. Can a filter sold to be used at wavelength λ_0 be used at a wavelength λ close to λ_0? Explain your answer.

3. Another type of filter (the so-called Christiansen filter) consists of a vessel containing a liquid (1) in which there are crushed pieces of a solid (2). Solid and liquid have the same refractive indexes at $\lambda = \lambda_0$. For example, a mixture of 56% of benzene and 44% of ethanol has the same refractive index as fused silica at λ = 350 nm. What happens when $\lambda \neq \lambda_0$? What physical phenomenon is at play here?

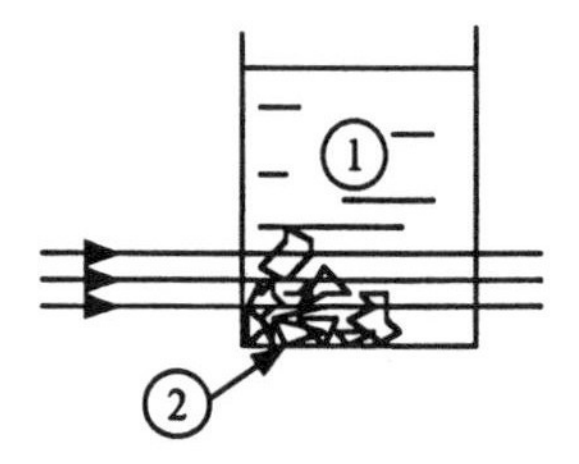

A. Dispersion in everyday life.

1. Decomposition can be due (i) to refraction (rainbow, glass or quartz prisms in lamps or chandeliers) or (ii) to interference (color of the thin objects like soap bubbles or a thin film of oil on a puddle of water). You can often see beautiful rainbows at the Niagara falls on the U.S.A side.

2.a. The angle at the center measures 4r (r is the refraction angle). Deviation is therefore $2\pi - 4r + 2\alpha$. We set the derivative of D with respect to α equal to zero and we solve for the α corresponding to the minimal deviation D_{min}. We obtain

$$\alpha_{min,1} = \text{Arc}\cos\left(\sqrt{n^2 - \sin^2\alpha_{min,1}}\,/2\right)$$

Or $\alpha_{min,1} = \text{Arc}\sin\sqrt{\left(4-n^2\right)/3}$. So D_{min} takes the value

$$D_{min,1} = 2\pi + 2\alpha_{min,1} - 4r_{min,1}$$

NA: With n = 1.33 we find $\alpha_{min,1}$ = 59.58 ° and $D_{min,1}$ = 317.47°. The rainbow is seen from the earth with an angle of 360 – 317.47 = 42.53°. The deviation decreases when the wavelength increases. The bow is purple at its concave lower part and red at its convex upper part.

b. The second bow is due to a second reflection inside the water drop. The second deviation can be calculated starting from the first. We must add angles $(\pi - r - \alpha)$ and $(\alpha - r)$. We find $D_2 = 3\pi + 2\alpha - 6r$. The minimum is obtained for $\alpha_{min,2} = \text{Arc}\sin\sqrt{(9-n^2)/8}$ and its value is $D_{min,2} = 3\pi + 2\alpha_{min,2} - 6r_{min,2}$. With $n = 1.33$ we find $\alpha_{min,2} = 71.94°$ and $D_{min,2} = 410.11°$. The second bow is seen from the earth with an angle of 50.11°. It lies above the first one. Its colors are the other way around: Red in the concave part and purple in the convex part.

c. Bows of higher orders have much weaker intensities and they are practically invisible.

B. The dispersion inside the laboratory.

1. and 2. The following figure gives a possible realization:

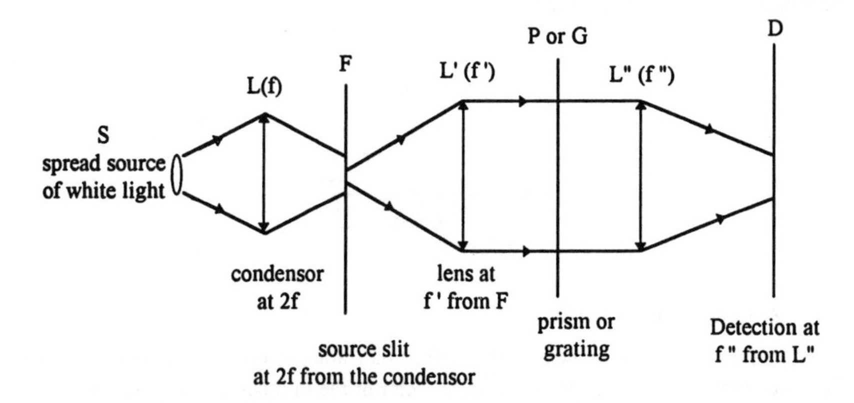

3. Experimental set up: It is identical to the one described in last question, but the prism and the grating are put in place simultaneously in such a way that their deviations are orthogonal. The resulting curves are indicated on the figure below.

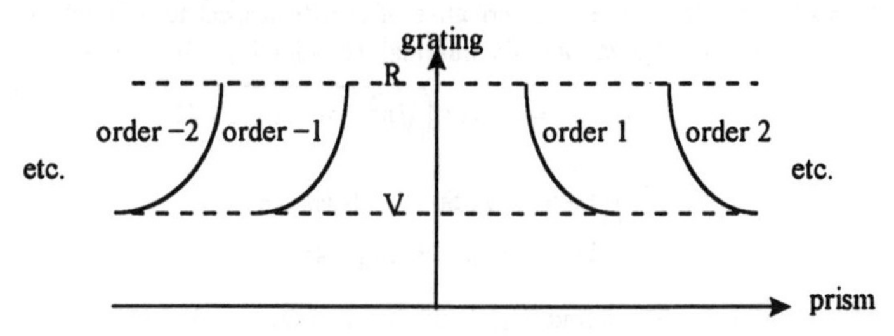

This experiment can be used as well to illustrate the scattering of a prism as that of a grating. The prism and the grating should have comparable dispersing powers so as to get a clearly understandable picture.

The grating deviates the red more than the purple. The prism deviates the purple more than the red.

4.a. For a prism with angle A we can write

$$\sin i = n \sin r \qquad \sin i' = n \sin r' \qquad r + r' = A \qquad D = i + i' - A$$

At the deviation minimum, we have $r = r' = A/2$,

$$D_{min} = 2\,\text{Arc}\sin\left[n \sin\left(A/2\right)\right] - A$$

At the minimum deviation, the geometrical aberrations are minimal.

b. Taking the derivative of the prism equations, we get

$$di'/dn = \sin A / \cos r \cos i' \qquad dD/d\lambda = \left(di'/dn\right)\left(dn/d\lambda\right) = \left(\sin A / \cos r \cos i'\right) dn/d\lambda$$

c. By writing A as a function of a and b, we get

$$\sin^2(A/2) = 1/(n^2 + 4a^2/b^2) \qquad dD_{min}/d\lambda = (b/a)\,(dn/d\lambda)$$

d. If the spectrum contains only two wavelengths which do not differ too much, we should obtain two slit images of width e, separated by $f\,(dD/d\lambda)\,\Delta\lambda$. We obtain

$$f\,(dD/d\lambda)\,\Delta\lambda = e \qquad R = \lambda f\,(dD/d\lambda)/e = \lambda b f\,(dn/d\lambda)/ae$$

Resolution is limited by the diffraction ($\lambda f/a$) induced by the finite transverse dimension (a) of the prism.

NA: We find $R = 1\,000$; $L\,R = b\,(dn/d\lambda) = 5\,000$.

5.a. The old grating manufacturing process consisted in engraving grooves on a support (with a hard tool) which was then used as a mold to obtain duplicates, for example by casting. Currently, gratings are manufactured by using interference processes on photosensitive supports before depositing reflective and protective substances under vacuum.

b. Let i and r be the incident and the refractive angles, respectively. We write the interference condition $\sin i \pm \sin r = p\lambda$. p is a whole number defining the interference order. + and – are used, respectively, for reflective and transmission gratings. Angles are defined to be positive in the direction of direct trigonometric rotation.

c. The condition $(a/\lambda)(\sin i - 1) \leq p \leq (a/\lambda)(\sin i + 1)$ must be satisfied. For normal incidence, p must lie between $-a/\lambda$ and $+a/\lambda$.

For 570 grooves per mm, $a/\lambda = 2.92$ and we have 5 orders corresponding to the numbers: – 2 ; – 1 ; 0 ; 1 ; 2. For 50 grooves per mm, $a/\lambda = 33.28$ and we have 67 orders (all the integers from – 33 to + 33).

d. The resulting electric field is given by

$$\underline{E} = E_0 \sum_{p=1}^{N} \exp\left(ip\varphi\right) \quad \text{where } \varphi = 2\pi a\,(\sin i + \sin r)/\lambda.$$

This is a limited geometrical progression of ratio ($i\varphi$). It is equal to

$$\{\exp(i\varphi) - \exp[i(N+1)\varphi]\} / [1 - \exp(i\varphi)]$$
$$= \exp[i(N+1)\varphi/2][\sin(N\varphi/2)/\sin(\varphi/2)]$$

so that the radiated intensity can be written as

$$I = \underline{E}\,\underline{E}^* = E_0^2 \sin^2(N\varphi/2)/\sin^2(\varphi/2)$$

The maxima of the function ($N^2E_0^2$) are obtained for $\varphi = 2p\pi$. Secondary zeros are obtained for $\varphi = 2p'\pi / N$, where p' is an integer (see figure below).

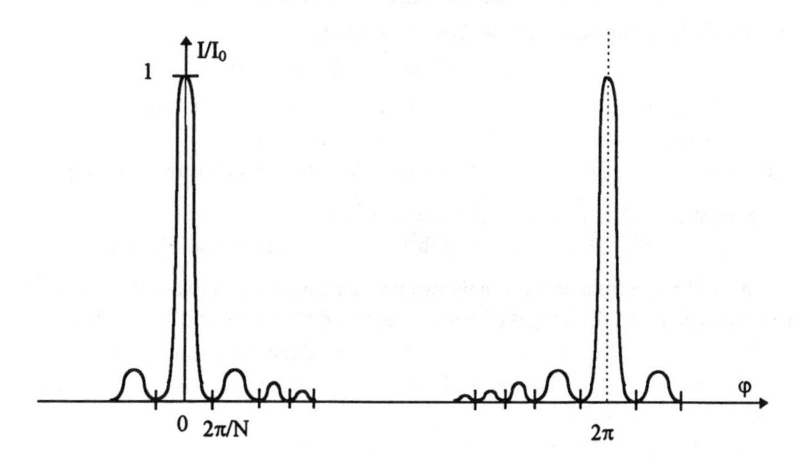

The LR is defined when the intensity maximum for wavelength $\lambda + \delta\lambda$ coincides with an intensity minimum for wavelength λ. We write

$$\varphi(\lambda + \delta\lambda) - \varphi(\lambda) \approx 2\pi a (\sin i \pm \sin r)\, \delta\lambda / \lambda^2 = 2\pi / N$$

And, remembering that a (sin i ± sin r) = pλ, we find that LR = pN.

We have (sin i ± sin r) = cτ / L = pλN/L = (λ/L) LR and therefore:

$$\tau^{-1} = (c/\lambda)\,(1/LR)$$

To avoid the disappearance of the interference fringes, this maximal value of the time path-difference must be smaller than the coherence time of the analyzed light. LR should be smaller than $c\tau_c / \lambda$. For instance, for the usual bands of high pressure discharge lamps the coherence time is about 10^{-9} s. LR should not be greater than 5×10^5. By assuming pure Doppler broadening (which is far from being the case in high pressure discharge lamps), LR varies as the square root of the mass/temperature ratio. For a grating with 5 000 grooves per mm (middle range quality), L should not be greater than 10 cm.

e. The resolution grows proportionally to the interference order, but the brightness decreases very quickly with p. By manufacturing échelette gratings, it is possible to concentrate all the light in a given order. The grating is said to be *blazed.* Let us write the conditions for the intensity maxima and minima for the three indicated mean wavelengths 600 p ≈ 350 (p + $p_{i,1}$ /2) ≈ 850 (p + $p_{i,2}$ /2). p is the interference order integer; $p_{i,1}$ and $p_{i,2}$ are two odd integers. We can choose a solution by taking p = + 2; $p_{i,1}$ = + 3; $p_{i,2}$ = −1. The maximum would then be obtained at 625 nm and the minima at 357 and 867 nm. In the échelette grating, the optical delay from one step to the next is 1 250 nm for these values. If θ represents the angle of the échelette with respect to the grating support, the angle of reflection must be equal to (2θ − i) (see figure below). Blazed grating works in reflection and we have the relation

$$a\left[\sin i + \sin\left(2\theta - i\right)\right] = p\lambda_{max}$$

Average plane of the grating

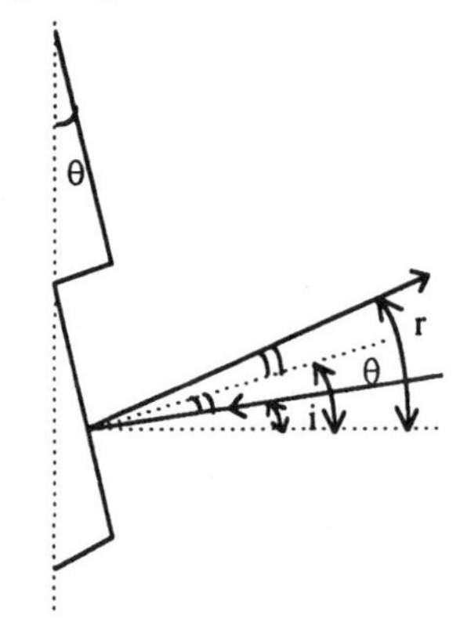

In other words, sin i + sin (2θ − i) = 0.712 . We can calculate the value of the blazing angle θ and define the geometry of the grating for any incident angle i.

f. Experimental set up. It is given by the following figure

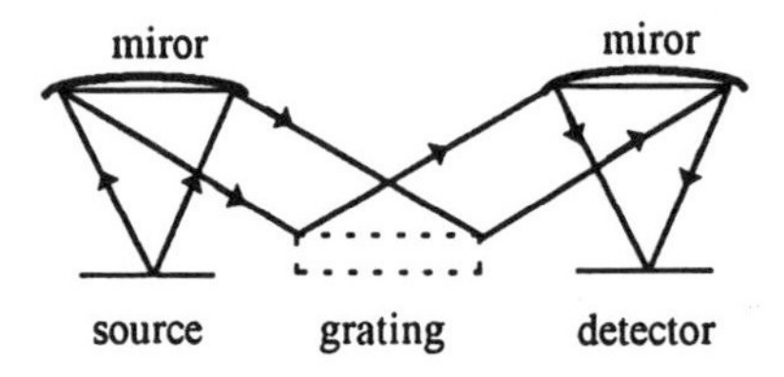

g. By taking the derivative of the relation sin i + sin r = mλ / a and remembering that dx = f dr, we obtain dx = fm dλ / a cos r. So that, using

$$\Delta x = e\,\Delta\lambda = ea\cos r / fm \text{ and } R = \lambda / \Delta\lambda + \lambda fm / \left[ea\sqrt{1-(m\lambda/a - \sin i)^2}\right]$$

we find R = 6 840 at order 1. R decreases when i increases. When i is vanishingly small, we have

$$R_{i=0}(m) = (\lambda f / ea)\left[m / \sqrt{1-(\lambda m / a)^2}\right]$$

h. When the phenomenon is not very luminous, the slits must be opened up. If there is a lot of light, we can close them. We can write:

$$LR = pN = (\lambda f / ea)\left[p / \sqrt{1-(p\lambda / a)^2}\right] \quad \text{or} \quad e/f = (\lambda / L)\left[1 / \sqrt{1-(p\lambda / a)^2}\right]$$

The numerical application gives $e / f = 6.4\ \mu m.m^{-1}$. The curve is drawn below.

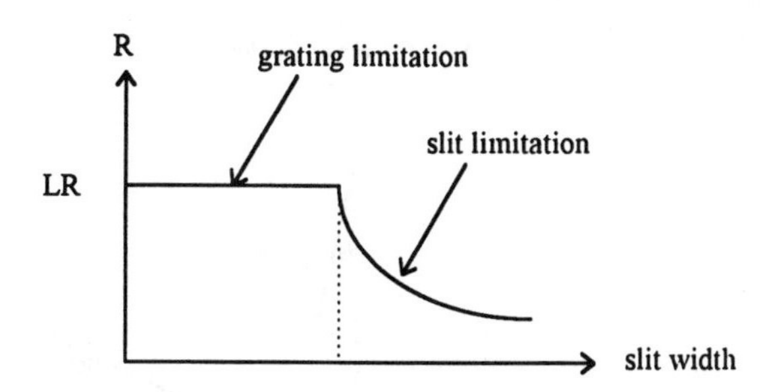

C. Filters.

1. Colored glasses absorb wavelengths selectively. Many kinds exist: Band-pass, low-frequency and high-frequency cutoff, etc. Their transmission obeys to the Beer-Lambert law for weak intensities and the optical density is proportional to the thickness of the glass. They can be nonlinear for high intensities so that their absorbance can saturate: In this case, they act like passive optical doors and are widely used in laser technology. The nature and the concentration of the colored doping is of course of fundamental importance.

2. Interference filters are based on the phenomenon of multiple interference. They are made from glass or silica supports, on which a variety of multi-dielectric coatings are deposited. The transmission and bandwidth calculations are performed numerically. If we simplify the problem by reducing it to a two-wave interference phenomenon, interference is constructive if $2ne/\lambda \cos r = 2p\pi$. Under normal incidence, the bandpass depends on the variations of the refractive index with wavelength. A bandpass of a few nanometer and a transmission of 80 % can be obtained with very advanced technology. The wavelength of the transmission maximum varies with the incident angle (up to a certain point).

3. If $\lambda \neq \lambda_0$, strong scattering due to the small pieces of silica occurs in all spatial directions so that the transmission of the parallel beam is greatly reduced.

IV.4.2. Fourier transform spectroscopy (from Agreg. P; A; 1992)

We will show that it is possible to obtain the profile of a source beam by using the interferogram recorded with a Michelson interferometer, and applying Fourier transformations.

1. Describe the Michelson interferometer and draw a picture of it. Give the expression of the amplitude A and of the intensity I obtained at the exit of the device in the direction perpendicular to the equivalent air plate, as a function of position x of the mobile mirror. Assume that the source is a perfectly monochromatic line of wavelength λ, i.e. of wave number $\sigma = 1 / \lambda$. The origin of the x coordinate is the position of the mirror corresponding to a vanishing thickness of the equivalent air plate. Draw I(x).

2. What is the value of I(x) when the source consists of two infinitely thin lines, with the same intensity, and with wave numbers σ_1 and σ_2? Using the results of this experiment, how would you characterize the source lines?

3. Assume now that the source band has a symmetrical wave number profile $g(\sigma)$. Show that I(x) includes both a flat line and a variable contribution linked to the Fourier transform of $g(\sigma)$. What is the shape of I(x) if $g(\sigma)$ is Gaussian; what is its width?

4. We measure the interferogram of the green line of mercury (546 nm). When using a high pressure spectral làmp, the interferogram disappears (it "scrambles") when the mobile mirror moves 3 mm to either side of the position corresponding to the flat color. For a low pressure mercury lamp, the scrambling occurs for a displacement which is ten times larger. Deduce the width $\Delta\sigma$ and the relative width $\Delta\sigma/\sigma$ of the profile $g(\sigma)$ in each case. How do these widths compare to the Doppler width? Assume that the temperature in the discharge is 500 K and the atomic mass of mercury will be taken to be 200.

1. The Michelson interferometer is described below:

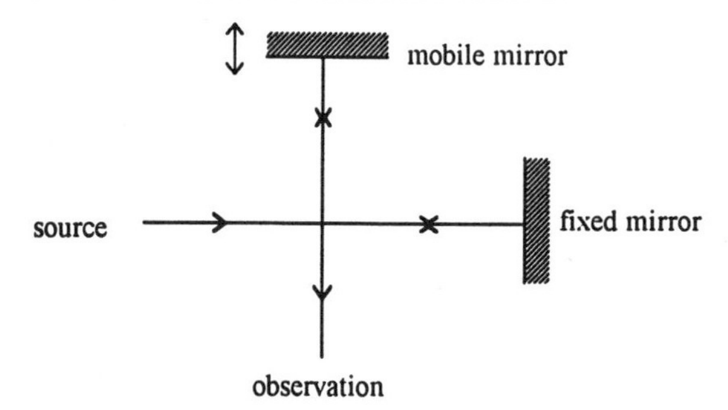

If E_0 is the amplitude of the incident field, the resulting field writes

$$E = \left(E_0 / \sqrt{2}\right)\left[\cos\omega t + \cos(\omega t + \varphi)\right] \text{ with } \varphi = 4\pi\sigma x$$

The light intensity writes $I = \left(E_0^2 / 2\right)(1 + \cos\varphi)$. The graph of I(x) is very simple. The first minimum is obtained for $x = \lambda / 4$.

2. The two lines are assumed to be incoherent so we add the intensities.

$$I = (I_0 / 2)\left[2 + \cos(4\pi\sigma_1 x) + \cos(4\pi\sigma_2 x)\right]$$
$$I = I_0\left[\cos 2\pi(\sigma_1 - \sigma_2)x . \cos 2\pi(\sigma_1 + \sigma_2)x\right]$$

We observe a beating phenomenon of wave number ($\sigma_1 - \sigma_2$), affecting an oscillation with wave number ($\sigma_1 + \sigma_2$). With this experiment, the sum and the difference of the two wave numbers can be measured, yielding the wave numbers themselves.

3. For a continuous line profile, we write

$$I = I_0 \frac{\int_{-\infty}^{+\infty}(1+\cos\varphi)\, g(\sigma)\, d\sigma}{\int_{-\infty}^{+\infty} g(\sigma)\, d\sigma} = I_0 \left[1 + \frac{\int_{-\infty}^{+\infty} \cos\varphi\, g(\sigma)\, d\sigma}{\int_{-\infty}^{+\infty} g(\sigma)\, d\sigma}\right]$$

The integral of g (σ) is assumed to be normalized to 1. By writing cos φ in complex form and taking into account the symmetry of g (σ), we find

$$I = I_0 \left[1 + T_F(X)\right] \text{ avec } T_F(X) = \int_{-\infty}^{+\infty} \exp(-i2\pi X\sigma)\, g(\sigma)\, d\sigma$$

$T_F(X)$ is the space-wave number Fourier transform of g (σ). The Fourier transform signal stands out against the flat background (I_0). For a Gaussian profile we get $g(\sigma) = \left(1/\sqrt{\pi}\sigma_0\right)\exp\left(-(\sigma/\sigma_0)^2\right)$, whose Fourier transform can be written in the form $\exp\left(-\pi^2 \sigma_0^2 X^2\right) = \exp\left(-4\pi^2 \sigma_0^2 x^2\right)$. This is a Gaussian line with a width of $1/2\pi\sigma_0$, rising against the background I_0 and which wraps a modulation whose frequency is twice the average central frequency of the line.

4. The numeric application gives the following results (HP and LP stand for high and low pressure, respectively):

	HP	LP
Δx (mm)	3	30
σ_0 (m^{-1})	53	5.3
$\Delta\sigma/\sigma$	2.9×10^{-5}	2.9×10^{-6}

The relative Doppler width is approximately equal to $\Delta\sigma/\sigma \approx \sqrt{3RT/Mc^2} \approx 10^{-6}$. Only the bandwidths of low pressure lamps are governed by the Doppler effect. In high pressure lamps, the width is mainly due to collisions between particles in the gas (pressure broadening).

IV.4.3. Microscopic classical theory of the absorption, part 1 (from CAPES; P; 1972)

A. Study of a free oscillator with one or two degrees of freedom

1. A solid of mass m, moving without friction on a horizontal plane, is fixed to the extremity M of a spring of constant K (constant ratio of the tension of the spring to its change in length, at compression as well as at extension) whose other extremity is fixed. The center of gravity G of the solid moves on the Ox axis, O being the equilibrium position of G; the motion of the solid is a straight translation. The force induced by the spring lies along the Ox axis.

a. Write the equation obtained by applying the fundamental principle of dynamics.

b. Give the angular frequency ω_0 and the period T_0 of the motion.

2. A diatomic molecule consists of two atoms M_1 and M_2 of masses m_1 and m_2.

a. When the distance between the two atoms is d, they are submitted to a force $f = k(d - a)$, attractive if $d > a$, repulsive if $d < a$. The formula is valid only for small values of $(d - a)$. Explain this result, specially by considering the potential energy.

b. Calculate the frequency and the period of the oscillations as functions of m_1, m_2 and k. Study the oscillations on the Gx axis of the molecule (G being the center of inertia), but do not take into account the possible rotations of this axis around G.

c. Hydrogen chloride absorbs electromagnetic radiation of wavelengths around $\lambda = 3.46\ \mu m$. This absorption band can be explained by a change of the vibrational state of the HCl molecule. Give a more detailed explanation.

Calculate all the possible values of the vibrational energy of this molecule in electron volts.

3. We now consider two solids of mass m_1 and m_2, fixed at the ends of two springs of constants K_1 and K_2, and linked by a spring of constant K'. x_1 and x_2 are the abscissa of the centers of gravity G_1 and G_2 of the two solids, measured from the equilibrium positions O_1 and O_2. At positions O_1 and O_2, the forces developed by the three springs vanish. O_1x_1 and O_2x_2 are oriented in the same direction.

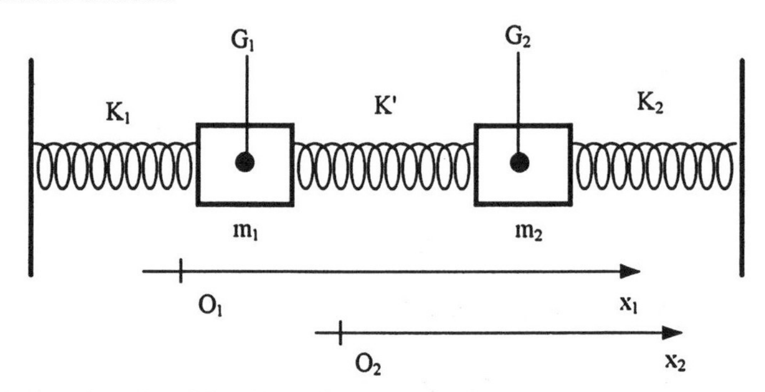

a. Write the equations describing the motion.

b. Solve these equations when the following conditions are true:

$$K_1 = K_2 = K\ ; \quad m_1 = m_2 = m$$

Hint: set X and Y such that

$$X = (x_1 + x_2)/2 \quad Y = (x_1 - x_2)/2$$

However, the reader is free to use any method of his choice. Show that x_1 and x_2 can be obtained by considering linear combinations of two sine functions, and give their angular frequencies ω_1 and ω_2.

c. If at $t = 0$, we have

$$x_1 = a,\ x_2 = 0,\ dx_1/dt = dx_2/dt = 0$$

Give x_1 and x_2 as functions of time.

d. x_1 and x_2 can be sine functions with the same frequency, either ω_1 or ω_2. For each of the two cases, give expressions of x_1 and x_2 as time dependent functions; indicate the positions and the velocities of the two solids at time $t = 0$ and make a diagram of these initial conditions. Give the details of the motion.

e. Study the case corresponding to $K' << K$. Consider especially the motions of the two solids when the initial conditions of question 3c are satisfied.

B. Study of a damped oscillator submitted to an external sinusoidal force

Again, we take up the system studied in question A1, but now the solid, of mass m, is submitted to a force $-Kx$ and also to two other forces lying along the Ox axis. The first one, due to friction, is proportional to the speed and equal to $(-2\mu m\, dx/dt)$. The second one is a driving force and is sinusoidal (with fixed frequency). It is equal to $F_0 \cos \omega t$.

1. Write the differential equation of the motion. Define parameter ω_0 as in question A.1.b and use it to eliminate parameter K.

2. After a damping period, characterized by the damping coefficient μ, the system settles in a permanent regime which is a sinusoidal motion of frequency ω. From here on, consider only this permanent regime corresponding to forced oscillations.

Write the solution of the differential equation in the form $x = X_1 \cos \omega t + X_2 \sin \omega t$.

Find the expressions of X_1 and of X_2. (hint: write $x = X_0 \cos (\omega t - \varphi)$ and then calculate X_0 and $\text{tg}\, \varphi$ by using a Fresnel construction or with the help of imaginary numbers; but a more rapid and more direct method also exists).

3. Consider only those cases where the damping is relatively small, so that terms in μ^2/ω_0^2 can always be neglected as compared to those in μ/ω_0.

a. Study the variations of X_1 as a function of ω. Give only the approximate form of the representative curve, but determine its extrema.

b. Study the variations of X_2 as a function of ω; give the approximate form of the representative curve; determine its extrema and the values of X_2 corresponding to the extrema of X_1.

Draw the two relative curves of the variations of X_1 and X_2 on the same graph.

c. What happens when the damping is extremely small, i.e. when μ tends to zero?

4. Give the expression of the average power consumed. Show that it can be expressed in a simple way as a function of F_0, X_2 and ω.

C. Application to the study of dispersion

We would like to study a schematic model explaining the origin of the dispersion of light in a transparent medium which is macroscopically neutral and consists of a collection of positive nuclei and negative electrons. In this problem, we assume that a rapidly changing electric field has no appreciable effect on the very heavy nuclei but that it induces a displacement of the electrons. In the absence of electric field, if you take a volume dV of this medium, the center of mass of the negative charges coincides with that of the positive charges. However, in the presence of an electric field E, this is no longer true: The material becomes polarized and the volume element acts like an electric dipole with moment **P**dv, where **P** is the polarization vector. Remember that in a perfect dielectric medium, the relation $\varepsilon\mathbf{E} = \varepsilon_0\mathbf{E} + \mathbf{P}$ holds, where ε_0 is the permittivity of vacuum and ε the absolute permittivity of the medium. The medium considered here is sufficiently dilute so that the field that acts on a particle is the electric field of the wave, i.e. we neglect the local field created by the possibly polarized surroundings.

1. A plane electromagnetic wave propagates in a homogeneous and isotropic insulating medium. Give the expression of the force exerted on an electron with charge $-$ e and velocity v. Show that the term due to the magnetic field is negligible when compared to that due to the electric field. From here on, assume that the force includes only the term induced by the electric field.

2. Let us considers a plasma, i.e. a completely ionized gas under low pressure. The effect of all other particles on a particular particle average out so that an electron can be considered as free. There are neither frictional nor damping forces.

a. Write the differential equation describing the motion of an electron (charge $-$ e, mass m_e) in an electric field $\mathbf{E} = \mathbf{E}_0 \cos \omega t$. The position of the electron is defined by its displacement vector **r**.

b. Integrate the equation and give the expression of **r** as a function of time.

c. Knowing that there are N free electrons per unit volume, give the expression of the polarization vector **P**. Derive the expression of the refractive index n as a function of the frequency ω of the wave. Set $\omega_p^2 = Ne^2 / m_e\varepsilon_0$.

d. Study the variations of n with ω. What happens when $\omega > \omega_p$? Explain the meaning of ω_p or of $\nu_p = \omega_p / 2\pi$. Is there a naturally occurring medium in which electromagnetic waves propagate in such a way? Numerical application: Calculate the number N of free electrons per cubic meter in a plasma for which ν_p = 20 MHz. Give the wavelengths of the waves that can propagate in this medium.

3. We now consider a gas consisting of non polar molecules. Admit the result that some of the electrons act as if they had an equilibrium position to which they are pulled back by a restoring force $f = - m_e \omega_0^2 \mathbf{r}$. Do not take account of friction nor of damping.

a. Write the differential equation describing the movement of such an electron under the action of an electric field $\mathbf{E} = \mathbf{E}_0 \cos \omega t$.

b. Consider the solution corresponding to forced oscillations and give the expression of **r** as a function of time.

c. Assuming that the medium contains N electrons per unit volume, calculate the polarization vector **P** and infer from it the refractive index n.

d. Draw the curve representing the variations of n^2 with frequency. Discuss the results according to the different values of n.

e. Let us now consider the case $\omega \neq \omega_0$. How should the above results be modified to take into account those obtained in question B ? What physical phenomenon intervenes in this case?

f. What do you know about the variation of the refractive index n with the frequency of the wave? What conditions should be satisfied to be able to use the model we developed here? How can it be improved? What do you know about the behavior of the refractive index for X rays?

A. Study of a free oscillator with one or two degrees of freedom.

1.a and b. The equation of the motion can be described by $m\ddot{x} + Kx = 0$. The angular frequency and the period are $\omega_0^2 = K/m$ and $T_0 = 2\pi/\omega_0$.

2.a. The force results from a linear approximation (Hooke's law). a is the equilibrium distance. The potential energy is equal to $(1/2)\,k\,(d - a)^2$.

b. By choosing a reference frame linked to the center of inertia and by neglecting the rotation, the motion is that of the reduced mass $m_1m_2/(m_1 + m_2)$. The angular frequency and the period are, respectively, $\omega_2 = K\,(m_1 + m_2)/\,m_1m_2$ and $T = 2\pi/\omega$.

c. This band is due to the absorption $v = 0 \rightarrow v = 1$. The corresponding vibrational energy is $E = ch/\lambda = 5.75 \times 10^{-20}\ \mathrm{J} = 0.36\ \mathrm{eV}$.

3.a. The equations of the motion can be written as

$$m\ddot{x}_1 + (K + K')x_1 - K'x_2 = 0 \quad \text{and} \quad m\ddot{x}_2 + (K + K')\,x_2 - K'x_1 = 0$$

b. The suggested change of variable yields:

$$m\ddot{X} + KX = 0 \quad \text{and} \quad m\ddot{Y} + (K + 2K')\,Y = 0$$

Returning to variables x_1 and x_2, we obtain a linear combinations of two sine functions with resonance frequencies ω_1 and ω_2 given by $\omega_1^2 = K/m$ and $\omega_2^2 = (K + 2K')/m$.

c. By expressing the initial conditions, we find

$$x_1 = (a/2)(\cos\omega_1 t + \cos\omega_2 t) \qquad x_2 = (a/2)(\cos\omega_1 t - \cos\omega_2 t)$$

d. The first case corresponds to $x_2 = x_1 = (a/2)\cos\omega_1 t$. In the second case, we have

$$x_2 = -\,x_1 = (a/2)\cos\omega_2 t$$

At time $t = 0$: $x_2 = x_1 = a/2$ and $|\dot{x}_2| = |\dot{x}_1| = \omega_1\,a/2$ (first case)

At time $t = 0$: $x_2 = -\,x_1 = a/2$ and $|\dot{x}_2| = -|\dot{x}_1| = \omega_2\,a/2$ (second case)

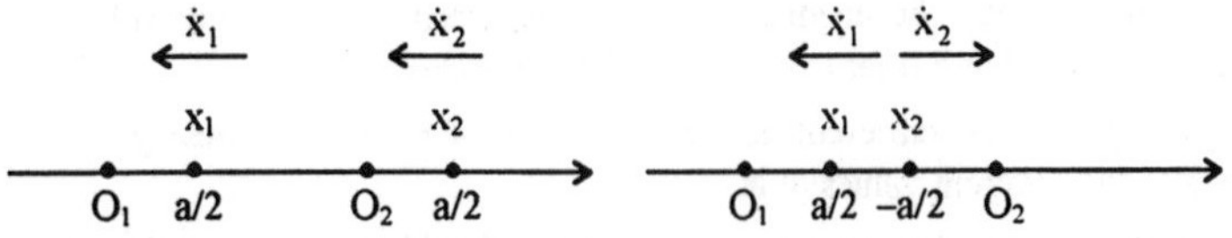

$\dot{x}_1$ $\dot{x}_2$ $\dot{x}_1$ $\dot{x}_2$

x_1 x_2 x_1 x_2

O_1 $a/2$ O_2 $a/2$ O_1 $a/2$ $-a/2$ O_2

First case (identical phases) Second case (opposite phases)

e. We have $\omega_1 = \omega_2 = \omega_0$ and $x_1 = a\cos\omega_0 t$ $x_2 = 0$. The motions are decorrelated. m_1 oscillates as in 1b. m_2 remains at rest.

B. Study of a damped oscillator submitted to an external sinusoidal force

1. The differential equation writes $m\ddot{x} + 2\mu m\dot{x} + m\omega_0^2 x = F_0 \cos\omega t$

2. After identification, we get

$$X_1 = (F_0/m)\left(\omega_0^2 - \omega^2\right)/\left[\left(\omega_0^2 - \omega^2\right)^2 + 4\mu^2\omega^2\right]$$

$$X_2 = (2\mu\omega F_0/m)/\left[\left(\omega_0^2 - \omega^2\right)^2 + 4\mu^2\omega^2\right]$$

3.a. By writing $\Delta\omega = (\omega - \omega_0)/\mu$, we find

$$X_1 \approx (F_0 / 2\omega_0\mu m)\,[\Delta\omega / (1 + \Delta\omega^2)]$$

The extrema $\pm (F_0)/(4m\mu\omega_0)$ are obtained for $\Delta\omega = \pm 1$, or $\omega = \omega_0 \pm \mu$.

b. In the same way, X_2 can be written $X_2 \approx (F_0 / 2\omega_0\mu m)\,[1/(1 + \Delta\omega^2)]$. The extremum $F_0 / (2\mu m\omega_0)$ is obtained for $\Delta\omega = 0$. For $\Delta\omega \pm 1$, we obtain $X_2 = F_0/(4m\mu\omega_0)$.

c. For a vanishing damping constant, the three extrema tend to infinity and they occur at $\omega = \omega_0$. Graphs are identical to those of Problem IV.4.5.

4. We can write $\overline{P} = \overline{\int_0^x F dx} = F_0 X_2 \omega / 2$. The displacement component with a phase at 90 degrees from the phase of the force contributes only to the average dissipated power (dissipative part of the response).

C. Application to the study of dispersion.

1. We have $\mathbf{F} = -e\mathbf{E} - ev \times \mathbf{B}$. For an optical wave propagating in vacuum $\varepsilon_0 E^2 = \mu_0 H^2$. Using the relation $\varepsilon_0\mu_0 c^2 = 1$, this reduces to $E = cB$. We see that, in the case of $v \ll c$, the magnetic force is very small compared to the electric force.

2.a. The differential equation writes $m\ddot{\mathbf{r}} = -e\,\mathbf{E}_0 \cos\omega t$.

b. We are not interested in the complete solution. We prefer to look for a particular solution which can be written as $\mathbf{r} = (e\mathbf{E}_0 / m_e\omega^2) \cos \omega t$.

c. We find $\mathbf{P} = -Ne\mathbf{r}$ and $n^2 = 1 - (\omega_p^2/\omega^2)$ with $\omega_p^2 = Ne^2/(m_e\varepsilon_0)$.

d. If $\omega = \omega_p$, $n = 0$. Only waves with frequencies greater than ω_p (cutoff frequency) can propagate. The refractive index is less than 1 and it tends to 1 when the frequency tends to infinity. The layers of the high atmosphere of the earth constitute a natural medium corresponding to the phenomena described here. We find $N = 4.973 \times 10^{12}\ m^{-3}$. Only waves whose wavelengths are smaller than 15 meters can propagate.

3.a. The differential equation is $m_e\,\ddot{\mathbf{r}} = -m_e\,\omega_0^2\,\mathbf{r} - e\mathbf{E}_0 \cos\omega t$.

b. The forced oscillations are described by the particular solution

$$\mathbf{r} = (e\mathbf{E}_0)/\left[m_e\left(\omega^2 - \omega_0^2\right)\right]\cos\omega t$$

c. We find $\mathbf{P} = -\,Ne\mathbf{r}$ and $n^2 - 1 = Ne^2/\varepsilon_0 m_e\left(\omega_0^2 - \omega^2\right) = \omega_p^2/\left(\omega_0^2 - \omega^2\right)$.

d. The following graph represents the variations of n^2 with ω. At resonance, it shows there is an anomaly in the refractive index.

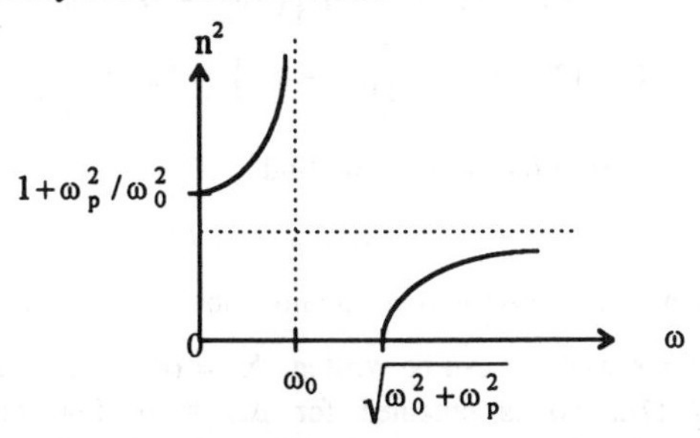

e. To describing the relaxation of the energy, we need to introduce a damping term (as in an "antenna" type emission, for example).

f. The model we just elaborated has two main drawbacks:

— It is a linear, and therefore a limited model (restoring force proportional to the elongation). It is useless for the understanding of nonlinear optics.

— It can describe the spontaneous absorption and emission phenomena, but it is not adapted to the description of the stimulated emission of radiation.

The refractive index can be less than 1 (phase velocity greater than the velocity of light).

IV.4.4. Microscopic classical theory of the absorption, part 2 (from X; M'; 2sd test; 1982)

In this problem, we are interest in the radiation absorbed by molecules of a gas assumed to be perfect and which is dilute enough so that its refractive index can be consider as equal to 1. We consider plane and linearly polarized light waves, propagating along the Oz axis of a direct and orthogonal reference frame Oxyz, whose Ox axis is parallel to the electric field vector of these radiations.

I

In this part, we study the absorption of a monochromatic wave of frequency ω by a molecule of mass M from a quantum point of view. When such a molecule absorbs a photon of energy E_γ, it goes from one state to a more excited one, and we call $\Delta\varepsilon$ the corresponding excitation energy. The laws of inelastic collisions apply to this absorption reaction.

1. In this question, we consider a molecule which is initially at rest at point O.

a. Using relativistic kinematics, find the relation between E, M and the rest mass M* of the molecule after absorption of the photon.

b. Derive the expression of $\Delta\varepsilon$ as a function of E_γ and M. Justify the inequality with physical arguments.

$\Delta\varepsilon < E_\gamma$. Calculate the fraction of the energy of the photon which is transformed in recoiling kinetic energy T of the molecule after the absorption.

c. Calculate the speed V* of the molecule after the absorption.

d. *Numerical application.* For $\omega = 2\times10^{15}$ s^{-1} and M = 41 GeV/c2, calculate E_γ, then calculate the relative error committed when confusing $\Delta\varepsilon$ and E_γ, and also calculate the recoiling speed V*. In what region of the spectrum does the wave of angular frequency ω lie?

2.a. Calculate the average kinetic energy W and the average quadratic speed V of gas molecules at temperature T = 300 K.

b. Does the recoil due to the inelastic collisions of photons on molecules notably modify W and V in the case of the above numerical application?

II

In the rest of the problem, we want to study the absorption of radiation using a classical model which describes the experimental results correctly in spite of its conventional character. This model, which does not take into account the real structure of the molecule, considers only one electron (mass m, electric charge – e) obeying to the laws of Newton mechanics. We admit that this electron moves on a line parallel to Ox passing through the center of inertia O′ of the molecule, and that it is submitted to three forces:

— The first force is the restoring force $-m\omega_0^2\mathbf{r}$, where **r** is the vector with its origin at O' and its end on the position of the electron, and ω_0 is a resonant angular frequency which depends on the molecule;

— The second force represents the effect of the electromagnetic field of the wave on the electron. We neglect the effect of the magnetic field because the speed of the electron is small when compared to the speed of light c. Last, we assume the electric field to take on the same value as at point O' in all the region considered since the molecular dimensions are small compared to the wavelength of the radiation.

— The third force is equal to $-(m/\tau)(d\mathbf{r}/dt)$ and it accounts for the finite value of the lifetime of the excited molecular state after absorption of the radiation. τ represents the mean lifetime of this state.

1. Study the motion of the electron, that is the variation of **r** as a function of t. Show that **r** tends to zero when t tends to the infinity in the absence of radiation, and explain how it tends to zero.

2. In the rest of part II, assume the wave to be monochromatic. The electric field at point z and at time t is expressed by

$$E_x = E_0 \cos \omega (t - z/c)$$

a. Write and solve the differential equation describing the motion of the electron in a path parallel to Ox. Show that, after transient relaxation phenomena, the motion of the electron on the Ox axis is a forced oscillation given by equation $x(t) = x_0 \cos(\omega t + \alpha)$.

b. Calculate the amplitude x_0 (assumed to be positive), sin α, and the power P absorbed by the molecule, i.e. the average power given to the electron by the field, as functions of E_0, ω, ω_0 and τ.

c. ω_0 and τ are determined by the nature of the molecule and E_0 by the field intensity. Study the variations of the positive values of P as a function of ω. Show that P reaches a maximum P_0 for a single value of ω. Determine this value and calculate P_0. Show that to a fixed value θP_0 of P (with $0 \leq \theta \leq 1$) correspond two values $\omega_+(\theta)$ and $\omega_-(\theta)$ of the angular frequency ω and calculate the gap $\Delta\omega(\theta) = |\omega_+(\theta) - \omega_-(\theta)|$ as a function of θ and τ.

d. When $\theta = 1/2$, the gap $\Delta\omega_{1/2}$ (noted $\Delta\omega$), is called the natural linewidth of the absorption line. Calculate $\Delta\omega$ and $\Delta\omega/\omega_0$ numerically for the values $\omega_0 = 2 \times 10^{15}\ s^{-1}$ and $\tau = 10^{-4}$ s. *NB:* For a molecule with many resonance frequencies but which lie far enough apart, these results remain valid in the vicinity of each resonance.

3. In this question, we assume that, due to the thermal agitation, the molecule undertakes a uniform motion along Oz with velocity v, with v slow enough so that the electric field measured in the reference frame of the inertia center O' is identical to the field measured at the same point in the laboratory frame; i.e. $|v_z| << c$.

a. Express the pulsation at which the molecule absorbs the maximum power as a function of ω_0, v_z and c.

b. The Doppler width $\delta\omega$ is defined as the difference between the angular absorption frequencies of molecules who all naturally absorb at ω_0, but who travel in opposite directions along the z axis at speeds $v_z = v_1$ and $v_z = -v_1$. Give the expression of $\delta\omega$ as a function of v_z.

c. Calculate $\delta\omega$ and $\Delta\omega/\omega_0$ numerically for the molecules considered in part I ($\omega_0 = 2 \times 10^{15}\ s^{-1}$, $M = 41\ GeV/c^2$), at temperature T = 300 K, and compare their Doppler and natural widths.

4. The mean lifetime of an excited state is an important physical parameter. How can this study lead to its determination? What is the main difficulty that affects this experimental determination?

I

1.a. Remembering that the molecule is initially at rest, we write the law of the conservation of energy as $\sqrt{p^2c^2 + M^{*2}c^4} = E_\gamma$ with $p = E_\gamma / c$. This gives

$$M^{*2}c^4 = M^2c^4 + 2E_\gamma\, Mc^2$$

b. The excitation energy is the difference between the energy of the molecule at rest after and before the absorption of the photon. It is written as

$$\Delta\varepsilon = \left(M^* - M\right)c^2 = Mc^2\left[\sqrt{1 + \left(2E_\gamma\right)/Mc^2} - 1\right]$$

The excitation energy is slightly smaller than the photon energy because of the molecular recoil (term p^2c^2 in the first equation of question 1.a). The kinetic energy writes $T = M^*c^2\left(1/\sqrt{1-v^2/c^2}-1\right)$ with $v = E_\gamma c/\sqrt{M^{*2}c^4+E_\gamma^2}$. By writing an expansion series in E_γ/Mc^2, limited to the second order, we find $T/E_\gamma = E_\gamma/\left(2Mc^2\right)$.

c. We find $V^* = \left(E_\gamma c/Mc^2\right)\left(1-E_\gamma/Mc^2\right)$.

d. Numerical application: $E_\gamma = 1.32$ eV; $R = (E_\gamma - \Delta\varepsilon)/E_\gamma = E_\gamma/Mc^2 \approx 3 \times 10^{-11}$; $V^* \approx Rc = 9$ mm.s^{-1}. We calculate a wavelength with a value close to 1 μm, at the limit between the red part of the visible spectrum and the infrared part.

2.a. $W = (3/2)kT = 3.87 \times 10^{-2}$ eV and $V = \sqrt{2W/M} \approx 412\ \mathrm{m.s^{-1}}$.

b. We have v << V, and the recoil effect, due to the inelastic collisions of the photons, disturbs V and W only within 10^{-5} and 10^{-10}.

II

1. The differential equation of the motion writes $\ddot{\mathbf{r}} + \dot{\mathbf{r}}/\tau + \omega_0{}^2\mathbf{r} = 0$. A particular solution is $\mathbf{r} = \mathbf{r}_0 \exp - (t/2\tau) \sin \omega'_0 t$ with $\omega'_0 = \sqrt{\omega_0^2 - (1/2\tau)^2}$. **r** tends to zero exponentially with a time constant 2τ when t tends to infinity.

2.a. The equation of the motion is

$$\ddot{x} + (1/\tau)\,\dot{x} + \omega_0^2\, x = \left(-eE_0/m\right)\cos\omega t$$

After relaxation, the motion of the electron is described by a particular solution of the whole equation. It is an oscillating motion, forced at the angular frequency of the field.

b. By substituting the suggested solution in the differential equation, we find

$$x_0 = eE_0\tau/m\sqrt{\tau^2\left(\omega_0^2-\omega^2\right)^2+\omega^2} \qquad \sin\alpha = \omega/\sqrt{\tau^2\left(\omega_0^2-\omega^2\right)^2+\omega^2}$$

And the mean power writes

$$P = \overline{dW(t)/dt} = \overline{-eE_x(t)\,dx(t)/dt} = \tau\left(e\omega E_0\right)^2/2m\left[\tau^2\left(\omega_0^2-\omega^2\right)^2+\omega^2\right]$$

c. The following figure shows P as a function of ω:

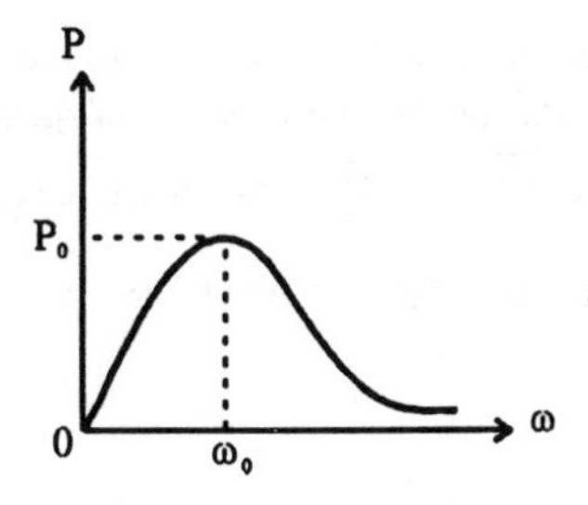

P reaches a maximum at $\omega = \omega_0$. $P_0 = \tau e^2 E_0^2 / 2m$.

Using the approximation $\omega_{+,-} \cong \omega_0 \quad \omega_0 - \omega_{+,-} = \pm(1/2\tau)\sqrt{(1-\theta)/\theta}$.

We can write the gap as $\Delta\omega(\theta) = (1/\tau)\sqrt{(1-\theta)/\theta}$

d. $\Delta\omega_{1/2} = 1/\tau$ and $\Delta\omega_{1/2} / \omega_0 = 10^6 / 2 \times 10^{15} = 5 \times 10^{-10}$.

3.a. We have $\omega_0 = \omega - v_z k = \omega (1 - v_z / c)$. The absorbed power is maximum for $\omega = \omega_0 / (1 - v_z / c)$.

b. The Doppler width is $\delta\omega = \omega_0 (2v_l/c) / [1 - (v_l /c)^2]$.

c. $v_z^2 = V^2/3 \quad v_z = 137.4 \, m.s^{-1} \quad \delta\omega / \omega_0 \approx 10^{-6}$. The Doppler width is about 2 000 times larger than the natural width of the line.

4. The natural width gives a direct determination of τ, but this natural width is generally hidden by the Doppler width. We must therefore develop so-called Doppler free methods (*cf.* Section VII.3).

IV.4.5. Microscopic classical theory of the absorption, part 3 (from Agreg. P; A; 1992)

1. Consider a dilute, linear, homogeneous and isotropic medium. This material is submitted to a sinusoidal electric field E(ω). The electrons of the gas are assumed to be elastically linked to the atoms. In this model, what is the equation of the motion of an electron, and give the solution for the forced regime.

2. Derive the complex polarization vector $\underline{P}(\omega)$ and the complex dielectric susceptibility $\underline{\chi}(\omega)$ of the material. Determine the complex dielectric constant. What characteristic angular frequencies intervene in these quantities? Give the real and imaginary parts of $\underline{\chi}(\omega)$ and the orders of magnitude of the heights and of the widths of these curves. Explain in a few sentences how the results of questions 1 and 2 should be modified if the medium were not dilute?

3. The electric field **E**(ω) belongs to a monochromatic plane wave which propagates in the said dilute medium, also assumed to be nonmagnetic. Recall the equation, derived from Maxwell's equations, which the complex expression of the electric field **E**(ω) of the wave must satisfy. Define the complex dispersion of the medium and its complex

refractive index (distinguish between the refractive and the extinction indices). Find a relation between these indices and $\underline{\varepsilon}(\omega)$.

4. For a glass, calculate the refractive index calculated in question 3 as a function of the wavelength λ for the visible part of the spectrum. What approximate form(s) usually describe(s) this function?

5. What is meant by abnormal dispersion? In what energy domain(s) does this phenomenon occur? What exactly is the anomaly? What are the microscopic phenomena which cause this effect?

6. What relation exists between the absorption and the attenuation coefficients of a plane wave?

7. In quantum mechanics, what does one call absorption, spontaneous emission, stimulated emission? Between what states do the absorption and the emission of light by an atom occur? What device makes use of stimulated emission? What parallel can be made between the classical treatment of the model based on the elastically linked electron, and the quantum model? What is meant by a selection rule? Give an example.

1. *Cf.* Problem IV.4.5.

2. We simply find

$$\underline{P}(\omega)=\sum_i N_i\, e x_i(\omega)=\left[e^2 E(\omega)/m\right]\sum_i N_i\left\{1/\left[\left(\omega_i^2-\omega^2\right)+i\gamma\omega\right]\right\}$$

And the susceptibility can be written as

$$\underline{\chi}(\omega)=\chi'(\omega)+i\chi''(\omega)=\left[\underline{P}(\omega)/\varepsilon_0 E(\omega)\right]=\left[e^2/\varepsilon_0 m\right]\sum_i N_i\left\{1/\left[\left(\omega_i^2-\omega^2\right)+i\gamma_i\omega\right]\right\}$$

The dielectric constant is $\underline{\varepsilon}(\omega)=\left[1+\underline{\chi}(\omega)\right]$. The resonant angular frequencies are those of the electronic oscillators. The real and imaginary parts of the susceptibility are

$$\chi'(\omega)=\left(e^2/\varepsilon_0 m\right)\sum_i N_i\left(\omega_i^2-\omega^2\right)/\left[\left(\omega_i^2-\omega^2\right)^2+\gamma_i^2\omega^2\right]$$

$$\chi''(\omega)=\left(e^2/\varepsilon_0 m\right)\sum_i N_i\,\gamma_i\omega/\left[\left(\omega_i^2-\omega^2\right)^2+\gamma_i^2\omega^2\right]$$

The graph is plotted on the following figure. We drew the functions $\Delta\omega/(1+\Delta\omega^2)$ and $1/(1+\Delta\omega^2)$, which are proportional to χ' and to χ'', respectively. $\Delta\omega$ is defined by $\Delta\omega = 2(\omega-\omega_i)/\gamma_i$.

If the medium is not dilute, the homogeneous broadening should also take into account the following phenomena:

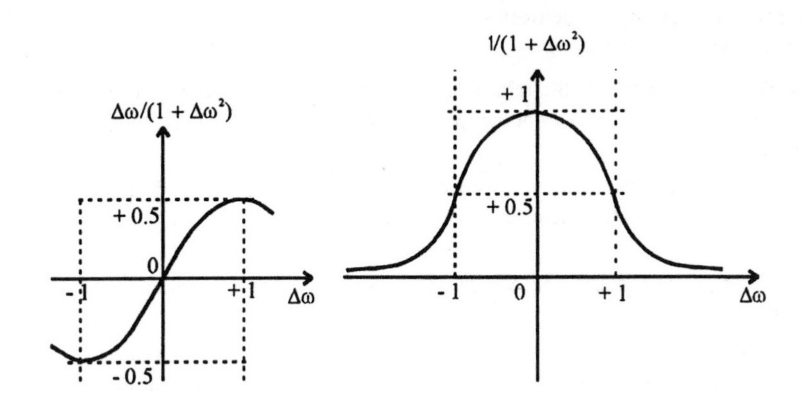

— A loss of phase, due to elastic collisions. Here one should introduce the correlation time T_2.

— The non radiating (nr) energy transfer to the surroundings, due to the inelastic collisions (dissipative coupling between the dipoles and their environment). This leads to setting $\gamma = T_1^{-1} + T_2^{-1} + \gamma_{nr}$.

— Inhomogeneous broadening due, for instance, to the Doppler effect (not very dilute gas and liquids) and to defects in a solid medium.

3. The electric field satisfies the wave equation

$$\Delta \underline{E}(\omega) - \left(1/c^2\right)\left[\partial^2 \underline{E}(\omega)/\partial t^2\right] = 0$$

The dispersion relation can be written as (where D is a constant)

$$\underline{n}^2 = 1 + \sum D/\left[1 + j\gamma/\omega_0 - (\omega/\omega_0)^2\right]$$

The refractive index is defined by $\underline{n}^2 = \underline{\varepsilon}_r = 1 + \chi' + i\chi''$. By setting $\underline{n} = n' + in''$, we find $n' \approx 1 + \chi'/2 \quad n'' \approx \varepsilon''_r/2 \approx \chi''/2$.

4. Far from the electronic bands, which are assumed to lie in the UV part of the spectrum, we get: $n'^2 - 1 = (e^2/\varepsilon_0 m)\sum_i N_i/(\omega_i^2 - \omega^2) \approx \sum_i A_i\lambda^2/(\lambda^2 - \lambda_i^2)$. This is the expression proposed by Sellmeier. It gives rise to the Cauchy series $n'^2 \approx A + B/\lambda^2 + C/\lambda^4 + \cdots$.

5. The so-called abnormal dispersion concerns the region where $\omega \approx \omega_i$, in which the refractive index decreases rapidly with the wavelength. Far from the electronic absorption bands, the refractive index decreases slowly with the wavelength and the dispersion is normal.

6. Ignoring the time-dependent term and not taking account of the amplitude, the electric field of the wave writes

$$\exp ikr \;\propto\; \exp(i2\pi nr/\lambda) \;\propto\; \exp\left[(i2\pi r/\lambda)(1+\chi'/2)\right]\exp(-\pi\chi''/\lambda).$$

The intensity of the plane wave varies as

$$\exp(-\alpha r) \text{ with } \alpha = 2\pi\chi''/\lambda = 4\pi n''/\lambda$$

7. *Cf.* Sections VII.1 et IV.3.2.

IV.4.6. Linewidth (from Agreg. P; A; 1992)

1. What is the cause of the natural width of an atomic line? Give an order of magnitude of this width, expressed in eV, and of the corresponding lifetime for an excited state.

2. What is the cause of the Doppler width $\Delta\nu$ of a line emitted by a gas at temperature T? Calculate the order of magnitude of $\Delta\nu$, and of the corresponding $\Delta\lambda$, for the line at $\lambda = 514.5$ nm of argon (atomic mass M = 40) at the temperature T = 500 K.

3. The line at 514.5 nm is the most intense of the lines of the commercial argon-ion laser. This line has a measured width of 10 GHz. Is this result in agreement with the evaluation of the Doppler width calculated in question 2? Explain.

A laser which functions on this line has a length of 1 m. How many longitudinal laser modes are found inside the emission line? What data determine the width of each mode?

4. In what experimental conditions is the width of a line emitted by a spectral lamp Doppler limited?

5. Explain why the Doppler effect perturbs the absorption process in the reference frame of a moving atom. So-called Doppler free techniques have been thought out so as to avoid this effect and to obtain very fine atomic data. Explain the two main existing Doppler free techniques.

1. The natural width of an atomic line is due to the relaxation of the energy, which classical microscopic theory explains by an antenna-type radiation, macroscopic theory by spontaneous emission, and semi-classical theory by the relaxation from a state of higher to one of lower energy. The following table indicates a few typical values of this width:

dyes in liquids	4×10^{-7} eV	10^{-8} s
low pressure gases	4×10^{-9} eV	10^{-6} s
doped solids	2×10^{-11} eV	5×10^{-4} s

2. The Doppler broadening is induced by the distribution of the molecular velocities in a gas at temperature T. The most probable speed v is such that

$$(1/2)\,Mv^2 = (3/2)\,RT$$

The frequency and wavelength broadening are given by $\Delta\nu = v / \lambda \approx 1.08$ GHz; $\Delta\lambda = \lambda^2\,\Delta\nu/c \approx 1$ pm.

3. The line at 514.5 nm has a width between 5 and 10 GHz (but note that the most intense line of the argon-ion laser lies at 488 nm). It is slightly wider therefore than the Doppler broadening at 500 K. In fact, the temperature of the tube of this laser is higher than 500 K. This explains the greater broadening since it is proportional to the square root of the temperature. On the other hand, the magnetic induction of about 1 KG which is applied to the tube to keep the ionic and electronic discharges on the axis of the capillary, gives rise to high ionic and electronic densities. Due to the nature and the density of the microscopic systems, the Doppler conditions are not satisfied. The broadening is affected by collisions (Holzmark effect). The longitudinal modes of the cavity are separated by $c / 2L = 150$ MHz. Therefore, there are 66 modes in the 10 GHz Doppler envelope. The width of each mode depends on both the finesse of the Pérot-Fabry interferometer and on the amplification by stimulated emission which causes an important decrease of the bandwidth. Because of the experimental imperfections, the bandwidth reduction has a finite (i.e. non zero) limit.

4. To limit collision-induced broadening, very low pressure lamps should be used.

5. In the reference frame of a moving atom, the frequency of an optical wave detected by this atom depends on its velocity (except in the case where the speed vector is normal to the wave vector of the optical wave). This effect gives rise to inhomogeneous broadening, which hides the natural widths of the lines. We could try to cool the sample and use molecular beams (where the mean speed vector is normal to the wave vector). Doppler free spectroscopies are described later in this book (*cf.* Section VII.3).

IV.4.7. Electric dipole approximation

Let us consider an atom. It consists of a nucleus with an electric charge + Ze, surrounded by Z electrons (charge: – e; rest mass: m_e) identified by subscript j (j going from 1 to Z). Assume that the center of mass is located at the nucleus and let it be the origin of a spatial reference frame. The atom is illuminated by a laser wave whose electric (**E**) and magnetic (**H**) fields derive from a potential vector **A** by $\mathbf{E} = -\dot{\mathbf{A}}$; $\mathbf{H} = (1/\mu_0)\,\nabla \times \mathbf{A}$. μ_0 is the magnetic permittivity of vacuum.

The static electric field **F**, induced by the electric charges of the atom, derives from a scalar potential ϕ by $\mathbf{F} = -\nabla\phi$.

1. Give a qualitative explanation for the following expression of the Lagrangian function L:

$$L\left(\mathbf{r}_j, \dot{\mathbf{r}}_j, t\right) = \sum_j \left[m_e\,\dot{\mathbf{r}}_j^2 / 2 - e\,\dot{\mathbf{r}}_j\,\mathbf{A}\left(\mathbf{r}_j, t\right) + e\,\phi\left(\mathbf{r}_j\right) / 2\right] - Z_e\,\phi(0) / 2$$

2. Compare the characteristic dimensions of the atom with those of the optical wave and show that it is justified to replace $\mathbf{A}(\mathbf{r}_j, t)$ by $\mathbf{A}(0, t)$.

3. Show that the Lagrange equations are not modified by adding the term $d[e\sum_j \mathbf{r}_j \mathbf{A}(0,t)]/dt$ to the Lagrange function. Give the new expression of the Lagrangian.

4. Calculate the expression of the Hamilton function, using the expression $\mathbf{p} = -\sum_j e\,\mathbf{r}_j$.

1. The reader may report to specialized books on analytical mechanics where the Lagrangian and Hamiltonian functions are defined and analyzed. To carry out the calculation, the mass of the electron can be neglected as compared to that of the nucleus (which is assumed to be the center of mass of the atom). In this case, the kinetic energy of the nucleus vanishes and the expression $\sum_j \left(m_e\,\dot{\mathbf{r}}_j^2/2\right)$ represents the non relativistic kinetic energy of the electrons. The expressions $\sum_j e\,\phi(\mathbf{r}_j)/2$ and $-Ze\,\phi(0)/2$ describe the electronic and nuclear parts, respectively, of the potential energy V induced by the static electric field as they appear in the Lagrangian (except for a change of sign).

The term $\sum_j e\,\dot{\mathbf{r}}_j \mathbf{A}(\mathbf{r}_j, t)$ describes a potential energy due to the interactions of the electrons with the electric field driven by the laser wave. Starting from these terms of the Lagrangian, the reader can easily show that the Lagrange equations lead to the Coulomb and Lorentz forces, whose existence is proved experimentally. Such an identity between the Lagrange equations and the fundamental law of dynamics justifies the form chosen for the Lagrangian.

2. Let us write a power expansion of $\mathbf{A}(\mathbf{r}_j, t)$, as a function of $\mathbf{r}_j$, around $\mathbf{r}_j = 0$:

$$\mathbf{A}(\mathbf{r}_j, t) = \mathbf{A}(0, t) + \mathbf{r}_j \left[\nabla_{\mathbf{r}_j} \mathbf{A}(\mathbf{r}_j, t)\right]_{\mathbf{r}_j = 0} + \dots$$

The second term of the expansion includes two characteristic lengths: The first one is an atomic characteristic length of the order of the angstrom ($\mathbf{r}_j$ and $\nabla_{\mathbf{r}_j}$), and the second one, an optical length of the order of the micron, is characteristic of the potential vector $\mathbf{A}$. The large difference (three to four orders of magnitude) between these two lengths implies that the potential vector $\mathbf{A}$ can be considered as constant over the atomic dimensions, yielding $\mathbf{A}(\mathbf{r}_j, t) \approx \mathbf{A}(0, t)$.

3. The Lagrange function is not defined uniquely. If a term is added to it which is the total derivative with respect to time of a function of the spatial coordinate $\mathbf{r}$, its properties do not change. We can therefore built a second Lagrange function L' which writes

$$L' = L + d\,[e\sum_j \mathbf{r}_j \mathbf{A}(0,t)]/dt = L + e\sum_j \mathbf{r}_j \dot{\mathbf{A}}(0,t) + e\sum_j \dot{\mathbf{r}}_j \mathbf{A}(0,t)$$

(the reader can check that this new Lagrange function indeed gives rise to the same Lagrange equations as the first one). So now we have

$$L'(\mathbf{r}_j, \dot{\mathbf{r}}_j, t) \approx \sum_j \left[m_e \dot{\mathbf{r}}_j^2/2 + e\mathbf{r}_j \dot{\mathbf{A}}(0,t) + e\phi(\mathbf{r}_j)/2\right] - Ze\phi(0)/2$$

4. The Hamilton function is $H = -L' + \sum_j m_e \dot{\mathbf{r}}_j^2$, yielding

$$H = \sum_j \left[m_e \dot{\mathbf{r}}_j^2/2 - e\phi(\mathbf{r}_j)/2 + Ze\phi(0)/2\right] - e\sum \mathbf{r}_j \dot{\mathbf{A}}(0,t)$$

The expression in square brackets represents the energy of the atom in the absence of radiation. Let us call it H_0. The last term can be written as

$$-e\sum_j \mathbf{r}_j \dot{\mathbf{A}}(0,t) = e\sum_j \mathbf{r}_j \mathbf{E}(0,t) = -\mathbf{pE}(0,t)$$

It represents the atom-radiation interaction energy in the EDA approximation.

IV.4.8. Fundamental laws of absorption (from Agreg. C; A; 1978)

We would like to derive the expression of the absorption coefficient $\varepsilon(\nu)$ of a substance as a function of the frequency ν. Let us consider a cylindrical cell, of section S, of length l, containing N molecules per unit volume in their ground state. The cell is crossed by a parallel beam of light of intensity I at the point of abscissa x ($I = I_0$ if $x = 0$). Each molecule which absorbs a photon $h\nu$ is lifted from its ground state i to its excited state j at a (1st order) rate whose constant is $k_{ij} = 8\pi^3 \rho |p_{ij}|^2 / 3h^2$, where ρ and p_{ij} are, respectively, the energy density and the transition moment.

a. Calculate the number dn of molecules that reach the excited state during one second per length dx.

b. Calculate the variation dI of the intensity between the abscissa x and x + dx during the same time.

c. Recall the differential expression of the Beer law at a fixed frequency, and use it to derive the expression of $\varepsilon(\nu)$.

d. The half-life time T of a chemical compound in an excited state is defined from the coefficient of spontaneous emission A_{ji} by $T = \ln 2/A_{ji}$ and the ratio A_{ji}/k_{ij} takes the value $8\pi h\nu^3/\rho c^3$. Show that it is possible to relate the half-life time T to the absorption coefficient $\varepsilon(\nu)$ in a simple way.

a. It is best to introduce a shape factor $\Phi(\nu)$ such that $\int_\nu \Phi(\nu)\,d\nu = 1$ and leading to

$$d^2n/dt = k_{ij}\,\Phi(\nu)\,NSdx$$

b. $dI(\nu) = -(h\nu/S)(d^2n/dt)$ with $I(\nu) = \rho c$.

c. $dI(\nu) = -\varepsilon(\nu)\,NI(\nu)\,dx$ and, by identification, we obtain

$$\int_\nu \varepsilon(\nu)\,d\nu = \left(8\pi^3 |p_{ij}|^2 / 3hc\right) \int_\nu \nu\Phi(\nu)\,d\nu$$

And, if the shape factor varies rapidly when compared to the frequency, i.e. by defining an average value of the frequency, we obtain

$$\int_\nu \varepsilon(\nu)\,d\nu = 8\pi^3 |p_{ij}|^2\,\bar{\nu} / 3hc$$

d. Finally we obtain $T = (\ln 2)c^2 / 8\pi\nu^2 \int_\nu \varepsilon(\nu)\,d\nu$

IV.4. 9. Electronic spectra of transition metals (from Agreg. C; C; 1983)

Study of the coordination of exchangeable ions in a zeolithe.

1. Atomic spectroscopy

1.1. *Main interaction energies in a many-electron atom.*

The energy levels of the electrons of the external shells depend on:

— The interaction of these electrons with the nucleus from which they are not completely protected by the internal shells;

— The interelectronic repulsion between the electrons of the external shell;

— The coupling of the orbital kinetic moments with the electron spin.

Classify the interactions affecting the electrons of a many-electron atom by order of decreasing importance.

1.2. *Russel-Saunders classification*

a. What is the Russel-Saunders LS coupling?

b. Give the spectroscopic terms associated to configurations d^1 and d^9.

c. The terms ^3P, ^3F, ^1D, ^1G and ^1S can be associated to configuration d^2. Give the Hund rules and derive the ground term of configuration d^2.

d. The gaps between terms ^3F and ^3P, ^3F and ^1D, ^3F and 1G, ^3F and ^1S are, respectively, equal to 15 B, 5 B + 2 C, 12 B + 2 C and 22 B + 7 C (B and C are the Racah parameters describing the interelectronic repulsion). Draw the energy diagram of the ion terms d^2 and d^8 knowing that $C \approx 4.7$ B.

1.3.a. Describe the influence of spin-orbit coupling on the ground term of a d^2 ion.

b. Let λ be the constant describing the spin-orbit coupling. Knowing that the energy gap between two successive J states is equal to $\Delta E_{J, J+1} = \lambda(J + 1)$, draw the energy-level diagram of the states of d^2 and of d^8 ions. On the same diagram, indicate the values of the spin-orbit coupling energies associated to each state (take the energy of the degenerate term as the origin).

2. Electronic spectra of complex ions and crystal field theory.

2.1. On the classification diagram of paragraph 1.1, indicate the position of the interactions with the crystal field in the following cases:

— A weak crystal field,

— A strong crystal field.

2.2. Weak crystal field.

2.2.1. Using a diagram, show the splitting of the spectroscopic terms associated to d^1 and d^9 ions in case of:

a. A weak octahedral field (symbolized by O_h),

b. A weak tetrahedral field (symbolized by T_d).

In each case, indicate the value of the energy associated to each term (in units Dq of the crystal field energy, choosing the origin of the energy at the degenerate term).

2.2.2. In a weak O_h field, the spectroscopic term ^3F of a d^2 ion splits according to:

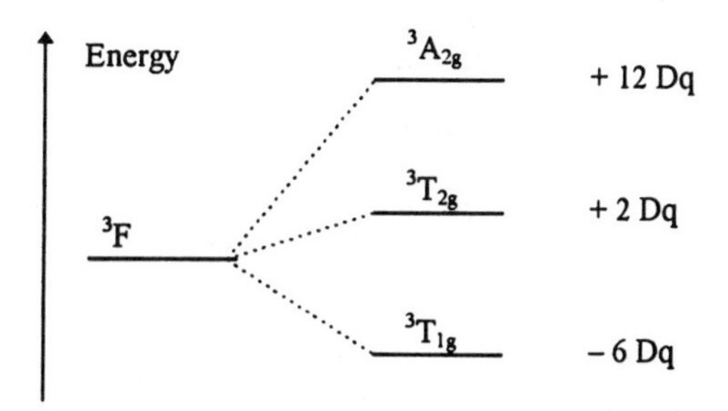

The level of term ^3P is unchanged and corresponds to level $^3T_{1g}$.

Derive the energy distribution of the triplet terms of a d^8 configuration, first in a weak O_h field, then in a weak T_d field.

2.2.3. Tetragonal distortion.

Let us consider a tetragonal bipyramid:

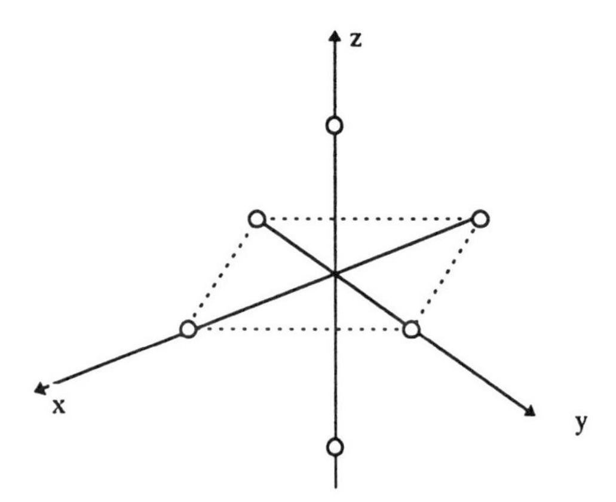

It belongs to symmetry group D_{4h}.

a. What are the symmetry elements and operations of this group?

b. By looking at the way orbitals d_{z^2}, $d_{x^2-y^2}$, d_{xy}, d_{yz} and d_{zx} are transformed by the symmetry operations of the D_{4h} group, establish the character table. Derive the correspondence between these d orbitals and the A_{1g}, B_{1g}, B_{2g} and E_g irreducible representations of symmetry group D_{4h}.

c. What are the energy distributions of terms d^1 and d^9 in an octahedral crystal field with tetragonal distortions due to an elongation along the Oz axis (D_{4h} symmetry)?

2.2.4. Crystal field with D_{3h} symmetry.

a. Give the symmetry elements and the symmetry operations of group D_{3h}.

b. Represent the influence of a crystal field with D_{3h} symmetry, induced by ligands placed at the tops of a trigonal bipyramid compressed along the Oz axis (Oz axis lies along the C_3 symmetry axis), on the energy diagram of the d orbitals.

c. The irreducible representations of symmetry group D_{3h} are A'_1 , E' and E ''. Using the results obtained in question 2.2.4.b, give the correspondence between the molecular d orbitals and the irreducible representations of group D_{3h} for d^1 and d^9 ions.

2.2.5. Electronic transitions.

a. State the Beer-Lambert law.

b. What are the selection rules concerning allowed electronic transitions?

c. Notwithstanding these, what mechanisms can explain the existence of d-d transitions?

d. How many transitions can be observed in the cases of d^2 and d^8 ions in O_h and T_d crystal fields?

e. The triplet terms of a d^2 ion are split by a D_{4h} tetragonal distortion induced by an O_h field according to:

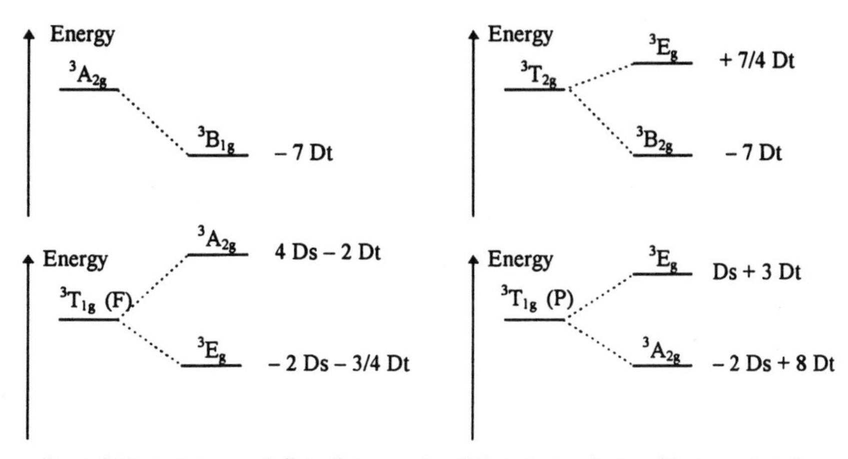

Ds and Dt are tetragonal distortion energies. What electronic transitions can be observed in the cases of d^2 and d^8 ions?

2.2.6. Electronic and stereochemical properties of single-nucleus complexes of copper (II) and of nickel (II).

a. Give the electronic configurations of copper (II) and of nickel (II).

b. Aqueous solutions of copper (II) are characterized by a single weak intensity absorption band whose maximum occurs at wavelength λ = 833 nm. The molar absorption coefficient is 11 $mol^{-1}.l.cm^{-1}$. What is the stereochemical formula of the hydrated complex? Explain the light blue color of the solution. In a highly concentrated ammonia solution, the absorption band shifts to λ = 606 nm. What are the chemical formula and the stereochemical formula of the complex ion? Indicate the electronic transition involved and explain the intense blue color of the solution.

c. The absorption band of copper (II) in faujasite-type hydrated zeolithes varies with the ratio Si/Al, that is, with the relative quantities of silicium and aluminium.

λ = 805 nm for copper containing X zeolithes in which the ratio Si/Al = 1.2.

λ = 823 nm for copper containing Y zeolithes in which the ratio Si/Al = 2.4.

What is the coordination number of copper (II) in the super-clusters of the zeolithes? Comment the differences between the absorption wavelengths. After dehydration, the absorption band splits into three bands whose maxima occur at 980, 800 and 667 nm in the case of copper containing zeolithes. Complete dehydration leads to a two-band spectrum with maxima at λ = 926 nm and λ = 690 nm. By examining the energy diagrams of questions 2.2.1, 2.2.3 and 2.2.4, deduce the coordination number of copper (II) in X anhydrous zeolithes. Identify the electronic transitions.

d. The aqueous solutions of nickel (II) absorbs at wavelengths λ_1 = 1 176 nm, λ_2 = 741 nm and λ_3 = 395 nm. Look at the energy diagrams of questions 2.2.2 and 2.2.5 and indicate what coordination number can be inferred from the existence of these three bands. The molar extinction coefficients are ε_1 = 1.6, ε_2 = 2.0 and ε_3 = 4.6 $mol^{-1}.\ l.\ cm^{-1}$, respectively. Find the coordination number of nickel (II) ions in aqueous solution. What electronic transitions can be associated to these absorption bands? Explain the green color of these solutions. What is the value of the energy of the crystal field induced by the water molecules? What is the value of the interelectronic repulsion parameter B?

e. A solution of nickel (II) ions in concentrated ammonia reveals three absorption bands. What are the coordination numbers for this complex ion? Knowing that two of the absorption maxima are located at $\lambda_1 = 926$ nm and at $\lambda_2 = 571$ nm, determine the coordination number of nickel (II) in this complex. By assuming that the value of B remains unchanged, at what wavelength would the third maximum lie? What is the color of the solution?

f. An aqueous solution of tetrachloronickelate (II) ion absorbs at the wavelengths $\lambda_1 = 2\ 500$ nm, $\lambda_2 = 1\ 282$ nm and $\lambda_3 = 667$ nm. The molar extinction coefficient is of the order of 17 $mol^{-1}.\ cm^{-1}$ for these three bands. Find the stereochemical formula of this complex ion. Identify the electronic transitions and calculate the crystal field energy and the interelectronic repulsion parameter B. Compare this value of B to that obtained in question b. Give a few comments.

2.3. Strong crystal field.

2.3.1. Give all the possible electronic configurations for ions in a strong octahedral field. Give the energies of the levels corresponding to each configuration (the electrons should be place in the e_g and in the t_{2g} orbitals)?

2.3.2. What electronic transitions can be predicted?

3. Molecular orbital theory.

Let us consider a complex ion of the form ML_6 in which M is a transition metal.

3.1. The bond between the metal and the six ligands is assumed to be a σ-type bond. Using a qualitative diagram of the molecular energy levels, describe the complex ion ML_6 in the frame of molecular orbitals.

3.2. Draw the same diagram for the following cases:

a. The ligands are π-electron donors.

b. The ligands are π-electron acceptors.

3.3. Infer what electronic transitions can occur for a d^1 ion in each of the previously described cases. Classify these transitions by order of increasing energy.

1.1. The classification is as follows:

Electron-core interaction > interelectronic repulsion > spin-orbit coupling

1.2.a. Since the energies are negative, the strongest coupling is that corresponding to the largest absolute value of the energy (stabilization by energy minimization). When the energy of the spin-orbit coupling (which increases as the fourth power of the atomic number Z) has a very small absolute value as compared to the interelectronic repulsion energy (in case of light atoms), the orbital kinetic moment and the spin moment are not coupled together, i.e. $\Sigma \mathbf{l} = \mathbf{L}$; $\Sigma \mathbf{s} = \mathbf{S}$; $\mathbf{J} = \mathbf{L} + \mathbf{S}$. The modulus of **J** can take on all values, in steps of one, lying between $(L+S)$ and $|L-S|$.

This is called the Russel-Saunders (RS) coupling. If the spin-orbit coupling is not so small, we must calculate **J** by adding the **j** vectors: $\mathbf{J} = \Sigma\mathbf{j} = \Sigma\,(\mathbf{l} + \mathbf{s})$. This second coupling is called the j – j coupling (*cf.* Section III.3).

b. d electrons correspond to l = 2, d^1 et d^9 are two single electron configurations with s = (1/2). This is a case of RS coupling and the two configurations write $^2D_{5/2}$ and $^2D_{3/2}$.

c. Hund rules predict that the lowest lying energy term is the one which (i) has maximal multiplicity and (ii) corresponds to the largest values of L. In this case, it is the triplet ^3F.

d. The diagram is given below:

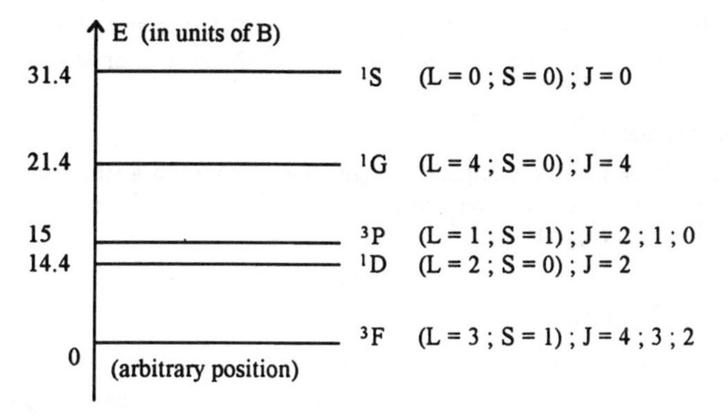

1.3.a. The spin-orbit coupling breaks the energy degeneracy of the two triplets ^3F and ^3P. Level ^3F generates levels $^3F_2$; $^3F_3$ and $^3F_4$.

b. Energy diagrams are given below. The two ions d^2 and d^8 are electronic images with respect to the half full d level. The energy distributions will be symmetrical with respect to the degenerate level.

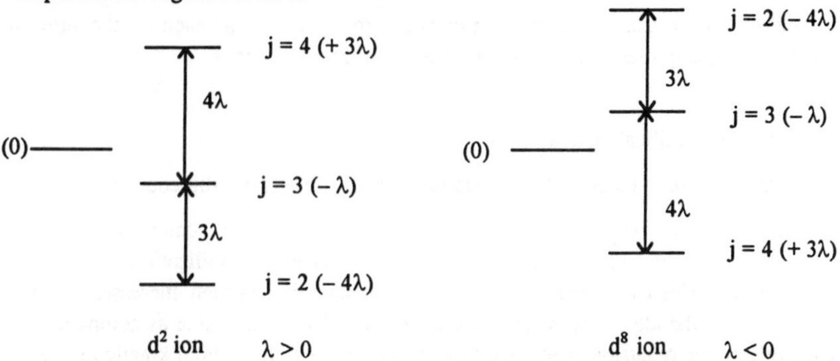

To find the energy position with respect to level ^3F which is the energy reference, we write that the center of mass of the distribution is zero. This is only true for a weak spin-orbit interaction. Taking into account that each level has a multiplicity of (2J + 1), we obtain the values indicated above.

2.1. In the case of a weak crystal field, the energies of the interactions with the crystal field lie between the interelectronic repulsion and the spin-orbit coupling. For a strong crystal field, interactions with the crystal field lie between the electron-core interaction and the interelectronic repulsion (or at the same level as the latter).

2.2.1.a. O_h symmetry. The diagram of the splitting is given below.

E_g : + 6 Dq

T_{2g} : – 4 Dq

d^1 ion

T_{2g} : + 4 Dq

E_g : – 6 Dq

d^9 ion

b. T_d symmetry. The diagram of the splitting is given below.

T_2 : + 4 Dq

E : – 6 Dq

d^1 ion

E : + 6 Dq

T_2 : – 4 Dq

d^9 ion

2.2.2. The triplet levels of a d^8 ion are ^3F and ^3P (*cf.* question 1.2.d).

In question 2.2.1. we were reminded that O_h and T_d distributions are symmetrical. Therefore we can construct the following diagram (remember that distributions of d^2 and d^8 are the other way around):

O_h symmetry:

^3P → $^3T_{1g,P}$ (unchanged)

^3F → $^3T_{1g,F}$ (+ 6 Dq)

^3F → $^3T_{2g,F}$ (– 2 Dq)

^3F → $^3A_{2g,F}$ (– 12 Dq)

T_d symmetry:

^3P → $^3T_{1g,P}$ (unchanged)

^3F → $^3A_{2,F}$ (+ 12 Dq)

^3F → $^3T_{2,F}$ (+ 2 Dq)

^3F → $^3T_{1,F}$ (– 6 Dq)

2.2.3.a. Symmetry elements and operations of the D_{4h} group are listed below on the first line of the table of question b.

b. The following table indicates the transformation of the d orbitals, and shows the irreducible representations for which they form 1- and 2-order bases.

D orbital	E	$2C_4$	C_2	$2C'_2$	$2C''_2$	i	$2S_4$	σ_h	$2\sigma_v$	$2\sigma_d$	Irreducible representations
d_{z^2}	1	1	1	1	1	1	1	1	1	1	A_{1g}
$d_{x^2} - d_{y^2}$	1	−1	1	1	−1	1	−1	1	1	−1	B_{1g}
d_{xy}	1	−1	1	−1	1	1	−1	1	−1	1	B_{2g}
$\{d_{yz}, d_{zx}\}$	2	0	−2	0	0	2	0	−2	0	0	E_g

For the E_g representation, you have to write the 2 × 2 matrices representing the operators on the (d_{yz} , d_{zx}) base. Their traces constitute the characters indicated on the table.

c. The diagram is given below.

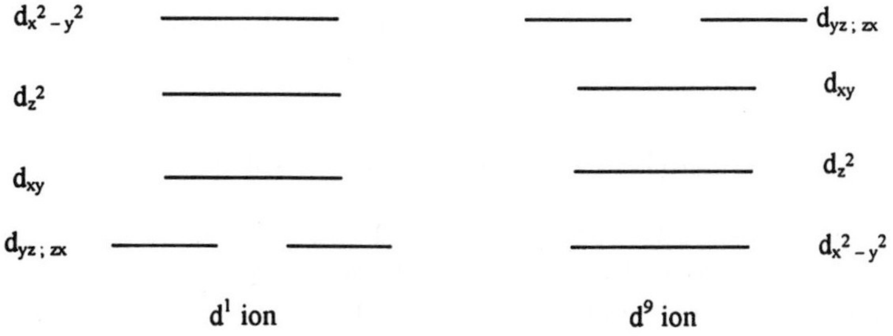

d¹ ion		d⁹ ion	
$d_{x^2-y^2}$	———	——— ———	$d_{yz\,;\,zx}$
d_{z^2}	———	———	d_{xy}
d_{xy}	———	———	d_{z^2}
$d_{yz\,;\,zx}$	——— ———	———	$d_{x^2-y^2}$

2.2.4.a. The symmetry elements and operations are E; $2C_3$; $3C_2$; σ_h; $2S_3$ and $3\sigma_v$.

b. The energy diagram is as follows:

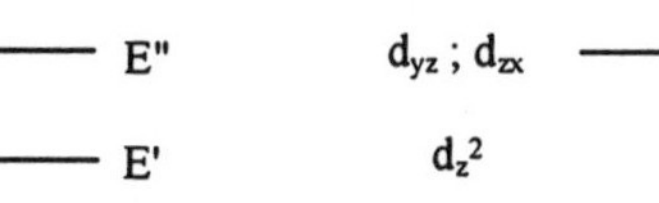

———	d_{z^2}
——— ———	(d_{yz}; d_{zx})
——— ———	($d_{x^2-y^2}$; d_{xy})

c. The relation is indicated below.

d¹ ion			d⁹ ion		
d_{z^2}	———	A'_1	$d_{x^2-y^2}$	——— ———	E'
d_{yz} ; d_{zx}	——— ———	E"	d_{yz} ; d_{zx}	——— ———	E"
$d_{x^2-y^2}$	——— ———	E'	d_{z^2}	———	A'_1

2.2.5.a. The Beer-Lambert law writes: $I_t = I_0 \exp(-\varepsilon lc)$. It indicates an exponential attenuation of the transmitted intensity I_t (l is the length of the absorbing sample, c the concentration of the absorbing species and ε the absorption coefficient). It is a limit law, only valid for low light intensities and weak optical densities of the medium.

b. The selection rules require $\Delta L = \pm 1$; $\Delta S = 0$.

c. There are other specific rules which can restrict some types of transitions. For example, when the two state vectors involved in a transition have a very definite parity, electronic transitions, as described by EDA, are limited to transitions between state vectors with different spatial parity by the so-called Laporte rules (*cf.* Section IV.3.2.3). This rule can be transgressed when the molecular vibrations modify the parity of the state vector. So-called d-d transitions constitute an example of the violation of the Laporte rules.

d. The following table indicates the main observable transitions.

	O_h	T_d
d^2	${}^3T_{2g,F} \leftarrow {}^3T_{1g,F}$ ${}^3A_{2g,F} \leftarrow {}^3T_{1g,F}$ ${}^3T_{1g,F} \leftarrow {}^3T_{1g,F}$	${}^3T_{2,F} \leftarrow {}^3A_{2,F}$ ${}^3T_{1,F} \leftarrow {}^3A_{2,F}$ ${}^3T_{1,P} \leftarrow {}^3A_{2,F}$
d^8	${}^3T_{2g,F} \leftarrow {}^3A_{2g,F}$ ${}^3T_{1g,F} \leftarrow {}^3A_{2g,F}$ ${}^3T_{1g,F} \leftarrow {}^3A_{2g,F}$	${}^3T_{2,F} \leftarrow {}^3T_{1,F}$ ${}^3A_{2,F} \leftarrow {}^3T_{1,F}$ ${}^3T_{1,P} \leftarrow {}^3F_{1,F}$

These transitions can be forbidden by Laporte rules (in the O_h symmetry group, all states considered are symmetrical with respect to spatial inversion). The values of the absorption coefficient are in fact relatively small. The three lines are not always observed and aqueous solutions containing the O_h d^2 complex show only two lines. The same complex in solid state shows all three lines.

e. The six observed lines are indicated on the next diagram (do not forget the level inversions between d^2 and d^8).

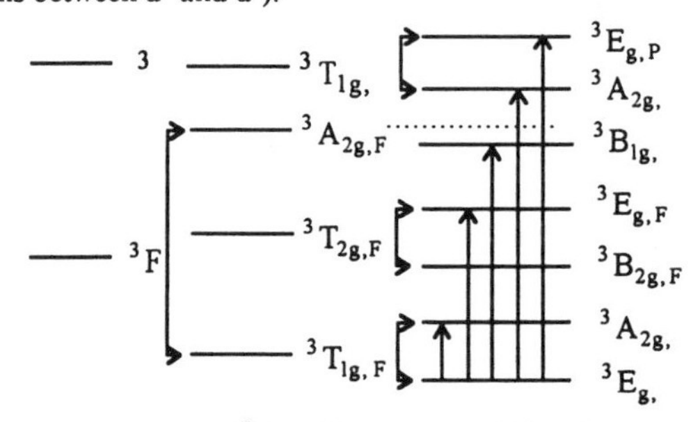

d^2 ion; O_h; tetragonal distortion

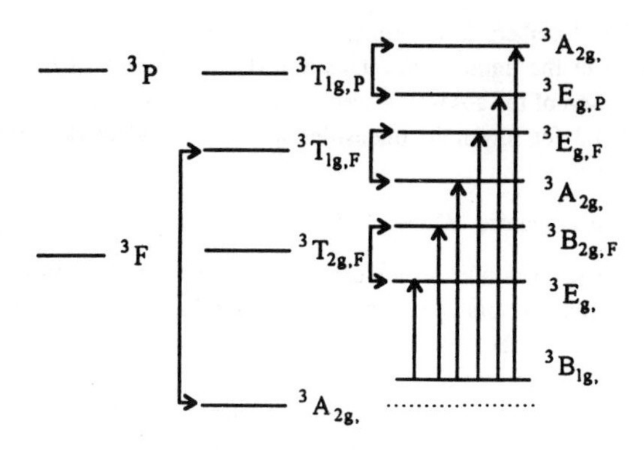

d^8 ion; O_h; tetragonal distortion

2.2.6.a. The electronic configurations are:

Cu (II): $1s^2\ 2s^2\ 2p^6\ 3s^2\ 3p^6\ 3d^9\ 4s^0$

Ni (II): $1s^2\ 2s^2\ 2p^6\ 3s^2\ 3p^6\ 3d^8\ 4s^0$

b. The small value of the absorption coefficient means we are dealing with a Laporte forbidden transition and indicates an O_h structure of the $\left[Cu(H_2O)_6\right]^{2+}$ ion. The light blue color is due to the fact that light is absorbed in the far red part of the spectrum. In a concentrated ammonia solution, an $\left[Cu(NH_3)_4(H_2O)_2\right]^{2+}$ ion is generated (tetragonal distortion O_h). The intense blue color is due to the hypsochromic effect of ammonia (the absorption maximum shifts from 833 nm to 606 nm). In aqueous solutions, the electronic transition corresponds to ${}^3T_{2g} \leftarrow {}^2E_g$. To identify the transition occurring in the combined complex, one must know the effect of the tetragonal splitting.

c. The structure of the absorption spectrum is very similar to those studied in last question so we opt for a O_h symmetry and a coordination number 6. When the aluminum ratio increases, a hypsochromic effect is observed due to a tetragonal distortion of the octahedron (stronger crystal field).

Partial dehydration results in a tetragonal distortion $O_h \rightarrow D_{4h}$. The three lines correspond to transitions ${}^3A_{1g} \leftarrow {}^3B_{1g}$; ${}^3B_{2g} \leftarrow {}^3B_{1g}$ and ${}^3E_g \leftarrow {}^3B_{1g}$ (by decreasing wavelengths). Complete dehydration results in a trigonal distortion $O_h \rightarrow D_{3h}$. By order of decreasing wavelengths, the two observed lines correspond to transitions ${}^3E'' \leftarrow {}^3A'_1$ and ${}^3E' \leftarrow {}^3A'_1$. The coordination number of copper in zeolithe is 5.

d. The small absorption coefficients suggest a coordination number 6 and an O_h symmetry. The only absorbance in the visible range occurs at wavelength 741 nm, i.e. in the red part of the spectrum. It is responsible for the green color associated to the hexapro nickel II ion. The answer to question 2.2.2, relative to the d^8 ion, is presented in the following table:

λ(nm)	ε ($mol^{-1}.l.cm^{-1}$)	Transition	Transition energy
1 176	1.6	${}^3T_{2g,F} \leftarrow {}^3A_{2g,F}$	10 Dq
741	2.0	${}^3T_{1g,F} \leftarrow {}^3A_{2g,F}$	18 Dq
395	4.6	${}^3T_{1g,P} \leftarrow {}^3A_{2g,F}$	15 B + 12 Dq

The two first lines give $\overline{Dq} \approx 800\ cm^{-1}$. By inserting this value in the third line of the table, we calculate a value of $1\ 000\ cm^{-1}$ for constant $\overline{B}$.

e. Question 2.2.2. results in coordination numbers of either 6 (O_h) or 4 (T_d). For coordination number 6, and with the values just obtained for Dq, we find $10\overline{Dq} \approx 8\ 000\ cm^{-1}$. In the same way, we find $8\overline{Dq} \approx 6\ 400\ cm^{-1}$ for coordination number 4. Since the experimental value is $\lambda_1^{-1} \approx 10\ 800\ cm^{-1}$, we choose coordination number 6. The third wavelength would correspond to an energy of $15\overline{B}+12\overline{Dq} \approx 27\ 600\ cm^{-1}$ giving $\lambda_3 \approx 370$ nm. The only absorption occurring in the visible range lies in the yellow-green part of the spectrum so that the solution is blue.

The larger value of the absorption coefficient for the three lines suggests a coordination number 4 and a T_d symmetry. The following table shows the energy values of the transitions.

λ (nm)	ε ($mol^{-1}.l.cm^{-1}$)	Transition	Transition energy
2 500	~ 17	${}^3T_{2,F} \leftarrow {}^3T_{1,F}$	8Dq
1 282	~ 17	${}^3A_{2,F} \leftarrow {}^3T_{1,F}$	18 Dq
667	~ 17	${}^3T_{2,P} \leftarrow {}^3T_{1,F}$	15 B + 6 Dq

We find $\overline{Dq} \approx 470\,cm^{-1}$; $B \approx 800\,cm^{-1}$. The decrease of $\overline{B}$ is due to the decrease of the interelectronic repulsion.

2.3.1. For the d^2 ion, we can use three possible stong field configurations, $t_{2g}^2 e_g^0$; $t_{2g}^1 e_g^1$ and $t_{2g}^0 e_g^2$ in order of decreasing energy. The energy are listed below.

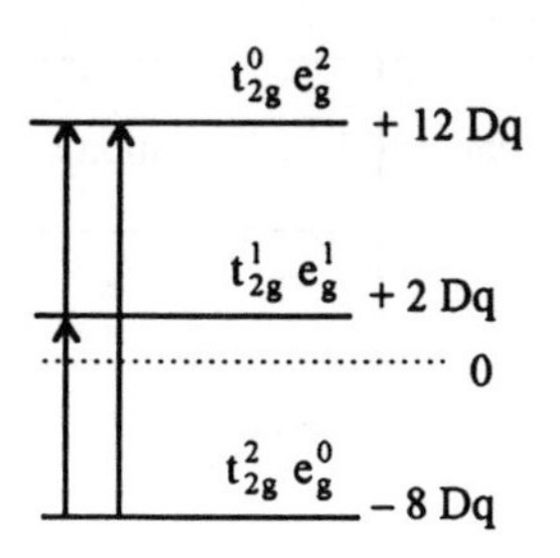

For the d^8 configuration, we must remember the d^2 configuration in a T_d symmetry and inverse it. We obtain the following diagram:

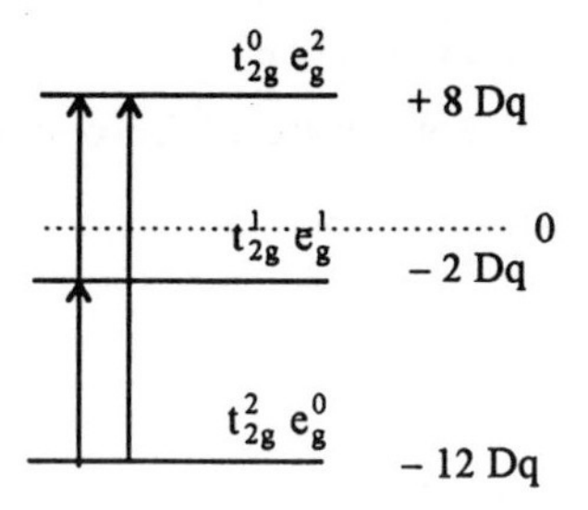

2.3.2. For the d^2 ion, configuration $t_{2g}^2 e_g^0$ gives levels with A_{1g}, E_g, T_{1g} and T_{2g} symmetries (take the direct product $T_{2g} \oplus T_{2g}$ and expand it according to the rule of Section III.2.4.2, Equation 3.2.4). In a same way, $t_{2g}^1 e_g^1$ gives T_{1g}; T_{2g} and $t_{2g}^0 e_g^2$ gives A_{1g}, A_{2g} and E_g. Transitions are possible between states with the same multiplicity, such as

$$^3T_{1g}\left(t_{2g}^1 e_g^1\right) \leftarrow {}^3T_{1g}\left(t_{2g}^2 e_g^0\right) \qquad {}^3T_{2g}\left(t_{2g}^2 e_g^0\right) \leftarrow {}^3T_{1g}\left(t_{2g}^2 e_g^0\right) \; (\sim 10\ \text{Dq})$$

$$^3A_{2g} \leftarrow {}^3T_{1g}\left(t_{2g}^2 e_g^0\right) \; (\sim 20\ \text{Dq}).$$

For the d^8 ion, configuration $t_{2g}^6 e_g^2$ gives levels with A_{1g}, A_{2g} and E_g symmetries (take the direct product $E_g \oplus E_g$). In the same way, $t_{2g}^5 e_g^3$ gives T_{1g}; T_{2g}, and t_{2g}^4 and e_g^4 give A_{1g}, E_g, T_{1g}, and T_{2g}. The possible transitions are

$$^3T_{1g}\left(t_{2g}^5 e_g^3\right) \leftarrow {}^3A_{2g} \qquad {}^3T_{2g}\left(t_{2g}^5 e_g^3\right) \leftarrow {}^3A_{2g} \; (\sim 10\ \text{Dq})$$

$$^3T_{1g}\left(t_{2g}^4 e_g^4\right) \leftarrow {}^3A_{2g} \; (\sim 20\ \text{Dq}).$$

The transition diagrams of the d^1 ion are listed qualitatively below (indicated in parentheses is a possible level occupation number).

3.1 and 3.3 (partial results)

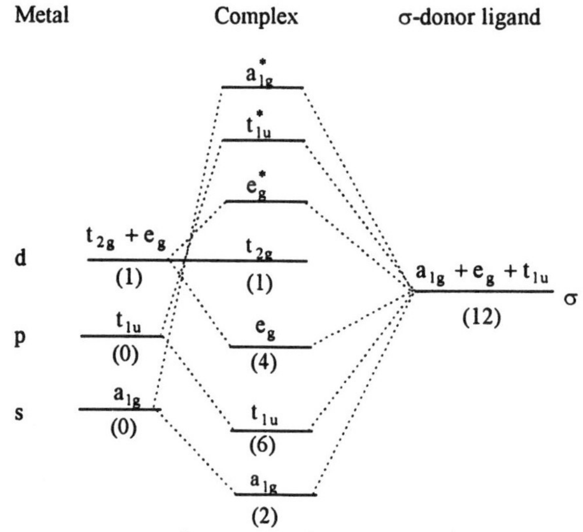

Possible transitions: $e_g^* \leftarrow t_{2g}$; $t_{1u}^* \leftarrow t_{2g}$; $a_{1g}^* \leftarrow t_{2g}$.

3.2 and 3.3 (partial result)

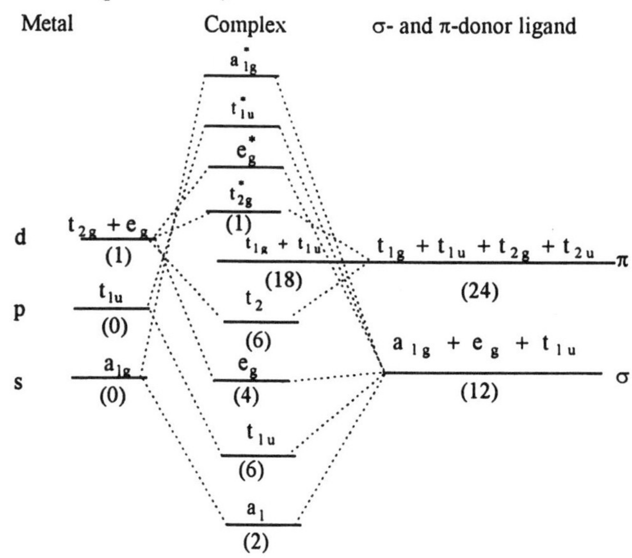

Possible transitions: $e_g^* \leftarrow t_{2g}^*$; $t_{1u}^* \leftarrow t_{2g}^*$; $a_{1g}^* \leftarrow t_{2g}^*$.

3.2.b and 3.3 (partial results)

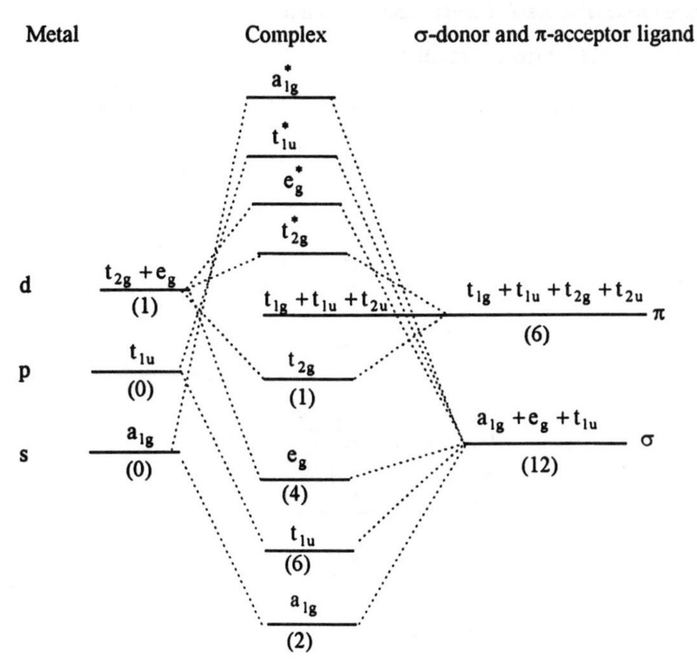

Possible transitions: $t_{1g} + t_{1u} + t_{2u} \leftarrow t_{2g}$; $t_{2g}^* \leftarrow t_{2g}$; $e_g^* \leftarrow t_{2g}$.

IV.4.10. Optical transitions in tetrahedral complexes

Consider a positive ion in a tetrahedral complex.

1. Show that the seven orbital given by

$$f_{z^3} = z\,(5z^2 - 3r^2)\,f(r)$$

$$f_{y^3} = y\,(5y^2 - 3r^2)\,f(r)$$

$$f_{x^3} = y\,(5y^2 - 3r^2)\,f(r)$$

$$f_{xyz} = xyz\,f(r)$$

$$f_{z(y^2-x^2)} = z\,(y^2 - x^2)\,f(r)$$

$$f_{y(x^2-z^2)} = y\,(x^2 - z^2)\,f(r)$$

$$f_{x(z^2-y^2)} = x\,(z^2 - y^2)\,f(r)$$

belong to three irreducible representations. What are these and give their degeneracy.

2. Is the $d_{xy} \rightarrow f_{xyz}$ transition allowed in a tetrahedral complex?

1. The following figure indicates the positions of a few symmetry elements of the T_d group (the others are easy to find).

The symmetry operators transform the orbitals in the same way as the corresponding product of the spatial coordinates. Let us first determine the action of the symmetry operators (shown to the right) on the x, y and z coordinates (it helps to look for the new positions occupied by atoms 1 and 2, then to represent the middle of the segment joining them so as to find the new position of the Ox axis).

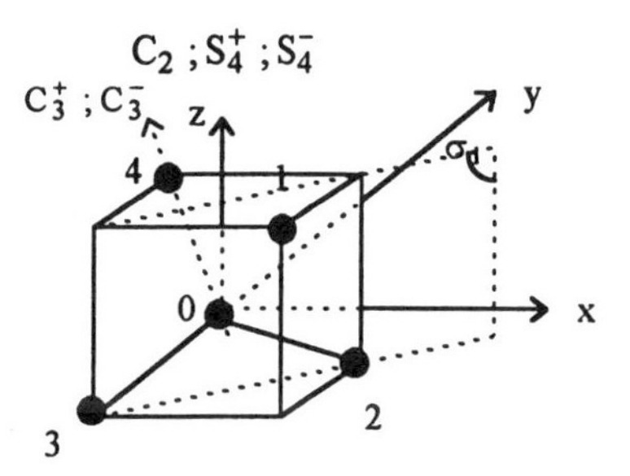

	E	C_3^+	C_3^-	C_2	S_4^+	S_4^-	σ_d
x	x	− z	− y	− x	y	− y	y
y	y	− x	z	− y	− x	x	x
z	z	y	− x	z	− z	− z	z

Using the f orbitals as a basis, it is easy to construct the matrices representing the operators and to calculate their traces. Next, we identify the associated irreducible representations.

We can thus show that:

— f_{xyz} forms a (1 dimensional) base for the irreducible representation A_1 (order 1; degeneracy 1).

— (f_{x^3} , f_{y^3} , f_{z^3}) forms a (3 dimensional) base for the irreducible representation T_2 (order 3; degeneracy 3).

— ($f_{x(z^2-y^2)}$, $f_{y(x^2-z^2)}$, $f_{z(y^2-x^2)}$) forms a (3 dimensional) base for the irreducible representation T_1 (order 3; degeneracy 3).

2. f_{xyz} belongs to the A_1 symmetry group. The three dipole moment components and d_{xy} belong to the T_2 symmetry group (please refer to the symmetry table of the T_d group). The double direct product contains the A_1 representation once and therefore the transition is allowed within EDA.

IV.4.11. Optical transitions in octahedral complexes

Let us consider an ion in an octahedral complex.

1.Show that the seven f orbitals given by

$$f_{z^3} = z(5z^2 - 3r^2)\,f(r)$$
$$f_{y^3} = y(5y^2 - 3r^2)\,f(r)$$
$$f_{x^3} = y(5y^2 - 3r^2)\,f(r)$$
$$f_{xyz} = xyz\,f(r)$$
$$f_{z(y^2-x^2)} = z(y^2 - x^2)\,f(r)$$
$$f_{y(x^2-z^2)} = y(x^2 - z^2)\,f(r)$$
$$f_{x(z^2-y^2)} = x(z^2 - y^2)\,f(r)$$

belong to three irreducible representations. Name them and give their degeneracy.

2. Are the transitions $d_{xy} \rightarrow d_{z^2}$ and $d_{xy} \rightarrow f_{xyz}$ allowed in an octahedral complex?

1. The demonstration is identical to that of Problem IV.4.10. The following figure indicates a few of the symmetry elements of the O_h group (the reader will easily find the others).

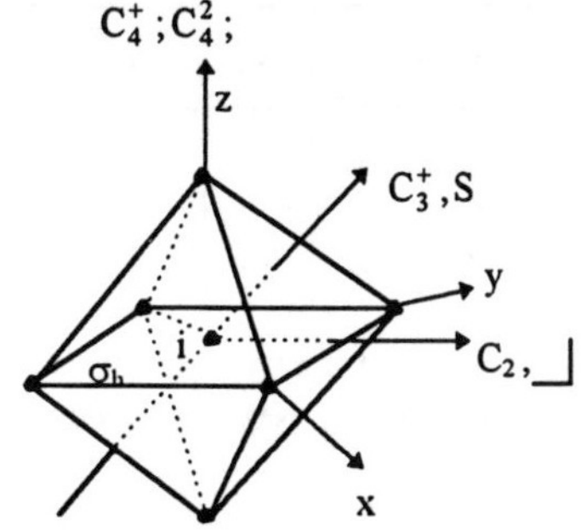

The transformation of the coordinates is indicated in the following table:

	E	C_3^+	C_2	C_4^+	C_4^2	i	S_4^+	S_6^+	σ_h	σ_d
x	x	y	y	y	−x	−x	y	−z	x	y
y	y	z	x	−x	−y	−y	−x	−x	y	x
z	z	x	−z	z	z	−z	−z	−y	−z	z

These results are quite similar to those found for the T_d group (Problem IV.4.10), but A_1, T_2 and T_1 are replaced by A_{2u}, T_{1u} and T_{2u}, respectively.

2. The components of the transition moment have a T_{1u} symmetry. d_{xy} and d_{z^2} have T_{2g} and E_g symmetries, respectively. The double direct product $E_g \oplus T_{1u} \oplus T_{2g}$ does not include the totally symmetrical irreducible representation A_1, so that the transition is forbidden within EDA.

f_{xyz} belongs to A_{2u}. The direct product $A_{2u} \oplus T_{1u} \oplus T_{2g}$ includes the irreducible representation A_{1g} once. The transition is therefore allowed within EDA.

IV.4.12. Symmetry of spin-orbitals and excited electronic states

The ground state of the ClO_2^- ion has a B_1 symmetry (only one electron in a p_x orbital with symmetry B_1). The ion interacts with a laser wave, which is polarized along Oy, and at resonance with the ion.

1. What is the symmetry of the excited spin-orbital?
2. What is the symmetry of the excited state?

1. In the C_{2v} symmetry group, y belongs to the irreducible representation B_2. The same is true for transitions with a non-vanishing Oy component in their transition moment. So we try to identify a irreducible representation X such that the reducible representation $B_1 \oplus B_2 \oplus X$ includes A_1. The only possibility is when $X = A_2$. Therefore the laser excited spin-orbital must have A_2 symmetry.

2. Since the ClO_2^- ion only has a single electron, the excited state must have A_2 symmetry.

IV.4.13. UV spectroscopy of carbon monoxide (from Agreg. C; A; 1978)

In the UV absorption spectrum of a CO molecule there are many lines. In particular, there is a set of lines corresponding to transitions from the electronic ground state to an excited electronic state for which v' = 0. For these lines, the wave numbers can be written in the form

$$\overline{\nu} = \overline{\nu_0} - 2\,166.34\left[v + (1/2)\right] + 12.69\left[v + (1/2)\right]^2 \text{cm}^{-1}$$

a. Calculate the photochemical dissociation energy D of a CO molecule in $kJ.mol^{-1}$.

b. Assuming that intensities are proportional to the population of the initial state, calculate the ratio of the first two line intensities at 500 K. Discuss the result. What are the consequences concerning the accuracy of the determination of constant D?

a. We look for a value v_M of v such that $\overline{\nu}_{v_M} = \overline{\nu}_{v_{M+1}}$. We find $v_M = 84.35 \approx 84$. The corresponding dissociation energy is $D \approx 9.13654 \times 10^4$ cm^{-1}, which is of the order of 1 092.3 $kJ.mol^{-1}$.

b. The ratio between intensities is given by the function exp $(-\Delta E/RT)$, where ΔE represents the energy gap between the two states considered first. We find $\Delta E \approx 2\,140.96$ cm^{-1}, of the order of 25.6 $kJ.mol^{-1}$. At 500 K, the ratio of the intensities is about 1/473. At 500 K, only the state corresponding to v = 0 is significantly populated. It is therefore difficult to get information about transitions starting from states v = 1, 2,.... Therefore the proposed method for measuring the dissociation energy is not very accurate.

IV.4.14. UV spectroscopy of methanal

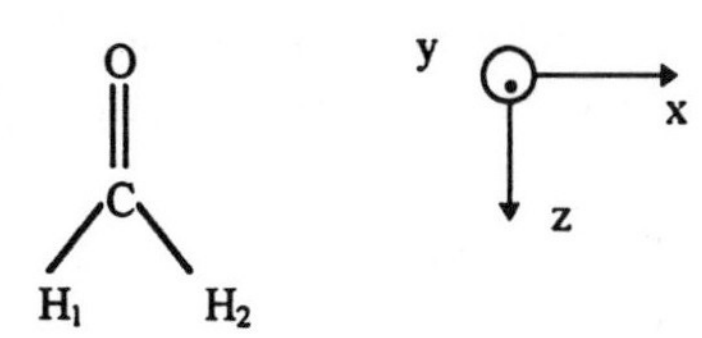

The molecule has sixteen electrons. Ignore the four core electrons $1s_C^2$ and $1s_O^2$ and solve the remaining twelve-electron problem.

1 - Electronic structure:

a. Find the symmetry elements and show them on a diagram.

b. Find the symmetry group.

c. Choose the 10-order base formed by the ten following AO's: [$1s_{H1}$; $1s_{H2}$; $2s_c$; $2p_{xc}$; $2p_{yc}$; $2p_{zc}$; $2s_o$; $2p_{xo}$; $2p_{yo}$; $2p_{zo}$]. Indicate the action of the group operators on the ten vectors of the chosen base in a table.

d. With the same base, indicate the different AO's (or their combinations) which can couple together, and give the irreducible representations of the group to which they belong.

e. Indicate the three general shapes of the MO's (obtained by LCAO) of the methanal molecule and the three irreducible representations for which they form a 1-order base.

f. In fact, HF - LCAO calculations lead to 10 delocalized MO's. The first eight (classified by increasing energy) are given in the following figure with their name ( ), their symmetry O, and their constitution [ ]. Complete the description. Place the 12 electrons in the ground state. Why do four empty levels exist? What do they represent?

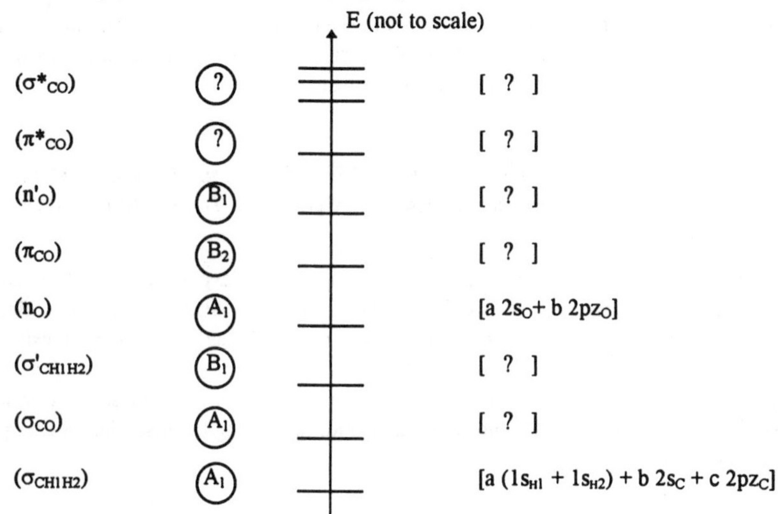

g. It is possible to treat methanal like a diatomic heteronuclear molecule (so-called fragment method), by considering it as if it were obtained from coupled fragments. The following diagram indicates the construction of the MO's as obtained from FO's.

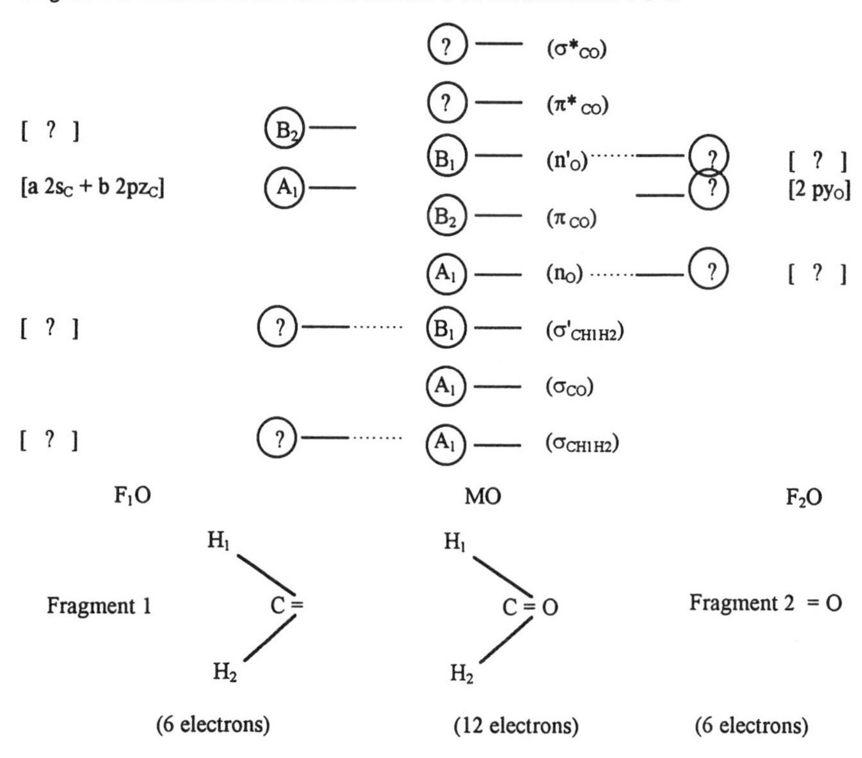

Place the twelve electrons in the ground state and fill in the blanks.

The four orbitals σ_{CO} ; π_{CO} ; n_{CO} ; σ_{CO}^* are obtained by coupling OF_1 and OF_2. Indicate the couplings by using dashed lines.

2 - UV absorption of methanal:

a. Write the ground state.

b. Give the symmetry of the ground state.

c. Consider the first four transitions between singlet states (within EDA). Draw a figure giving the electronic structure, the symmetry of the state, the nature and the properties of the electronic transitions of methanal.

1.a. The following figure gives the symmetry elements:

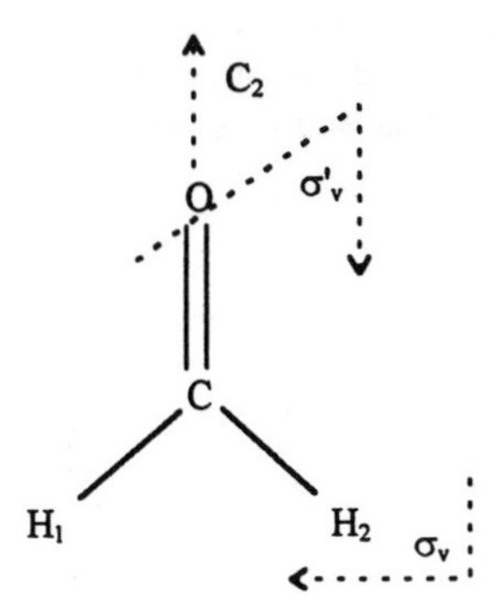

b. The molecule belongs to the C_{2v} group.

c. The following table indicates the action of the operators of the group:

	$1s_{H_1}$	$1s_{H_2}$	$2s_c$	$2p_{xc}$	$2p_{yc}$	$2p_{zc}$	$2s_o$	$2p_{xo}$	$2p_{yo}$	$2p_{zo}$
E	$1s_{H_1}$	$1s_{H_2}$	$2s_c$	$2p_{xc}$	$2p_{yc}$	$2p_{zc}$	$2s_o$	$2p_{xo}$	$2p_{yo}$	$2p_{zo}$
C_2	$1s_{H_2}$	$1s_{H_1}$	$2s_c$	$-2p_{xc}$	$-2p_{yc}$	$2p_{zc}$	$2s_o$	$-2p_{xo}$	$-2p_{yo}$	$2p_{zo}$
σ_v	$1s_{H_1}$	$1s_{H_2}$	$2s_c$	$2p_{xc}$	$-2p_{yc}$	$2p_{zc}$	$2s_o$	$2p_{xo}$	$-2p_{yo}$	$2p_{zo}$
σ'_v	$1s_{H_2}$	$1s_{H_1}$	$2s_c$	$-2p_{xc}$	$2p_{yc}$	$2p_{zc}$	$2s_o$	$-2p_{xo}$	$2p_{yo}$	$2p_{zo}$

1.d. By multiplying, column by column, the terms of the previous table by the characters of each of the irreducible representations of the C_{2v} group, then taking a vertical sum, and finally dividing by the number of elements (the so-called projector method), we get

A_1	$\frac{1}{2}(1s_{H_1}+1s_{H_2})$	$\frac{1}{2}(1s_{H_1}+1s_{H_2})$	$2s_c$	0	0	$2p_{zc}$	$2s_o$	0	0	$2p_{zo}$
A_2	0	0	0	0	0	0	0	0	0	0
B_1	$\frac{1}{2}(1s_{H_1}-1s_{H_2})$	$\frac{1}{2}(1s_{H_1}-1s_{H_2})$	0	$2p_{xc}$	0	0	0	$2p_{xo}$	0	0
B_2	0	0	0	0	$2p_{yc}$	0	0	0	$2p_{yo}$	0

Irreducible representation	AO (or their combinations which can couple together)
A_1	$(1s_{H_1}+1s_{H_2});2s_C;2p_{zC};2s_O;2p_{zO}$
A_2	no coupling
B_1	$(1s_{H_1}-1s_{H_2});2p_{xC};2p_{xO}$
B_2	$2p_{yC}$; $2p_{yO}$

e. So the general form of the MO's can be written as

$$A_1 \rightarrow a\left(1s_{H_1}+1s_{H_2}\right)+b2s_C+c2p_{zC}+d2s_O+e2p_{zO}$$
$$B_1 \rightarrow a\left(1s_{H_1}-1s_{H_2}\right)+b2p_{xC}+c2p_{xO}$$
$$B_2 \rightarrow a\,2p_{yC}+2p_{yO}$$

f. This description can be easily refined

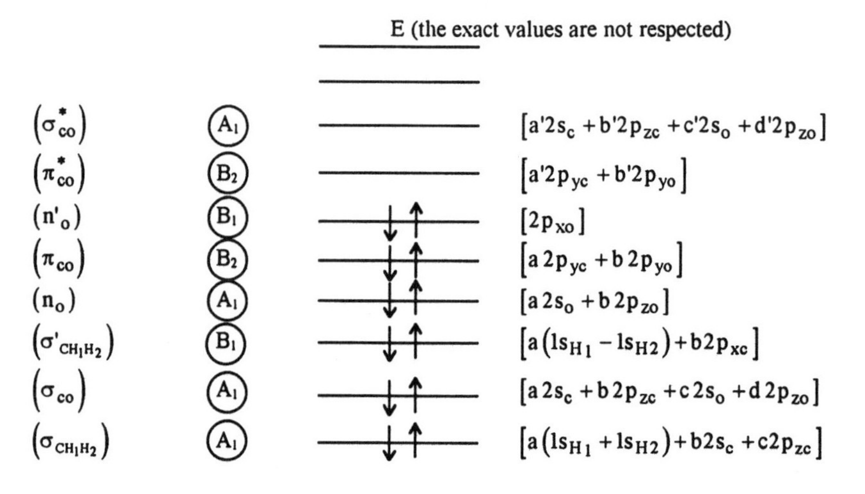

— Symmetry of the (π^*_{CO}) and (σ^*_{CO}) orbitals: They are two antibonding orbitals with the same symmetries as the corresponding bonding orbitals (B_2 et A_1, respectively).

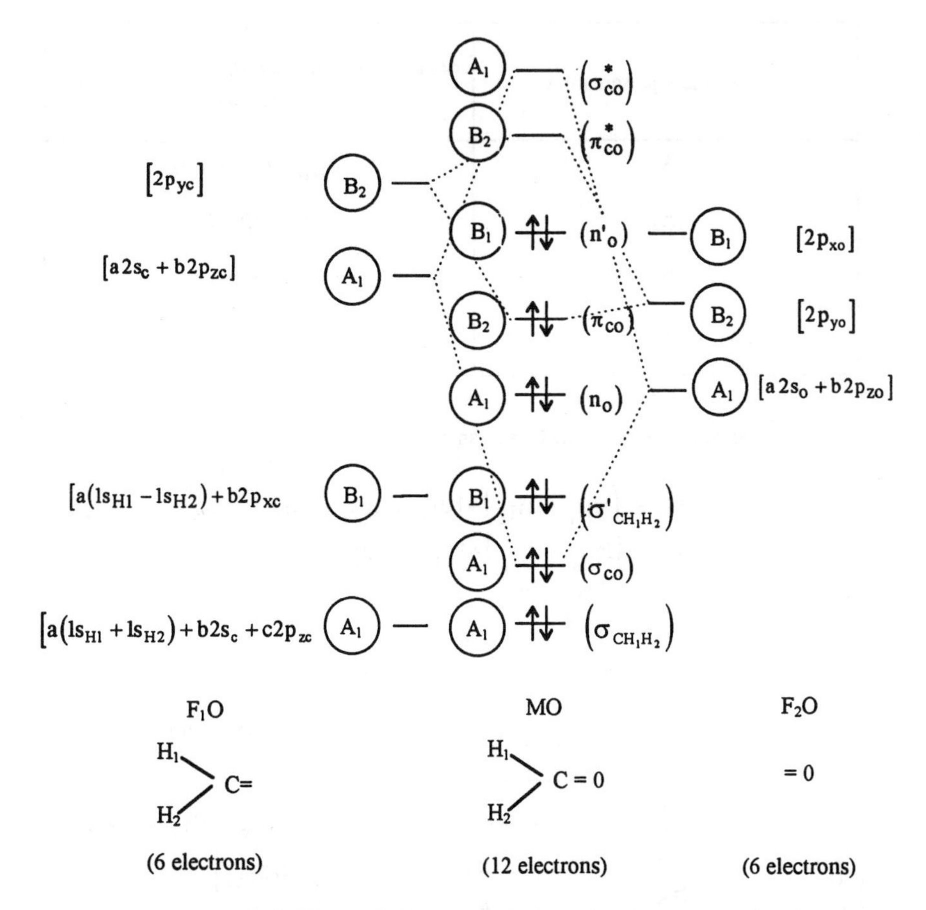

— Form of the MO: To take into account the symmetry when starting from the general form determined in question 1.e, we limit the LCAO expansion to the relevant atoms. Thus the σ_{CO} orbital (A_1 symmetry) can be written:

$$a\ 2s_C + b\ 2p_{ZC} + c\ 2s_O + d\ 2p_{ZO}.$$

So now we can fill in the empty parentheses.

— The 12 electrons occupy 6 levels. However, the base we chose is 10 dimensional, so that four levels remain vacant in the ground state. These will be used to describe the excited electronic state.

The molecular orbitals (σ_{CH}) and (σ'_{CH}) have the same energies as the two F_1O of lowest energy with which they can be identified. In the same way, (n_O) and (n'_O) have the same energies as two F_2O with which they can also be identified. The atomic orbital $2p_{yo}$ of the OF_2 fragment has a B_2 symmetry. The highest F_1O belongs to B_2 so it must be identical to $2p_{yc}$. The four molecular orbitals σ_{CO}, π_{CO}, π^*_{CO}, and σ^*_{CO} are obtained by coupling $OF_1 + OF_2$. Remembering that coupling can occur only between FO's with the same symmetry, it is easy to draw the dashed lines on the figure.

2.a. The ground state is $\sigma^2_{CH_1H_2}\ \sigma^2_{CO}\ \sigma'^2_{CH_1H_2}\ n^2_O\ \pi^2_{CO}\ n'^2_O$.

b. The ground state has A_1 symmetry.

c. The following table indicates the nature and the properties of the electric dipole transitions.

Name of the state	Electronic structure (the four core levels are not indicated)	Symmetry of the state	Nature of the transition	Transition (allowed or forbidden)	Polarization of the electric dipole moment of the transition
Ground	$\pi^2_{co}\ n'^2_o\ \pi^{*0}_{co}\ \pi^{*0}_{co}$	A_1	/	/	/
1st excited	$\pi^2_{co}\ n'^1_o\ \pi^{*1}_{co}\ \sigma^{*0}_{co}$	$B_1 \oplus B_2 = A_2$	$n \to \pi^*$	forbidden	/
2nd excited	$\pi^1_{co}\ n'^2_o\ \pi^{*1}_{co}\ \sigma^{*0}_{co}$	$B_2 \oplus B_2 = A_1$	$\pi \to \pi^*$	allowed	z
3rd excited	$\pi^1_{co}\ n'^2_o\ \pi^{*0}_{co}\ \sigma^{*1}_{co}$	$B_2 \oplus B_1 = A_2$	$\pi \to \sigma^*$	allowed	y
4th excited	$\pi^2_{co}\ n'^1_o\ \pi^{*0}_{co}\ \sigma^{*1}_{co}$	$B_1 \oplus B_1 = A_1$	$n \to \sigma^*$	allowed	x

IV.4.15. UV spectroscopy of cyclobutadiene

Cyclobutadiene (restricted to the four $2p_z$ electrons and treated in the Hückel approximation).

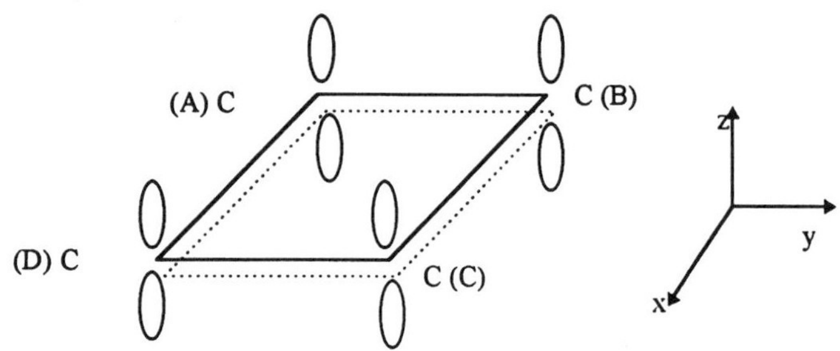

Basis: $|2p_{zA}\rangle; |2p_{zB}\rangle; |2p_{zC}\rangle; |2p_{zD}\rangle$ $\quad |\psi\rangle = c_a|2p_{zA}\rangle + c_b|2p_{zB}\rangle + c_c|2p_{zC}\rangle + c_d|2p_{zD}\rangle$

1. Write the equation which gives the energies of the cyclobutadiene MO's in the Hückel approximation.

2. Calculate these energies [to simplify the calculation, set $x = (\alpha - E)/\beta$].

3. Calculate the coefficients c_a, c_b, c_c and c_d for each MO, and express the vectors associated to these MO's.

4. Give the complete energy diagram of this molecule (classify the energy levels, the associated energies, and the corresponding MO's).

5. Give the symmetry group of the molecule, assumed to be plane and reduced to the four $2p_z$ atomic orbitals.

6. Give the symmetries of the MO's.

7. Indicate the electronic structure and the magnetic properties of the electronic ground state.

8. Can the molecule have electric dipole transitions polarized along the Oz axis?

1. The equation yielding the eigenvalues of the energy can be written (*cf.* Section III.3.3.2)

$$\begin{vmatrix} \alpha - \overline{E} & \beta & 0 & \beta \\ \beta & \alpha - \overline{E} & \beta & 0 \\ 0 & \beta & \alpha - \overline{E} & \beta \\ \beta & 0 & \beta & \alpha - \overline{E} \end{vmatrix} = 0$$

2. By expanding this determinant, then dividing by β^4 and setting $x = \left(\alpha - \overline{E}\right)/\beta$, we obtain a fourth-degree equation whose roots are -2 ; 0 (a double root) and $+2$. So the energies are $\alpha + 2\beta$; α ; α and $\alpha - 2\beta$ (do not forget that α and β are negative parameters).

3. By replacing E by its values in the above equation, we get a system of three equation with four unknowns that we solve. Using the normalization relation, we then find

Energies	Vectors
$\alpha + 2\beta$	$\|\psi_1\rangle = (1/2)(\|2p_{zA}\rangle + \|2p_{zB}\rangle + \|2p_{zC}\rangle + \|2p_{zD}\rangle)$
α (twice degenerate)	$\|\psi_2\rangle = (1/2)[(\|2p_{zA}\rangle - \|2p_{zC}\rangle) + (\|2p_{zB}\rangle - \|2p_{zD}\rangle)]$ $\|\psi'_2\rangle = (1/2)[(\|2p_{zA}\rangle - \|2p_{zC}\rangle) - (\|2p_{zB}\rangle - \|2p_{zD}\rangle)]$
$\alpha - 2\beta$	$\|\psi_3\rangle = (1/2)(\|2p_{zA}\rangle - \|2p_{zB}\rangle + \|2p_{zC}\rangle - \|2p_{zD}\rangle)$

4. The energy diagram is

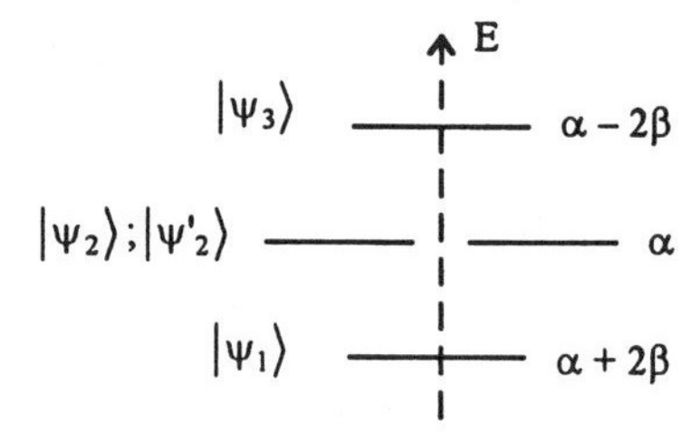

5. The symmetry elements are indicated on the next figure.

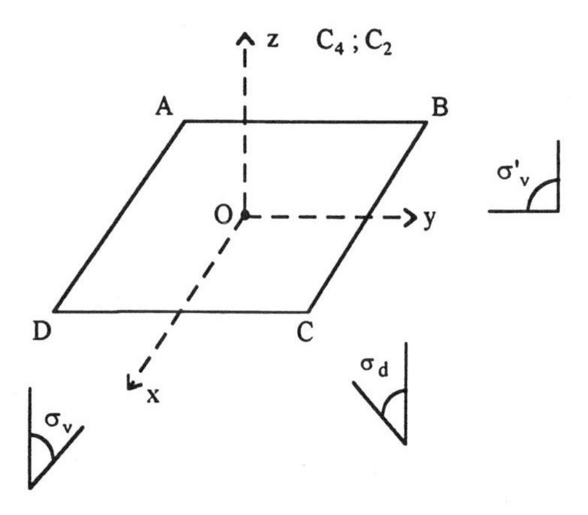

A horizontal symmetry plane cannot exist because of the different signs affecting the mathematical representations of the lobes of the $2p_z$ atomic orbitals. The Hückel representation belongs to the C_{4v} group.

6. $|\psi_1\rangle$ forms a 1-order base for the irreducible representation A_1.

$|\psi_2\rangle$ and $|\psi'_2\rangle$ form a 2-order base for the irreducible representation E.

$|\psi_3\rangle$ forms a 1-order base for the irreducible representation B_2.

(Write the matrices representing the action of each operator successively on the chosen base, then determine each of the successive traces so as to identify the corresponding representations).

7. The system consists of four electrons occupying the $|\psi_1\rangle$ orbital (two electrons) and the $|\psi_2\rangle$ and $|\psi'_2\rangle$ orbitals (one electron in each). Thus the symmetry of the ground state is $A_1 \oplus E\ B_1 \oplus B_2 = A_2\ E = A_1$. It is easy to check that each product of 2×2 matrices representing the operators on the $|\psi_2\rangle, |\psi'_2\rangle$ base gives the 2×2 unit matrix (for the C_4 operator, one must take the product of the matrices representing

the C_4^+ and C_4^- operators). The ground state belongs to the completely symmetrical representation. It is a paramagnetic state because it has two electrons with unpaired spins.

8. The z coordinate forms a 1-order base for the irreducible representation A_1. The electronic transition which could occur would be the $|\psi_2\rangle$ or $|\psi'_2\rangle \rightarrow |\psi_3\rangle$ transition. The product of the characters of the E, A_1 and B_2 representations yields a representation with characters 2; 0; – 2; 0 and 0, i.e. the E representation. The contemplated transition is therefore forbidden (*cf.* Section IV.3.2.2).

IV.4.16. UV Spectroscopy of conjugated linear polyenes

This problem concerns the π-electron system of a trans-hexatriene molecule $H-(CH=CH)_n-H$ (with n = 3) represented below.

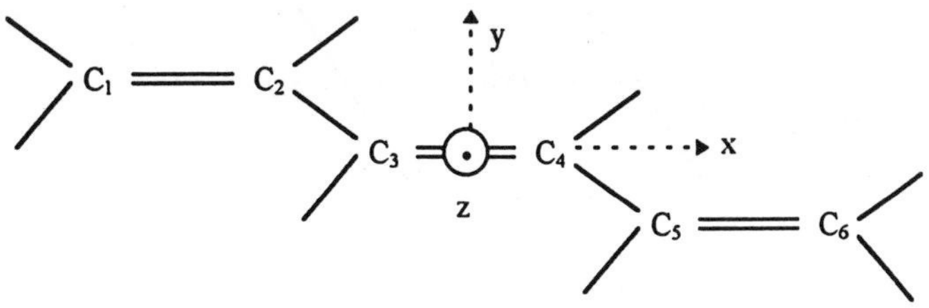

C_1 C_2 C_3 C_4 C_5 C_6 x y z

The molecule is assumed to be plane (Oxy plane). It absorbs in the UV part of the spectrum and the wavelength corresponding to the absorption maximum is approximately 268 nm.

A. Calculation of the average bond length by the so-called Free Electron Molecular Orbital (FEMO) model

In this model, the electrons of a conjugated molecule are treated as independent particles. The potential energy is zero inside a one-dimensional box of length L = 2nR (R is the average length of a C—C bond in the studied molecule).

1. Solve the problem of a particle in an infinite square potential well and draw the energy diagram representing the first four levels and their associated energies. Place the six electrons of the molecule in the ground electronic state, then in the first excited singlet state.

2. Express the energy associated to the electronic transition corresponding to the longest wavelength.

3. Calculate R and the energy associated to this transition (expressed in eV) by comparing theoretical and experimental values. Does the value of R seem reasonable?

B. Calculation of the resonance integral using the Hückel method.

The MO's are represented by linear combinations $|\varphi\rangle = \sum_i c_i |2p_i\rangle$

Summation (i) is performed over the six carbon atoms. 2p is the $2p_z$ atomic orbital of the carbon atom.

We set $\langle 2p_i|H|2p_i\rangle = \alpha$ (negative Coulomb integral)

$\langle 2p_i|H|2p_{i+1}\rangle = \langle 2p_{i-1}|H|2p_i\rangle = \beta$ (negative resonance integral)

The electronic repulsion, the overlap of the AO's and the resonance between the AO's of non-adjacent atoms are not taken into account.

1. Determine the symmetry group of the molecule. Determine the characters of the reducible representation constructed on the base of the six $2p_z$ AO's of the carbon atoms. Reduce it to six irreducible representations of the symmetry group of the molecule. Determine the six energy levels (labeled from 1 to 6 in order of increasing energy) of the MO diagram and the corresponding irreducible representations. The three real roots of the two following third-degree equations are

$x^3 - x^2 - 2x + 1 = 0$ + 1.80 ; − 1.25 ; + 0.44

$x^3 + x^2 - 2x - 1 = 0$ − 1.80 ; + 1.25 ; − 0.44

Place the six electrons of the molecule in the electronic ground state.

2. Show that the transition $3 \rightarrow 4$ is allowed. What is its polarization? Calculate the energy associated to the electronic transition corresponding to the largest absorption wavelength as a function of parameter β.

3. Calculate β by comparing theoretical and experimental values. Comment the value obtained.

A.1. The energies are given by

$$E_n = n^2\, h^2/8m_eL^2 = n^2\, E_0 \text{ with } n = 1, 2, 3, \ldots \text{ and } E_0 = h^2/8m_eL^2$$

We obtain the following diagram:

Level	Ground state	1st excited state	Energy
n = 4	—	↓	$E_4 = 16\,E_0$
		7 E_0	
n = 3	↑↓	↑	$E_3 = 9\,E_0$
n = 2	↑↓	↑↓	$E_2 = 4\,E_0$
n = 1	↑↓	↑↓	$E_1 = E_0$

Ground state 1st excited state

2. $\Delta E = E_4 - E_3 = 7E_0 = 7h^2/8m_e L^2$.

3. The energy of the transition is

$h\nu = hc/\lambda = 6.6 \times 10^{-34} \times 3 \times 10^{8}/(0.268 \times 10^{-6} \times 1.6 \times 10^{-19}) = 4.62$ eV

and the value of R is

$$R = L/6 = (1/6)\sqrt{\frac{7 \times (6.6)^2 \times 10^{-68}}{8 \times 9.1 \times 10^{-31} \times 4.62 \times 1.6 \times 10^{-19}}} = 0.13 \text{ nm}$$

This value is close to those corresponding to single and double C-C bonds.

B.1. The molecule belongs to the C_{2h} group whose characters are indicated in the following table:

	E	C_2	i	σ_h	
A_g	1	1	1	1	
B_g	1	–1	1	–1	
A_u	1	1	–1	–1	z
B_u	1	–1	–1	1	x; y

We choose the six $2p_i$ atomic orbitals of the carbon atom as a base. The characters of the corresponding reducible representation are 6 ; 0 ; 0 and – 6. The reduction yields the sum 3 B_g + 3 A_u. Let us now calculate the energies associated to the MO's $|\psi\rangle = \sum_i C_i \, 2p_i$.

OM B_g:

$$i|2p_1\rangle = -|2p_6\rangle ; i|2p_2\rangle = -|2p_5\rangle ; i|2p_3\rangle = -|2p_4\rangle$$

with $\chi_{(i)} = +1$ and therefore $C_1 = -C_6$; $C_2 = -C_5$; $C_3 = -C_4$

Setting $k = (\alpha - E)/\beta$, the system to be solved takes the form

$$\begin{cases} C_1 k + C_2 = 0 \\ C_1 + C_2 k + C_3 = 0 \\ C_2 + C_3(k-1) = 0 \end{cases} \text{ and compatibility requires } \begin{vmatrix} k & 1 & 0 \\ 1 & k & 1 \\ 0 & 1 & k-1 \end{vmatrix} = 0$$

we get the equation $k^3 - k^2 - 2k + 1 = 0$, whose roots are: + 1.80 ; – 1.25 ; + 0.44

OM A_u:

$\chi_{(i)} = -1$, therefore $C_1 = C_6$; $C_2 = C_5$; $C_3 = C_4$

The system to be solved is

$$\begin{cases} C_1 k + C_2 = 0 \\ C_1 + C_2 k + C_3 = 0 \\ C_2 + C_3(k+1) = 0 \end{cases} \text{ and compatibility requires } \begin{vmatrix} k & 1 & 0 \\ 1 & k & 1 \\ 0 & 1 & k+1 \end{vmatrix} = 0$$

we get the equation $k^3 + k^2 - 2k - 1 = 0$, whose roots are -1.80 ; $+1.25$; -0.44

Thus, we obtain the following energy diagram:

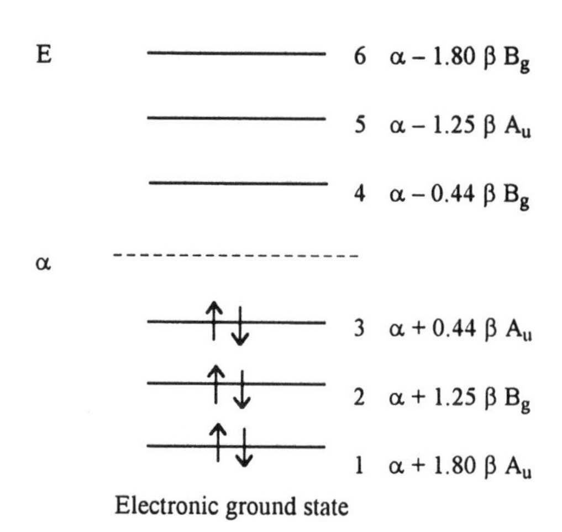

2. Transition $3 \rightarrow 4$ is allowed in EDA ($A_u \oplus B_u \oplus B_g = A_g$) and its polarization lies in the Oxy plane. Transition $3 \rightarrow 4$ corresponds to the absorption maximum. $\Delta E = -0.88\ \beta$.

3. $\beta = -4.62/0.88 = -5.25$ eV. We find a negative resonance energy about two times larger than that generally admitted as the value of the delocalization energy (which is about – 2.5 eV). Here, as for most methods used in quantum chemistry, the values of α and β dependent very much on the described property.

IV.4.17. Visible spectroscopy of colored acid-base indicators (from Agreg. C; B; 1989)

Determination of the pK_a of a colored indicator by using visible spectrophotometry.

1. Let us first consider an indicator whose two forms absorb in the visible range. The absorption spectra of these two forms are represented on the figure below.

a. What is the name of point C? What is its main characteristic?

b. In the rest of this problem, we study a few aqueous solutions in which the total concentration of the indicator (acid and basic forms) is maintained constant and equal to a $mol.l^{-1}$. At a fixed pH of the solution, the ionization coefficient of the acid form of the indicator is called α.

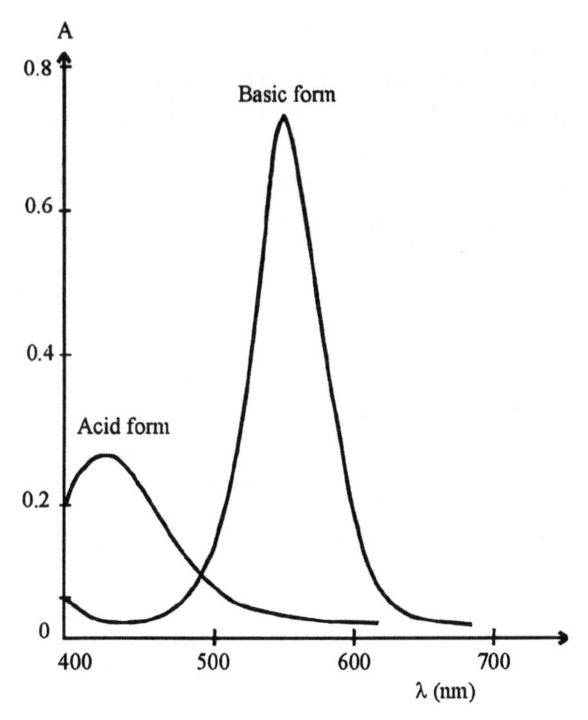

b.1. Express the difference $pH - pK_a$ as a function of α.

b.2. Give the expression of the absorbance $A(\lambda)$ of the solution as a function of a, α, the molar extinction coefficients ε_a and ε_b of, respectively, the acid and basic forms of the indicator, and l, the width of the solution crossed by the light beam of wavelength λ. Show that, in the cases of large values of the acidity and of the basicity, the values of $A(\lambda)$ tend to two limit values $A_a(\lambda)$ and $A_b(\lambda)$. Give these two limits

b.3. Show that a relation exists which links the difference $pH - pK_a$ to $A(\lambda)$, $A_a(\lambda)$ and $A_b(\lambda)$.

c. Imagine a spectrophotometric method to determine the pK_a of the indicator whose spectra are presented on the following figure. Discuss the choice of the wavelength.

2. We would now like to study an indicator for which only the basic form A^- absorbs in the visible range.

a. Give an example of an indicator of this type.

b. Show that the results obtained in part 1 (for a two-color indicator) can be transposed to fit the case of a one-color indicator. What becomes of the relation established question l.b.3 in this case?

c. Results.

Measurements of the absorbances of solutions which all had the same indicator concentration a gave the following results, recorded at a wavelength of 550 nm:

Number of the solution	1	2	3	4	5	6	7
pH	8.8	9.0	9.2	9.4	9.6	10.8	11.0
A(λ)	0.23	0.28	0.35	0.40	0.49	0.71	0.71

Use a graphic method and the relation established in question 1.b.3 to determine the pK_a of the indicator.

1.a. Point C is the isobestic point: The absorbances of the acid and basic forms are equal.

b.1. We get $pK_a = pH - \log\left[A^-\right]/[AH]$ so that $pH - pK_a = \log[\alpha/(1-\alpha)]$.

b.2. By summing the absorbances we obtain $A(\lambda) = [\varepsilon_a\, a(1-\alpha) + \varepsilon_b\, a\alpha]l$. The limit values of the absorbance in acid and basic media are, respectively: $A_a(\lambda) = \varepsilon_a al \quad A_b(\lambda) = \varepsilon_b al$.

b.3. We get

$$A = A_a(1-\alpha) + A_b\alpha \text{ and } pH - pK_a = \log[(A - A_a)/(A_b - A)]$$

c. We measure the transmitted intensity of a He-Ne laser wave (λ = 632.8 nm) as a function of pH. The acid form absorbs very little. We have previously determined A_b for this wavelength by increasing the pH to basic values. The graph of $\log[A/(A_b - A)] = f(pH)$ is a straight line. Its intersects the Ox axis at the value $pH = pK_a$. The experiment is easy to do and the measurement is both simple and accurate.

2.a. Phenolphtalein fulfills the condition.

b. The answer has already been given at question 1.c.

c. The following table can be exploited graphically ($A_b = 0.71$):

N°	1	2	3	4	5	6	7
pH	8.8	9.0	9.2	9.4	9.6	10.8	11.0
A	0.23	0.28	0.35	0.40	0.49	0.71	0.71
log ($A/A_b - A$)	– 0.32	– 0.19	– 0.012	0.11	0.35		

We find $pK_a = 9.2$.

IV.4.18. UV spectroscopy of aromatic compounds (from Agreg. C; A; 1994)

The energies of the π-levels of a cyclic conjugated polyene with n carbon atoms (n odd) is given by

$$E_j = \alpha + 2\beta \cos(2j\pi/n) \quad \text{with } j = 0, \pm 1, \pm 2, ..., \pm n/2 \quad \text{and} \quad \beta = -294 \text{ kJ.mol}^{-1}$$

1. Represent the energy levels of the π-orbitals of benzene. Find the absorption wavelength for the electronic transition corresponding to the lowest energy. The UV-visible spectrum of benzene has an absorption band with three maxima occurring at $\lambda = 180$ nm ($\varepsilon = 40000$), $\lambda = 203.5$ nm ($\varepsilon = 7400$) and $\lambda = 254$ nm ($\varepsilon = 204$). Comment this.

2. We want to study the effect of the substitution of a hydrogen atom by a virtual atom with the same electronegativity as that of a carbon atom and which has a nonbonding doublet. Represent the new distribution of the energy levels on the same diagram as that of question 1.

The electronegativity of the atom is increased progressively. Represent the evolution of the energy levels.

3. Next, we want to look at the displacement of the so-called primary band (203.5 nm). In the case of polysubstituted benzenes, the effects of the substituents are roughly additive and the following increments can be assigned:

OH: + 7 nm; NH_2: + 27 nm ; O^-: + 32 nm

Explain the observed relative values.

1. We can identify the following four energy levels:

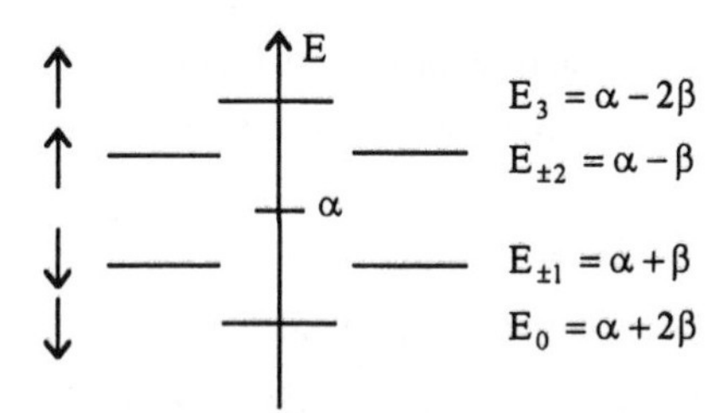

The two levels corresponding to $j = \pm 1$ and $j = \pm 2$ are degenerate. The electrons with the highest energy occupy level $E_{\pm 1}$. The electronic transition with the largest wavelength corresponds to the $E_{\pm 1} \rightarrow E_{\pm 2}$ transition with an energy of $-2\beta = 588$ KJ.mol^{-1}. The corresponding wavelength is $\lambda = hc/2\beta \approx 203.5$ nm. This model correctly describes one line (203.5 nm) of the observed spectrum.

2. An atom with the same electronegativity as that of a carbon atom, but which has a nonbonding electronic doublet, introduces an extra level n of energy α and with $n-\pi^*$ type transitions. The effect on the other energy levels is represented on the diagram above by arrows (→). If the electronegativity of the substituent increases, the

energy of level n decreases, thus perturbing levels E_0 and $E_{\pm 1}$ more strongly than levels $E_{\pm 2}$ and E_3.

3. The three substituents have free electron doublets and we observe $n \rightarrow \pi^*$ transitions. The electronegativity of the substituent decreases when the OH group is replaced successively by NH_2, and by O^-. The energy gap between level n and level $E_{\pm 2}$ decreases (because the energy of level $E_{\pm 2}$ increases less quickly than that of level n) and the wavelength of the observed transition increases.

IV.4.19. Determination of the pK_a of a compound in an excited electronic state (from Agreg. C; A; 1994)

We want to study the acid-base dissociation of napht-2-ol in its ground state, called AH (equilibrium 1, see below), and in its first excited state AH * (equilibrium 2, see below).

$$(1)\ AH + H_2O \leftrightarrow H_3O^+ + A^- : \text{acidity constant } K_a$$

$$(2)\ AH^* + H_2O \leftrightarrow H_3O^- + A^{*-} : \text{acidity constant } K_a^*$$

1. Compare the values of pK_a and of pK_a^* and explain your answer.

2. ΔE (AH) and $\Delta E(A^-)$ are the energy gaps between the ground and the excited states for the acid and the basic forms, respectively. Determine the relation linking $\Delta E(AH)$, $\Delta E(A^-)$, pK_a, pK_a^* and T to each other by operating a few reasonable approximations.

3. In a M/10 chlorhydric acid solution, the spectrum of napht-2-ol shows an absorption maximum at wave number $\sigma(AH) = 30.8 \times 10^3$ cm^{-1}. In the same conditions, the fluorescence spectrum has a band lying at $\sigma'(AH) = 28.3 \times 10^3$ cm^{-1}. With a M/10 sodium hydroxide solution, the two maxima lie at $\sigma(A^-) = 29.0 \times 10^3$ cm^{-1} and $\sigma'(A^-) = 24.1 \times 10^3$ cm^{-1}, respectively. Explain why $\Delta E(AH)$ can be estimated from the half sum of σ (AH) and σ'(AH). Knowing that the pK_a of napht-2-ol is equal to 9.5 at 25°C, find its pK_a^* at the same temperature.

1. In the excited electronic state, we can say, very qualitatively, that the oxygen atom of the phenol group carries a positive charge increment (the transition is of the $n \rightarrow \pi^*$ type and the spatial localization of the lone doublet of the oxygen atom is less marked). The acidity increases and we have the order relation $pK_a > pK_a^*$.

2. By expressing the two acidity constants, we find that their ratio is equal to $K_a / K_a^* = [A^-][AH^*]/[A^{-*}][AH]$. Mawwell-Boltzmann statistics then yield

$$\exp[-\Delta E(AH)/RT] = [AH^*]/([AH]+[AH^*])$$

$$\exp\left[-\Delta E\left(A^-\right)/RT\right] = \left[A^-*\right]/\left(\left[A^-\right]+\left[A^-*\right]\right)$$

And, assuming ΔE (AH) >> RT and ΔE (A^-) >> RT, we find

$$K_A / K_A^* \approx \exp[-\Delta E(AH)/RT] / \exp[-\Delta E(A^-)/RT]$$

or
$$pK_a^* \approx pK_a + (1/2{,}3RT)[\Delta E(A^-) - \Delta E(AH)]$$

3. The half sum $(\sigma + \sigma')/2$ indeed gives an approximate value of ΔE when the energy levels of the ground state and of the excited state have comparable widths ($\Delta E = [\sigma + \sigma']/2$ if the two widths are equal). Therefore, we find $\Delta E\ (AH) \approx 29.55 \times 10^3\ cm^{-1}$ and $\Delta E\ (A^-) \approx 26.55 \times 10^3\ cm^{-1}$. Numerical application yields a value of $pK_a^* \approx 3$.

IV.4.20. Radiation absorbed by matter (from Agreg. C; C; 1982)

1. Transition moments.

The absorption of light energy by matter occurs because of the interaction between the electric field **E** of the wave and the electric charges of the molecules. This absorption is proportional to the square of the transition moment **p** that can be defined by $p_{12} = \langle \psi_1 | \sum_i r_i | \psi_2 \rangle$, where ψ_1 and ψ_2 are the state vectors of the system in its initial and final states, respectively, and $\mathbf{r}_i$ is the vector defining the position of the ith electron.

a. Show that the transition is forbidden between states with different spectral multiplicities.

b. Make a qualitative comparison between the absorptions of the π–π^* and the n–π^* transition.

2. Absorption spectrum.

a. Express the Franck-Condon principle.

b. The absorption spectrum is in fact composed of bands which can be resolved in lines. Interpret this result using a Jablonski-type diagram.

3. Evolution of the excited state:

a. The system goes from the ground state S_0 to the first excited state S_1. Let us call the first triplet state T_1. With the help of a Jablonski diagram, indicate the different possible ways the system can go from state S_1 to state S_0.

b. Explain why a photochemically activated molecule reacts from its triplet state.

1.a. The structure of the spin function used to describe the singlet and the triplet state (*cf.* Problem III.3.4) and the orthogonality of the α and β functions make that the product of a function representing a singlet state by a function representing a triplet state vanishes. Within EDA, a singlet → triplet transition (or the inverse) is therefore forbidden.

b. The $\pi \to \pi^*$ absorption is more intense than the $n \to \pi^*$ absorption which is generally forbidden from symmetry considerations, or, if it is not, has a smaller transition moment because of the smaller overlap of the orbitals.

2.a. The Franck-Condon principle concerns the comparison between the duration of the photon impact and the characteristic period of interatomic vibrations . During the impact of a photon on a molecule, the interatomic distance is preserved. This means that the transition is represented by a vertical line in a diagram where the electronic energy is plotted as a function of the interatomic distance.

b. The bands correspond to the different $S_0 \to S^*$ transitions. Each electronic level exhibits a vibrational structure. The lines correspond to the various authorized transitions between these discrete levels.

3.a. Transitions are represented on the following diagram (Jablonski), where the energy is plotted on the vertical axis.

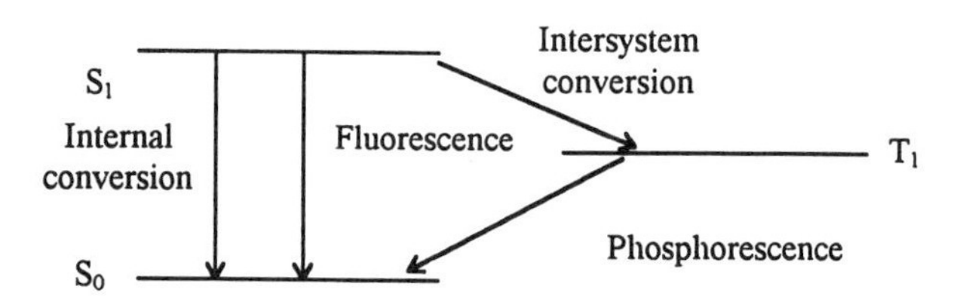

(The vibrational levels are not shown)

b. The lifetime of a triplet state is much longer then that of a singlet state so that the populations of these triplet states are often larger then the singlet populations.

IV.5. Appendix: From density operator to population rate operator

The population rate operator is defined from the average density operator (called ρ) by

$$\rho(t) = \sum_j \int_{-\infty}^{t} \rho_d(j, t, t_0) \lambda(t_0) dt_0$$

j represents one of the two states a and b. The average density operator describes the evolution from t_0 to t. The integral takes into account the whole past of the system, from the origin of time at $-\infty$ to time t. λ is the population rate of the state. Let us calculate the derivative with respect to time of the population rate operator.

$$
\begin{aligned}
d\rho / dt &= \lim(dt \to 0)\left[[\rho(t+dt) - \rho(t)\right] \\
&= \lim(dt \to 0)
\end{aligned}
$$

$$
\times\left\{(1/dt)\sum_j\left[\int_{-\infty}^{t+dt} dt_0 \lambda_j(t_0)\rho_d(j, t+dt, t_0) - \int_{-\infty}^{t} dt \lambda_j(t_0)\rho_d(j, t, t_0)\right]\right\}
$$

$$
= \lim(dt \to 0)
$$

$$
\times\left\{(1/dt)\sum_j\left[\int_{-\infty}^{t} dt_0 \lambda_j(t_0)\rho_d(j, t+dt, t_0) - \int_{-\infty}^{t} dt_0 \lambda_j(t_0)\rho_d(j, t, t_0)\right]\right\}
$$

$$
+ \lim(dt \to 0)\left[(1/dt)\sum_j \int_t^{t+dt} dt_0 \lambda_j(t)\rho_d(j, t+dt, t)\right]
$$

We have expressed the time derivative of the average density operator and the last expression can also be written as

$$
d\rho / dt = \sum_j \int_{-\infty}^{t} dt_0 \lambda_j(t_0)\, d\rho_d(j, t, t_0)/dt + \sum_j \lambda_j(t)\rho_d(j, t, t)
$$

The time derivative of the average density operator is given by the Liouville equation. $\rho_d(a, t, t)$ is the density operator relative to state "a" which is populated at time t and observed at the same instant t. In base (a, b) it takes the form $\begin{pmatrix} 1 & 0 \\ 0 & 0 \end{pmatrix}$. In a same way, $\rho_d(b, t, t) = \begin{pmatrix} 0 & 0 \\ 0 & 1 \end{pmatrix}$, and we obtain

$$
\begin{aligned}
d\rho / dt = &\sum_j \int_{-\infty}^{t} dt_0 \lambda_j(t_0)(2\pi / ih)\left[H_0, \rho_d(j, t, t_0)\right] \\
&+ \lambda_a(t)\begin{pmatrix} 1 & 0 \\ 0 & 0 \end{pmatrix} + \lambda_b(t)\begin{pmatrix} 0 & 0 \\ 0 & 1 \end{pmatrix}
\end{aligned}
$$

In the commutator, the integral does not apply to the Hamiltonian operator. We identify the population rate operator and get

$$
d\rho(t)/dt = (2\pi / ih)\left[H, \rho(t)\right] - (1/2)\left[\Gamma\rho(t) + \rho(t)\Gamma\right] + \chi
$$

with $$\chi = \begin{pmatrix} \lambda_a & 0 \\ 0 & \lambda_b \end{pmatrix}$$

IV.6. Bibliography

4.6.1. An introduction to theoretical UV-visible spectroscopy can be found in the articles:

JAFFE H.H., BEVERIDGE D.L. and ORCHIN M. — Understanding Ultraviolet Spectra of Organic Molecules, *J. of Chem. Educ.*, 44, 383 (1967).

THOMSEN V.B.E. — Why Do Spectral Lines Have a Linewidth ?, *J. of Chem. Educ.*, 72, 616 (1995).

And in the following books:

BANWELL C.N. — *Fundamentals of Molecular Spectroscopy*, Mc Graw (1983).

CHABANEL M. et GRESSIER P. — *Liaison chimique et spectroscopie*, Ellipses (1991).

HOLLAS J.M. — *Modern Spectroscopy*, Second Edition, J. Wiley (1992).

STRUVE W.S. — *Fundamentals of Molecular Spectroscopy*, J. Wiley (1989).

4.6.2. A more specialized book is:

SINCLAIR R. and DENNEY R. — *Visible and Ultraviolet Spectroscopy*, J. Wiley (1987).

4.6.3. Applications to organic and to inorganic chemistry are described in:

JAFFE H.H. and ORCHIN M. — *Theory and Applications of Ultraviolet Spectroscopy*, J. Wiley (1964).

4.6.4. The UV-visible spectroscopy of complex ions can be found in:

FIGGIS B.N. — *Introduction to Ligand Fields*, R. E. Krieger Publishing Company (1986).

Chapter V

Spectroscopy of Optical Activity

V.1. General features

By the *optical activity* of a system, whether macroscopic or microscopic, we mean the ability of this system to modify the propagation of a circularly-polarized light wave (i) by acting on its propagation velocity (through the *refractive index*, and therefore the *phase*) and (ii), in some cases, by acting both on its propagation velocity and on the amplitude of the transmitted wave (*absorption coefficient*). Therefore, this material system will fundamentally distinguish between *right and left circularly-polarized* waves. However, we already indicated (*cf.* Section II.2.2.2) how in a material without optical activity, a linearly-polarized and non absorbed wave can be considered as the superposition of two counter-rotating right and left circularly-polarized waves of the same amplitude which propagate without absorption at the same velocity. So we can see that optical activity is in fact about the three fundamental kinds of polarization, i.e. elliptic, linear and circular polarization. There are two kinds of optical activity. The first kind (i) is properly called *optical activity*. The second kind (ii) is called *circular dichroism*. Figure 5.1.1 illustrates these two optical activities for a linearly-polarized incident wave.

The phenomenon of *rotatory polarization* was discovered on a crystalline sample of quartz by the French physicist Arago in 1811. Four years later, Biot showed that this phenomenon can also occur in isotropic substances (for example in turpentine). Limonene — a liquid at room temperature — can induce a rotation of about 107° per decimeter in the yellow part of the spectrum! And for many years, nicotine held the record of optical activity (– 164° dm^{-1}), until the discovery of liquid crystals some of whose phases can present considerable optical activity.

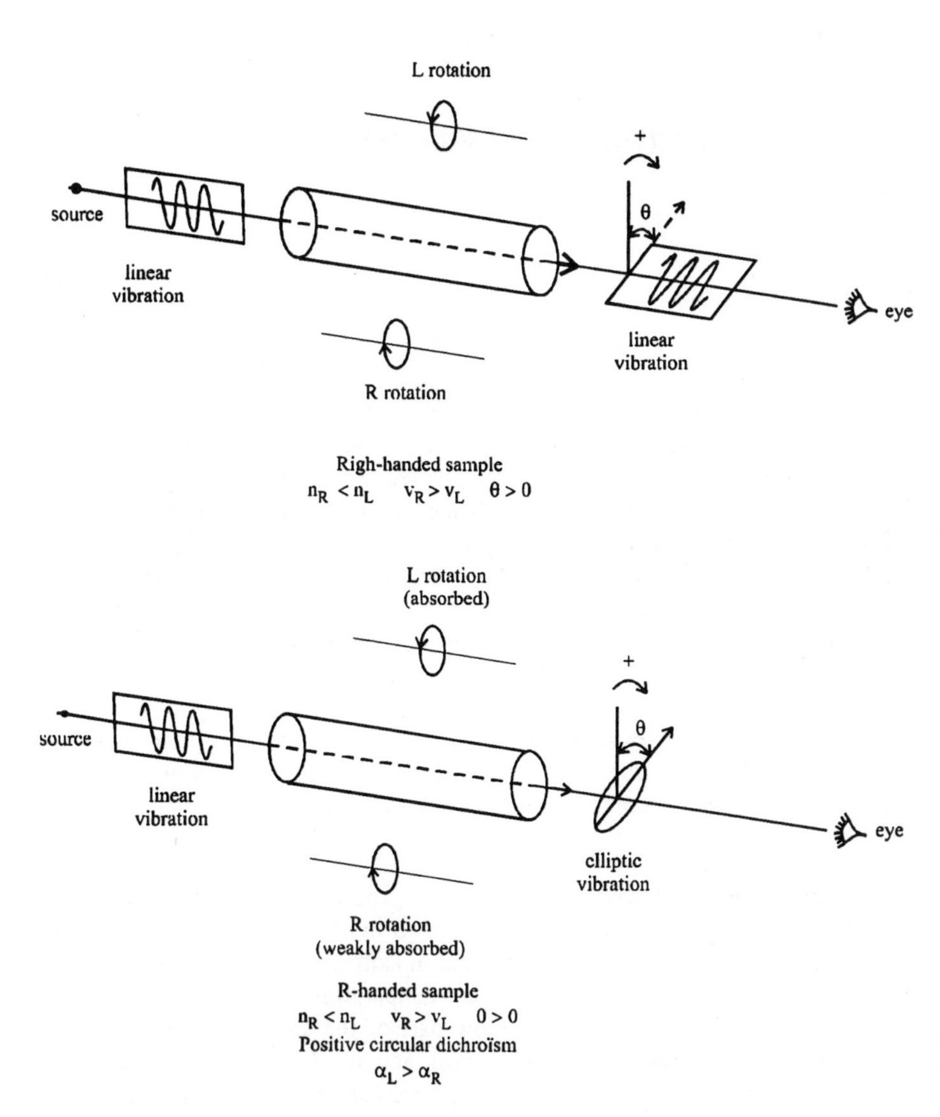

Figure 5.1.1: Illustration of natural optical activity in the case of a linearly-polarized incident vibration. (Top) optical activity: Rotation without change of amplitude of the polarization of the incident vibration. (Bottom) circular dichroism: The incident vibration becomes elliptic (the position of the axes of the ellipse defines the rotation and the ratio of the lengths of its axes — called the ellipticity — measures the difference between the left and the right absorptions).

Circular dichroism was put in evidence for the first time by Cotton during his study of the colored solutions of various tartrates.

To be complete, we would like to mention the existence of two similar effects:

— The *Faraday effect* (1846), in which a magnetic field induces the rotation of the polarization plane of an optical wave whose wave vector is parallel to the direction of the magnetic field.

— the *inverse Faraday effect*, more recently discovered by Pershan in 1966, according to which an intense circularly-polarized laser wave induces a static magnetic field along the direction of its wave vector.

V.2. Phenomenological descriptions, classical and semi-classical interpretations

V.2.1. Natural optical activity

V.2.1.1. Fresnel's kinetic theory

Fresnel's kinetic theory is a very simple phenomenological theory (extensively exploited in Problem V.5.1) which expresses the rotation induced by a linearly-polarized vibration, in experimental conditions and for wavelength λ, as a function of the thickness e of the sample, the wavelength λ of the radiation and the refractive indices n_L and n_R of the material.

Let us consider a vibration, linearly polarized along the Ox axis and propagating along the Oz axis of an orthonormed frame Oxyz. As said earlier, the adjective "*right*" or "*left*" is defined with respect to the observer. The adjective "*dextrorotatory*" (right) applies to a wave turning clockwise, while "*levorotatory*" (left) applies to a wave turning counter-clockwise. We represent the linear vibration at the entrance of the sample by the sum of two circularly-polarized vibrations (*cf.* Equations 2.2.30 and 2.2.31):

left: $x_L = (E_0/2)\cos\omega t \quad y_L = (E_0/2)\sin\omega t$

right: $x_R = (E_0/2)\cos\omega t \quad y_R = -(E_0/2)\sin\omega t$

Phase changes φ_L and φ_R have appeared at the exit of the sample, and the circular vibrations can be written as

left: $x'_L = (E_0/2)\cos(\omega t + \varphi_L) \quad y_L = (E_0/2)\sin(\omega t + \varphi_L)$

right: $x'_R = (E_0/2)\cos(\omega t + \varphi_R) \quad y_R = -(E_0/2)\sin(\omega t + \varphi_R)$

so that the components of the vibration along the Ox and the Oy axes at the exit of the sample are

$$x' = x'_L + x'_R = E_0 \cos[\omega t + (\varphi_L + \varphi_R)/2] \cos[(\varphi_L - \varphi_R)/2]$$
$$y' = y'_L + y'_R = E_0 \cos[\omega t + (\varphi_L + \varphi_R)/2] \sin[(\varphi_L - \varphi_R)/2] \quad (5.2.1)$$

with the time-dependent term shows that we are dealing with a linear vibration which rotated by an angle θ, defined by

$$\mathrm{tg}\theta = y'/x' = \mathrm{tg}[(\varphi_L - \varphi_R)/2]$$

and the rotation is equal to

$$\theta = \pi(n_L - n_R)e/\lambda \quad (5.2.2)$$

It is proportional to the difference between the refractive indices for the left and the right vibrations and inversely proportional to the wavelength of the light. This expression has been suggested by Fresnel. To be complete, it needs a sign convention. We shall admit that dextrorotatory substances induce a positive rotation and levorotatory substances induce a negative rotation. For a dextrorotatory substance, $n_R < n_L$ and the right circularly polarized vibration propagates faster than the left circularly polarized vibration.

In a levorotatory substance the situation is exactly opposite.

V.2.1.2. Physical origin of the Born effect and of Born's tensor theory

We showed in Chapter IV that the polarization **P** was proportional to the electric field **E** and we defined the susceptibility χ as the ratio between these two entities. However, in a 3D orthonormal reference frame, the ratio between two vectors is in fact a tensor of rank two, and we must write, in a general way

$$P_i(\omega) = \varepsilon_0 \chi^1{}_{ij}(-\omega, \omega) E_j(\omega) \quad (5.2.3)$$

We used (i) Einstein's sum convention over the repeated (mute) subscript j and (ii) a second convention consisting of including the frequency of the electric field, ω, in the frequency dependence of the susceptibility: The first ω in the parentheses defines the frequency of the polarization, while the second ω refers to the frequency of the electric field. Moreover, the sum of the frequencies inside the parentheses must be equal to zero which explains the minus sign in front of the first ω. This convention is very often used in optics. Writing $\chi^1{}_{ij}(-\omega, \omega)$ simply indicates that both the polarization and the electric field oscillate with the same frequency ω. $\chi^1{}_{ij}$ is a tensor of rank 2, characterized by the two indices i and j. Equation 5.2.3 defines a linear relation between the cause (E) and its effect (P). In this case, the susceptibility is a susceptibility of order 1 (indicated by the exponent 1 placed after χ). In fact, there is no reason for such a linear behavior. Though physical effects do not usually depend linearly on their causes, physicists very often *assume* their linearity in a first approximation! We could in fact have added the term $\chi^2{}_{ijk}(-2\omega,\omega,\omega)E_j(\omega)E_k(\omega)$ to the above expression of the polarization, thereby defining a susceptibility of order 2 which would give rise to a 2ω-frequency component of the polarization, proportional to the square of the electric field. This a term is of great importance in nonlinear optics. Thus, the term placed in the right hand member of Equation 5.2.3 turns out to be the first term of a development of the polarization in series of increasing power of the electric field. The order of the susceptibility increases by one unit with each term and, correlatively, the rank of the tensor describing this susceptibility increases by one.

But, on top of these terms involving the successive powers of the field, we can formally add other terms to the polarization by varying the physical nature of the effects described. For example, remembering that the optical wave propagates along the Oz axis, i.e. that the associated electric field possesses a spatial gradient along this axis, we can write

$$P_i(\omega) = \varepsilon_0[\chi^1{}_{ij}(-\omega,\omega)E_j(\omega) + \eta_{ijk}(-\omega,\omega,0)\nabla_k E_j(\omega)] \qquad (5.2.4)$$

The second term of the right hand member of the equation, starting from the kth component at zero-frequency (subscript 0) of the spatial gradient of the jth component of the electric field E, assumed to oscillate at frequency ω, indeed begets a component of the polarization **P** at frequency ω.

If we assume that a plane wave (*cf.* Section II.2.2.2) represented by

$$E_j(\omega) = E_j(0)\exp\left[i\left(\mathbf{kr} - \omega t\right)\right]$$

interacts with the system, the gradient of the field is

$$\nabla_k E_j(\omega) = i|\mathbf{k}|u_k E_j(\omega)$$

Vector **u** defines propagation direction of the wave. Thus we obtain

$$P_i(\omega)/\varepsilon_0 E_j(\omega) = \chi^1{}_{ij}(-\omega,\omega) + i|\mathbf{k}|\eta_{ijk}(-\omega,\omega,0)u_k = \chi_{ij}(-\omega,\omega) \qquad (5.2.5)$$

Let us look now for the consequences resulting from the insertion of the 2nd term in the expression of the susceptibility. We consider an isotropic material, i.e. one for which the vectorial properties do not depend on spatial direction. This is the case, for example, in a molecular solution in the absence of orientational constraints. The *macroscopic* tensor η_{ijk}, defined in the laboratory frame, is obtained from the *microscopic tensor* $\eta_{\alpha\beta\gamma}$ and can be written as

$$\eta_{ijk} = N < c_{i\alpha}c_{j\beta}c_{k\gamma} >_0 \eta_{\alpha\beta\gamma}$$

N is the density of the microscopic systems and the coefficients c are the directional cosines of the transformation microscopic frame → laboratory frame. The symbol $< >_0$ defines the statistic orientational average, taken in an isotropic way , i.e. on a macroscopic sample without any privileged orientation. Formally, it is a *true* symmetrical tensor of rank 6 (Appendix V.6.1 defines and compares true tensors and pseudotensors), built by using the direct product of two *Levi-Civita* pseudotensors ε (*cf.* Appendix V.6.2), and it writes as $< c_{i\alpha} c_{j\beta} c_{k\gamma} >_0 = a\varepsilon_{ijk}.\varepsilon_{\alpha\beta\gamma}$. The pseudoscalar a is determined by externally multiplying both members of this equation by ε_{ijk}. The left member yields $\varepsilon_{\alpha\beta\gamma}$ and the right member 6a $\varepsilon_{\alpha\beta\gamma}$ (*cf.* Appendix V.6.2). We therefore conclude that a = 1/6 and we obtain the value $< c_{i\alpha} c_{j\beta} c_{k\gamma} >_0 = (1/6)\varepsilon_{ijk}.\varepsilon_{\alpha\beta\gamma}$ for the isotropic average. It follows that

$$\eta_{ijk} = (N/6)\varepsilon_{ijk}.\varepsilon_{\alpha\beta\gamma}.\eta_{\alpha\beta\gamma}$$

The triple contraction performed on parameters α, β and γ yields a scalar number which we write as

$$\varepsilon_{\alpha\beta\gamma}\eta_{\alpha\beta\gamma}/6 = \eta$$

So we have

$$\eta_{ijk} = (\varepsilon_{\alpha\beta\gamma}.\eta_{\alpha\beta\gamma}/6)\varepsilon_{ijk} = \eta\varepsilon_{ijk} \qquad (5.2.6)$$

The tensor describing the optical activity of a fully isotropic material is therefore proportional to the Levi-Civita tensor and is antisymmetrical with respect to the permutation of the three indices.

And, remembering that the dielectric constant and the susceptibility are linked by the relation

$$\varepsilon_{ij}{}^{r} - \delta_{ij} = P_i/\varepsilon_0 E_j = \chi_{ij}/\varepsilon_0 \qquad (5.2.7)$$

we obtain

$$\varepsilon_{ij}{}^{r} - \delta_{ij} = \chi^1\delta_{ij}/\varepsilon_0 + i2\pi\eta\varepsilon_{ijk}u_k/\varepsilon_0\lambda \qquad (5.2.8)$$

With this equation, we can attribute a structure to the electrical permittivity which, in the orthonormed reference frame Oxyz and for an electric field with components E_x and E_y propagating along the Oz axis (labeled by the vector **u**), can thus be written as

$$\left[\varepsilon_{ij}{}^{r}\right] = \begin{bmatrix} \varepsilon^r & i\varepsilon'^r & 0 \\ -i\varepsilon'^r & \varepsilon^r & 0 \\ 0 & 0 & \varepsilon^r \end{bmatrix} \qquad (5.2.9)$$

So the term $\nabla_k E_j(\omega)$, which is due to the spatial dispersion along the propagation axis of the electric field, has resulted in the addition of two *imaginary nondiagonal terms* $\varepsilon'^r = |\mathbf{k}| u_z \eta$ to the electrical permittivity tensor. The tensor becomes *antisymmetrical* in this case.

By using the following relations, derived from the constitutive relations of Maxwell's equations (*cf.* Appendix II.4.2):

$$D_i = \varepsilon_0 \varepsilon_{ij}{}^r E_j = \varepsilon_0 E_i + P_i$$
$$n^2 - 1 = P_i / \varepsilon_0 E_i \qquad (5.2.10)$$

Equation 5.2.9 leads to

$$\left(\varepsilon^r - n^2\right) E_x + i\varepsilon'^r E_y = 0$$
$$-i\varepsilon'^r E_x + \left(\varepsilon^r - n^2\right) E_y = 0 \qquad (5.2.11)$$

This system has a nontrivial solution only if the following equation

$$\begin{vmatrix} \varepsilon^r - n^2 & i\varepsilon'^r \\ -i\varepsilon'^r & \varepsilon^r - n^2 \end{vmatrix} = 0$$

is satisfied, i.e. for the two values n_+ and n_- of the refractive index given by

$$n_-^2 = \varepsilon^r - \varepsilon'^r \quad n_+^2 = \varepsilon^r + \varepsilon'^r \qquad (5.2.12)$$

By successively inserting these two values in Equations 5.2.11, we obtain the two solutions

$$E_y / E_x = \exp(i\pi/2)$$
$$E_y / E_x = \exp(-i\pi/2)$$

They represent two circularly-polarized waves (E_x and E_y have a 90 degrees phase shift), a dextrorotatory and a levorotatory one, propagating in the material with different phase velocities, which results in a phase change written as

$$\varphi = 2\pi e \Delta n / \lambda = 2\pi e \varepsilon'^r / n\lambda \qquad (5.2.13)$$

and in a rotation of $\theta = \varphi/2$ for a linearly-polarized incident wave.

These results agree with Born's formalism (*cf.* Bibliographic reference 5.7.2) of the natural rotatory power (defined by $\rho = \theta / e$) which introduced a *gyration tensor* G_{kl} such that $\varepsilon_{ijl} G_{kl} = 2\pi\eta\varepsilon_{ijk} / \lambda$.

Comment: Since P is a true tensor, the term $\eta_{ijk}(-\omega,\omega,0)\nabla_k E_j(\omega)$ must be uneven for any spatial inversion. ∇_k and E_j are both uneven, so their product is even. Therefore η_{ijk} must be uneven and behave like $(-1)^3$, which makes it a true tensor of rank 3. Since ε_{ijk} is a pseudotensor, η must necessarily be a pseudoscalar. G_{kl} is a pseudotensor and θ is a pseudorotation. This has a particularly important physical consequence since the rotation must behave with spatial inversion as $(-1)^{0+1}$. *It is therefore uneven which means that when the propagation direction of the wave is reversed, the angle θ turns in the other direction.* Natural rotation therefore vanishes when the light wave is reflected, that is, when it goes through the active material two times, once in each direction. This property differentiates natural optical activity from the optical activity induced by a magnetic field applied along the propagation axis (Faraday effect).

V.2.1.3. Classical and semi-classical microscopic theories

The details of these theories can be found in Bibliographic reference 5.7.1.

To study electronic absorption, described in Chapter IV (*cf.* Section IV.3.1), classical microscopic physics uses a helical type mechanical model. An electron is constrained to move on a helix (radius r; pitch $2\pi a$) whose axis lies along the Oz axis of an orthonormed frame. The wavevector **k** of a plane optical wave (of frequency ω, with its electric field along Ox, and its magnetic induction along Oy) also lies along this axis. If s represents the curvilinear abscissa along the helix, the motion is described by the following differential equation

$$m\ddot{\mathbf{s}}_j(t) = -K\mathbf{s}_j(t) + \mathbf{F}_{em} \qquad (5.2.14)$$

F_{em} is an electric force (e), of magnetic origin (m), due to the magnetic induction B of the wave. Indeed, induction B creates an electromagnetic force (e.m.f) on the helix and therefore an electric field whose circulation over one turn of the helix is equal to this e.m.f. A simple calculation gives the following expression for the electric force associated to this field:

$$\mathbf{F}_{em} = \mathbf{s} r^2 e\dot{B}/|\mathbf{s}| 2\sqrt{r^2 + a^2} \qquad (5.2.15)$$

Constant K is an elastic restoring force. Neglecting the transitory regime and looking for a particular solution of Equation 5.2.14 in the form

$$\mathbf{s}_j = \mathbf{s}_{0j} \exp[i(\omega t + \varphi)]$$

we find $\mathbf{s} = -\mathbf{F}_{em} / \left[m\left(\omega_0^2 - \omega^2\right)\right]$. We define an eigenfrequency by setting $\omega_0^2 = K/m$. An electric dipole moment therefore appears, labeled $\mathbf{p}_{em}$ because of its magnetic origin, which lies parallel to B and along the Oy axis (unit vector **j**). It can be expressed as

$$\mathbf{p}_{em} = -e(\mathbf{sj})\mathbf{j} = \left[e^2 r^2 l / 2m\left(r^2 + l^2\right)\right]\left[\dot{\mathbf{B}}/\left(\omega_0^2 - \omega^2\right)\right] \qquad (5.2.16)$$

This electrical moment therefore (i) is carried by B, (ii) is proportional to the first derivative of B with respect to time, i.e. it has a modulus proportional to the modulus of the magnetic induction associated to the optical wave and its phase is delayed by 90 degrees with respect to this induction, and last, (iii) has a resonant nature.

This moment radiates an electrical field, carried by Oy, whose modulus is proportional to its own modulus, and which has a 90 degrees phase delay. After calculating this field and summing over all the electrons j of the sample (*cf.* Bibliographic reference 5.7.2), we find that the optical wave which carries an electric field **E** along Ox before crossing the sample, carries a extra electric field after crossing it. This extra field has a weak amplitude (because of its magnetic origin), it is carried by Oy, and it has a phase delay of π with respect to **E**. The resulting wave now carries an electric field which is still linearly polarized, but which has *rotated by an angle* $\theta = \left|\mathbf{E}'_{em}\right| / |\mathbf{E}|$ with respect to the Ox direction.

Comments:

— In this model, we did not considered the direct action of the induction **B** on the electron (the magnetic force is $\mathbf{F}_m = -e\mathbf{v} \times \mathbf{B}$) because $\mathbf{F}_m$ has no component along Oy.

— This microscopic model is in agreement with the phenomenological description presented in the previous section. Indeed, we showed that optical activity was due to the component of the spatial gradient of the electric field along the propagation axis. The term $\nabla_k E_j$ (*cf.* Equation 5.2.4) can be written $\nabla_z E_x$ and is the Oy component of the rotational $\nabla \times \mathbf{E}$. However, Maxwell's equations show that $(\nabla \times \mathbf{E})_y = -\dot{\mathbf{B}}_y$. In fact, the two origins of optical activity we discussed coincide in the phenomenological and in the microscopic classical treatments. The *spatial dispersion* of the electric field along the propagation axis is equivalent to the *time variations* of the magnetic induction carried by the wave.

— The contribution is directly connected to the absorption of the optical wave by the sample and diverges at resonance.

— We considered only one oscillator, with an eigenfrequency ω_0, and so we described a microscopic system with only one absorption band. As in Chapter IV, the generalization to systems with several absorption bands is obtained by adding up the contributions of each of the bands.

A strictly equivalent model can be obtained by positioning the helix axis along Ox and by considering the action of the oscillating electric field. This field induces an electric force on the electron which moves along the helix. An electric current results, which induces a magnetic induction along Ox. This induction creates a magnetic moment along Ox, of electric origin and noted μ_{me}, which in turn radiates a magnetic induction along the same axis. The resulting optical wave carries a linearly polarized magnetic induction which has *rotated by an angle* $\theta = |\mathbf{B}'_{me}| / |\mathbf{B}|$ with respect to the Oy direction. The Ox component of the force $\mathbf{F}_m$ is now different from zero but it can be neglected when compared to the electric force $\mathbf{F}_{me}$. This model gives a contribution to the optical activity which is strictly equal to that resulting from the first model.

In the literature, a certain number of other models can be found, which derive more or less from the two previous ones. Let us only mention Kuhn's model, developed in the nineteen thirties (*cf.* Bibliographic reference 5.7.2). It associates two coupled oscillators oscillating in the same plane and was widely used in chemistry.

In conclusion, we can say that:

— The microscopic classical models all use a helical description of the material, a fertile description which has extensively been used these last years for the theoretical treatment of the optical properties of certain phases of liquid crystals;

— We already noted a deep analogy between this treatment and the classical description of electronic absorption presented in Chapter IV. Let us note, however, a very fundamental difference: *The magnetic induction carried by the optical wave, which we did not consider previously, here plays the main role.*

Of course, this microscopic classical treatment presents a major drawback, already indicated in Chapter IV, namely it remains merely descriptive and as such does not allow any theoretical predictions related to the structure of the microscopic systems. To be able to do this, it needs some complementary semi-classical explanations.

To this end, we shall rely on the first classical helical model described in this chapter, and in order to calculate the contribution of one band, we shall use a procedure very similar to that described in detail in Section IV.4.2. With the same approximations as above, we shall calculate the quantum average value of the electric dipole moment induced by the magnetic induction **B.** The rate population operator we shall use is calculated from a Hamiltonian which is derived from a magnetic type interaction energy. The resulting two-level model can, if necessary be generalized by summing over all the bands of a molecule.

Thus, Equation 4.3.29 can be written as

$$H = H_0 - \mu\mathbf{B} \qquad (5.2.17)$$

μ is the magnetic dipole moment of the microscopic system and is equal to

$$\mu = (-e/2m)\sum_j \mathbf{r}_j \times \mathbf{p}_j \qquad (5.2.18)$$

where **r** and **p** are, respectively, the position and the linear momentum vectors (do not confuse the linear momentum with the electric polarization which in this book are represented by the same letter p!). The summation is performed over all the electrons j of the system.

Please observe that the electric dipole energy $-\mathbf{pE}$ has been replaced by the magnetic dipole energy $-\mu\mathbf{B}$. The diagonal element of the rate population operator (Equation 4.3.32) is proportional to μB_0 and the average electric polarization induced by the magnetic induction is given by a relation very similar to relation 4.3.35. But there are three differences: (i) the saturation S is completely negligible because the magnetic interaction energies are so small, (ii) the population difference D_0 is equal to the density of microscopic systems in the ground state (the excited state is always weakly populated) and (iii) the product p^2E_0 must be replaced by the expression $p\mu B_0$, where μ is the magnetic transition momentum $\langle b|\mu|a\rangle$. *The optical activity is therefore proportional to the product of the electric and the magnetic transition moments of the system.* In fact, in the phenomenological treatment we showed that the optical activity was described by the nondiagonal, strictly imaginary, component of the dielectric tensor. It is therefore the imaginary component of the product $p\mu$ which must be used in this description for the general case (and in any case, the linear momentum operator is a strictly imaginary operator). Last, summing over all the transitions would yield the most general result.

V.2.1.4. Dispersion of the optical activity

The dispersion of the optical activity is fundamentally the same as the dispersion of a susceptibility of order 2. In Bibliographical reference 7.6.3 for example, the reader can find the most general expression for the nth-order susceptibility. By setting $n = 2$ in the general expression and taking into account the frequency structure $(-\omega, \omega, 0)$ of the tensor $\eta_{\alpha\beta\gamma}$, we find that in the vicinity of resonance, the dispersion is the same as that established in the semi-classical theory and that it is expressed by a relation which is identical to that of the real part of the 1st order susceptibility (*cf.* Section IV.3.2.2). It varies with the shift to the resonance frequency as

$$\left[(\omega_0 - \omega)/\gamma\right] / \left[1 + (\omega_0 - \omega)^2/\gamma^2\right] \qquad (5.2.19)$$

ω_0 et γ are, respectively, the resonance frequency and its linewidth.

The dispersion is very similar to that of the real part of the refractive index, and it is not surprising that in *regions* which are *transparent* to the studied radiation, the dispersion law of classical microscopic theory is perfectly identical to Sellmeier's law of the dispersion of the refractive index (*cf.* Problem IV.4.6). This is the so-called Drude law and it can be expressed as (*cf.* above and Equation 5.2.16)

$$\rho \approx 1/(\lambda^2 - \lambda_0^2) \tag{5.2.20}$$

This law will be exploited in depth in Problem V.4.1.

V.2.2. Circular dichroism

V.2.2.1. Extension of Fresnel's theory

We must now take into account the fact that the two circularly-polarized waves propagating in the material have different absorption coefficients. As in Chapter IV, we call α_L and α_R the absorption coefficients (defined for the intensity) of the left and right waves arising from the decomposition of an linearly polarized incident vibration propagating along the Ox' axis of an orthonormed frame. At the exit of a plate of thickness e, the circular vibrations are expressed as

$$\begin{aligned} E_L &= \begin{pmatrix} 1 \\ i \end{pmatrix} (E_0/2)(\alpha_L/2)\exp[i(\omega t + \varphi_L)] \\ E_R &= \begin{pmatrix} 1 \\ -i \end{pmatrix} (E_0/2)(\alpha_R/2)\exp[i(\omega t + \varphi_R)] \end{aligned} \tag{5.2.20}$$

Here we used the complex notation introduced in Section II.2.2.2. By projecting on the Ox' and the Oy' axes, we obtain

$$\begin{aligned} E_{x'} &= E'_L \cos(\omega t + \varphi_L) + E'_D \cos(\omega t + \varphi_R) \\ E_{y'} &= E'_L \sin(\omega t + \varphi_L) - E'_D \sin(\omega t + \varphi_R) \end{aligned}$$

with $E'_L = E_0\alpha_L/2$ and $E'_R = E_0\alpha_R/2$.

This vibration is elliptical. $E_x \cos(\omega t + \varphi)$ and $E_y \sin(\omega t + \varphi)$ represent its components in the frame consisting of its eigenaxes. If θ denotes the angle between the eigenaxes of the ellipse and the Ox' and Oy' axes, we can write in the Ox', Oy' frame:

$$\begin{pmatrix} E_{x'} \\ E_{y'} \end{pmatrix} = \begin{pmatrix} \cos\theta & -\sin\theta \\ \sin\theta & \cos\theta \end{pmatrix} \begin{pmatrix} E_x \cos(\omega t + \varphi) \\ E_y \sin(\omega t + \varphi) \end{pmatrix}$$

then, by noting $\varphi = (\varphi_L + \varphi_R)/2$ and $\Delta\varphi = (\varphi_L - \varphi_R)/2$, or

$$\varphi_L = \varphi + \Delta\varphi \text{ and } \varphi_R = \varphi - \Delta\varphi$$

we find

$$E_{x'} = \left(E_R^{'} + E_R^{'}\right)\cos\Delta\varphi\cos(\omega t + \varphi) + \left(E_R^{'} - E_L^{'}\right)\sin\Delta\varphi\sin(\omega t + \varphi)$$
$$E_{y'} = -\left(E_R^{'} - E_L^{'}\right)\cos\Delta\varphi\sin(\omega t + \varphi) + \left(E_R^{'} + E_L^{'}\right)\sin\Delta\varphi\cos(\omega t + \varphi)$$

and, by identifying

$$E_x \cos\theta = \left(E_R^{'} + E_L^{'}\right)\cos\Delta\varphi$$
$$-E_y \sin\theta = \left(E_R^{'} - E_L^{'}\right)\sin\Delta\varphi$$
$$E_x \sin\theta = \left(E_R^{'} + E_L^{'}\right)\sin\Delta\varphi$$
$$E_y \cos\theta = -\left(E_R^{'} - E_L^{'}\right)\cos\Delta\varphi$$

These equations lead to

$$\theta = \Delta\varphi = \left(\varphi_L - \varphi_R\right)/2 \qquad (5.2.21)$$

$$E_x / E_y = (E_R^{'} + E_L^{'})/\left(E_L^{'} - E_R^{'}\right) \qquad (5.2.22)$$

Equation 5.2.21 is rigorously identical to Equation 5.2.2. The absorption of the circular wave does not modify the natural optical activity. Now it is the Ox axis of the ellipse which makes an angle θ with the incident linear vibration. The absorption only affects the ellipticity of the vibration at the exit of the plate. The ratio between the axes of the ellipse is given by Equation 5.2.22, which is also equal to

$$E_x / E_y = \left|\left(\alpha_R + \alpha_L\right)\left(\alpha_L^{'} - \alpha_R\right)\right| \qquad (5.2.23)$$

The circular dichroism is called positive if $\alpha_G > \alpha_D$ and negative in the opposite case.

Equations 5.2.21 and 5.2.23 are the leading equations describing this phenomenon, illustrated by Problem V.5.2.

V.2.2.2. Extension of Born's theory

We can also assume that the optical activity tensor $\eta_{\alpha\beta\gamma}$ has an imaginary part and include it into the mathematical reasoning of Section V.2.1.2. Again, we identify two circular waves, but now the refractive index has a complex value and Equation 5.2.12 takes the form

$$n^2_{-,+} = \varepsilon^r \pm \varepsilon'^r \pm i\varepsilon''^r \qquad (5.2.24)$$

Using the same reasoning as in Section IV.3.2.2, we can show that in this case the two waves propagate with different absorbances, and again we obtain the previously established results. In particular, the circular dichroism absorption curves are very comparable to those represented on Figure 4.3.6 (the same similarity exists between the dispersion of natural optical activity, represented by a Lorentzian, and the dispersion of the real part of the refractive index as defined by Equation 4.3.12).

V.3. Symmetry and structure of the optical activity tensor

— *In isotropic substances*

We have shown previously that in isotropic substances, the optical activity tensor is proportional to the Levi-Civita tensor. It is therefore an antisymmetrical tensor with respect to all three indices. In a 3D reference frame, there are 27 components in a tensor of rank 3. The 3 components with three identical indices and the 18 components with 2 identical indices all vanish. The tensor therefore contains only 6 nonzero components, one of which is independent.

— *More generally*

We can ask if this property of total antisymmetry, which characterizes the optical activity tensor in an isotropic microscopic material, also applies to the most general molecular microscopic tensor, i.e. the tensor of an anisotropic system. This is not so. In the most general case, the optical activity tensor is antisymmetrical only with respect to the first two indices (a demonstration of this, based on the thermodynamics of irreversible processes, can be found in Bibliographic reference 5.7.6). This is called *natural* reduction. As a consequence, the three components with identical indices, as well as the 6 components whose first two indices are identical, always vanish. The most general tensor (triclinic system and symmetry class C_1) therefore consist of 18 nonzero elements of which 9 are independent.

But we also know that the *symmetry* of the system plays a determining role in the structure of the optical activity tensor. In general, we can say that if a microscopic system — or a part of this system — presents a *helical* structure, its susceptibility (or polarizability) depends on the rotation direction of the applied circular wave. This rotation can occur in the same direction as, or in opposite direction to, that of the helix. We understand why some symmetry elements (symmetry center, plane mirror and improper rotation axis for example) are "helicophobic" and incompatible with *chirality*, defined as the property of optical activity. A convenient rule imposes that a chiral system (from the Latin word meaning hand) cannot be superposed on its mirror image.

In fact, the more symmetric the system is, the more structure the optical activity tensor has. Consequently, the number of independent and nonzero elements decreases with increasing symmetry. Of course, for certain nonchiral symmetry classes there are no nonzero elements at all. This second reduction is due to the symmetry and is due to the fact that each tensor component must remain invariant under the action of each symmetry operator of the group to which the system belongs. Thus, the described physical property must have at least the symmetry of the studied object. A very convenient method to determine the number of nonzero and independent components has been suggested by *Fumi* (*cf.* Bibliographic reference 5.7.4). Fumi noticed that the tensor components transform in the same way as the corresponding coordinate products. The tensor component is nonzero if the corresponding coordinate product remains invariant under the action of the symmetry operator and is zero in the opposite case. However, we must point out that this method can only be applied to symmetry classes for which we can define Cartesian axes which remain unchanged under the action of the symmetry operators. In the other cases, a more general analytic method must be used (*cf.* Bibliographic reference 5.7.4).

Appendix V.6.3 defines the structure of the optical activity tensor for all crystalline systems and for all chiral symmetry classes.

V.4. Experimental techniques and some examples of results

A traditional application concerning the inversion kinetics of saccharose is presented in Problem V.5.3. Here, we want to describe a few recent practical experimental setups concerning the properties of *liquid crystals* (*cf.* Bibliographic reference 5.7.5) along with some of their results. Liquid crystals are known for their very great polymorphism, and some of their phases (especially their cholesteric phases) show remarkable chiral properties. These present a helical structure whose spatial period, called "*pitch*", lies in the micron range. This structure results in a marked chirality in the visible part of the spectrum and they have a characteristic wavelength λ_0 which plays the same part as the eigenwavelength of the microscopic system described earlier. But here, there is no intrinsic absorption when the radiation wavelength λ is close to λ_0. Instead, a phenomenon of *selective reflection* occurs, and part of the optical wave sees its propagation modified both in direction and in polarization. These phenomena give rise to very important applications which have been used in numerous industrial realizations, for example in the domain of display. Moreover, these compounds are very often *thermotropic*, i.e. their properties (and in particular their pitch λ_0) vary with temperature. This offers widely exploited opportunities. Whereas in traditional experiments in optical activity and in circular dichroism dispersion, the wavelength λ of the incident radiation is varied in order to sweep the range including λ_0, in the case of thermotropic crystals liquids, λ is kept constant while the temperature (i.e. λ_0) is varied. We can therefore use a relatively unsophisticated laser source and use its coherence properties (*cf.* Section VII.2.2) to improve considerably the quality of the experiments which we are about to describe at least partially in the following section.

V.4.1. Measurement of the rotatory power

An example of an experimental setup is shown on Figure 5.4.1

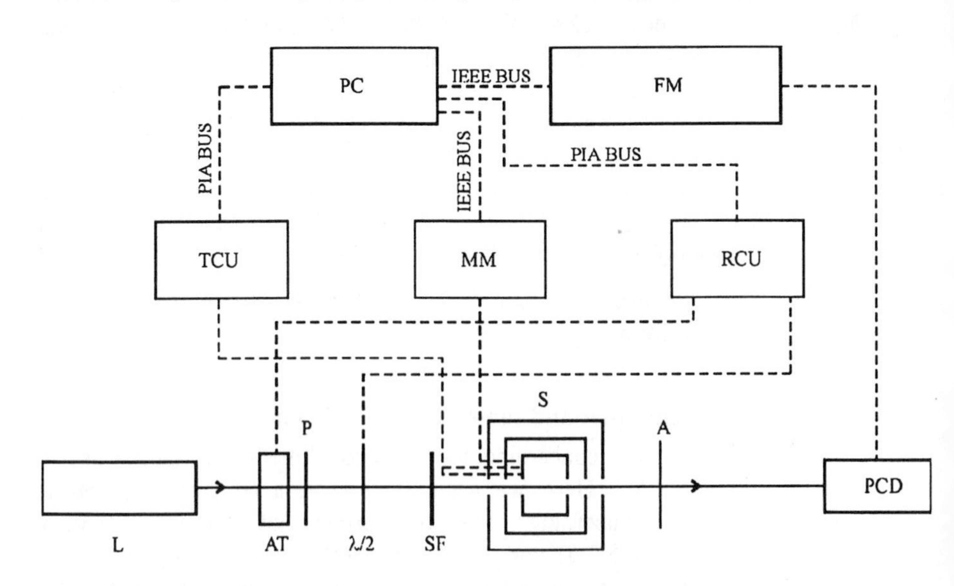

Figure 5.4.1: Rotatory power investigations on oriented liquid crystal samples [from J. R. Lalanne and coll., Phys. Rev. E, 52, 2, 1846(1995); reproduced with the kind authorization of the American Physical Society]. L: He-Ne laser, λ = 632.8 nm, P = 1 mW; AT: Motorized attenuator ; A, P: Linear polarizers; SF: Spatial filter ; S: Sample and its oven; PCD: Photon counting device; FM: Frequency meter; RCU: Rotatory control unit; MM: Multimeter; TCU: Temperature control unit; PC : Microcomputer IBM PS 8550.

The liquid crystal sample (MHPOBC) is 100 mm thick and it is uniaxial at high temperature. Its optical axis is parallel to the wavevector (Oz axis) of the incident laser wave. It is placed in a triple-wall oven which is temperature stabilized within a milli Kelvin, and illuminated by a wave of very weak intensity. The laser intensity is carefully stabilized. Measurements can be carried out step by step, the smallest interval between two successive temperatures being 5 mK. The polarizers P and A are crossed. At each temperature, the rotation of the linear polarization of the wave, induced by the optical activity which we want to measure, is compensated by the rotation of the half-wave plate. Twice the value of this rotation gives the optical activity of the sample. The experiment, which needs a few days to sweep a temperature range of a few degrees, is completely automated and controlled by a microcomputer.

Figure 5.4.2 gives an example of a recording.

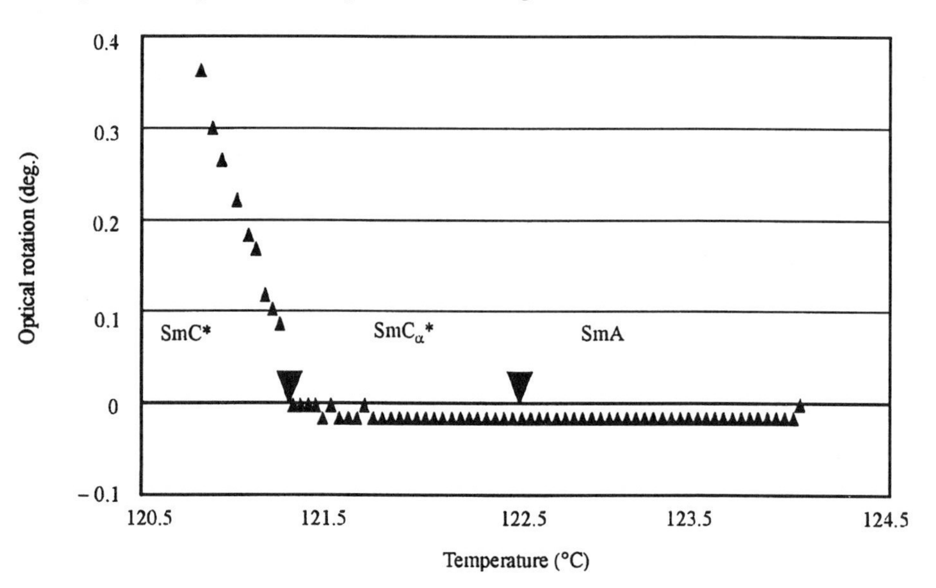

Figure 5.4.2: Optical activity of the smectic A (SmA), smectic C_α (Sm C_α*) and smectic C* phases of a 100 mm thick homeotropic sample of R-MHPOBC (for Sm C*, only the high temperature part of the phase is shown). [from J.R. Lalanne and coll., J. Phys.II, France, 4, 2149 (1994), reproduced with the kind authorization of the Editions de Physique].*

We do not detect any optical activity in the two SmA and SmC_α^* phases within the accuracy (0.01 degree) of the experiment. However, we observe that the so-called ferroelectric SmC* phase has a discontinuity in its optical activity at the phase transition and that its activity increases rapidly when the temperature decreases. This increase is greatest and it changes sign at a temperature between 119.90 and 119.95 °C.

This phenomenon corresponds to a *selective reflection* occurring in this phase. At this temperature, the corresponding length λ_0 is exactly equal to the wavelength of the incident radiation (632.8 nm). In fact, we record a dispersion curve very comparable to that given by Equation 5.2.15.

These results are useful to determine the very wide variety of structures, many of them almost unknown, presented by these substances.

V.4.2. Measurement of circular dichroism

To measure circular dichroism, the same experimental setup can be used. By inserting an electro-optical modulator on the laser path, right and left circularly-polarized waves can be generated in turn. For substances presenting slightly different absorptions for dextrorotatory and for levorotatory waves, the photomultiplier will detect both a c.w. component and a component modulated at the frequency of the modulator. This modulated signal can be detected separately with the help of a synchronous detection system capable of detecting only the chosen modulation frequency. More conventional devices can also be used. They yield results comparable to those presented on Figure 5.4.3.

We observe the presence of two bands located at about 290 nm and 480 nm. The first band indicates the presence of two broad bands (noted $L_{a(eff)}$ and $L_{b(eff)}$) due to intrinsic electronic absorptions of the molecule, which overlap in the vicinity of 290 nm. These bands arise from the π–electron system of the benzene core in the molecule. They have orthogonal transition moments and induce two bands whose ellipticities have opposite signs in the recorded circular dichroism spectrum.

This change of sign is very visible on the figure. On the other hand, the band with negative ellipticity which lies at 480 nm is not due to a molecular electronic absorption but is linked to a selective reflection of the laser wave by the film of TFMHPDOPB. This result seems to be very interesting and particularly revealing of the two effects contributing to the resonant character of circular dichroism in liquid crystals. Experiments on circular dichroism can also be undertaken by dissolving molecules of a dichroic dye in a liquid crystal phase. In this case, the liquid crystal phase induces a helical arrangement of the dye molecules in the material, thereby giving rise to circular dichroism.

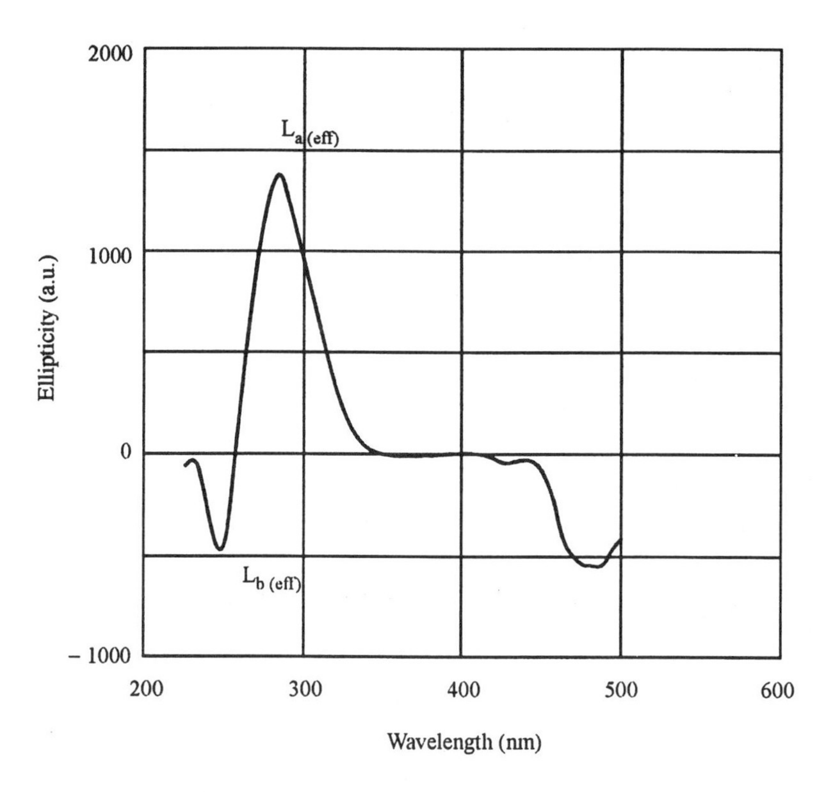

Figure 5.4.3: Circular dichroism of the liquid crystal R-TFMHPDOPB [from A. Fukuda and coll., Liquids Crystals, vol. 18, 2, 239 (1995), reproduced with the kind authorization of the editor].

V.5. Exercises and problems

V.5.1. Rotatory polarization (from Agreg. C; B; 1987)

Conventions concerning the orientation: In this section, the light always propagates along the Oz axis in the direction of increasing z. Rotations around Oz and the corresponding angles are counted positively if they take place in the positive direction around $\mathbf{u}_z$ (that is, for an observer who receives the light beam in his eye).

We call "left rotation" a rotation in the positive direction and "right rotation" a rotation in the negative rotation.

1. A quartz plate, of thickness e and whose parallel faces ($z = 0$; $z = e$) are cut perpendicularly to the optical axis, is placed perpendicularly to the Oz axis.

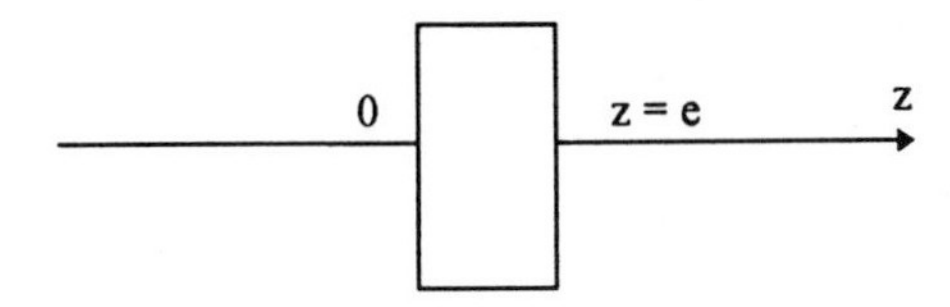

A monochromatic plane wave, propagating in the positive direction along the Oz axis, is linearly polarized along the Ox axis at the entrance of the plate:

$$z = 0 : \mathbf{E} = \mathbf{u}_x E_0 \cos \omega t$$

We make the following hypotheses using Fresnel's theory:

— Quartz, an optically active material, transmits right and left circularly-polarized vibrations without change;

— Quartz has refractive index n_R for right circular vibrations and refractive index n_L, different from n_R, for left circular vibrations.

a. Show that the wave is still linearly polarized after crossing the plate, but that the direction of the vibration has turned by an angle α around Oz.

b. By using the sign convention indicated in the beginning of this section, determine the algebraic values of the angle α and of the rotatory power $\rho = \alpha/e$ as functions of λ_0, e, n_R and n_L.

c. What circular vibration (right or left?) propagates more rapidly in a dextrorotatory quartz plate? If n'_R and n'_L are the refractive indices of a dextrorotatory quartz and n''_R and n''_L those of a levorotatory quartz, what are the relations linking n'_R, n'_L, n''_R and n''_L ?

2. Now we insert a plate cut perpendicularly to the optical axis in a quartz crystal and of thickness e = 6 mm between initially crossed Nicol polarizers P and A.

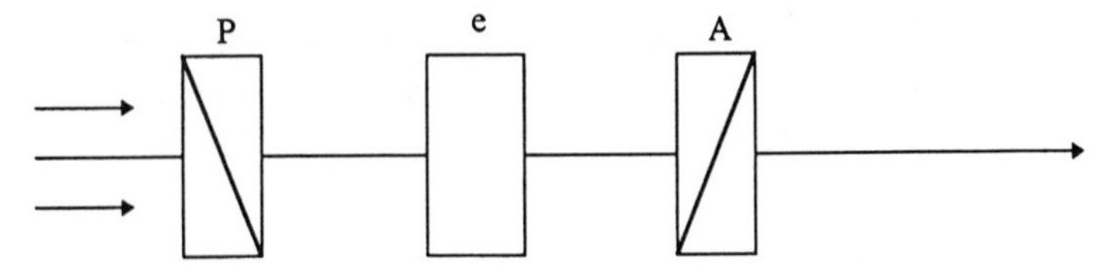

The whole is illuminated by a plane wave of wavelength $\lambda_0 = 0.589$ nm. It is observed that extinction can be restored by turning the analyzer by 49.7° towards the left (that is, in the positive direction around $\mathbf{u}_z$). Determine the value of α knowing that the quartz crystal is dextrorotatory and that the absolute value of angle α lies between 0 and π. Derive the value of the refractive index difference $\Delta n = n_R - n_L$ at wavelength λ_0 for a quartz crystal.

3. Let us now consider two coupled prisms.

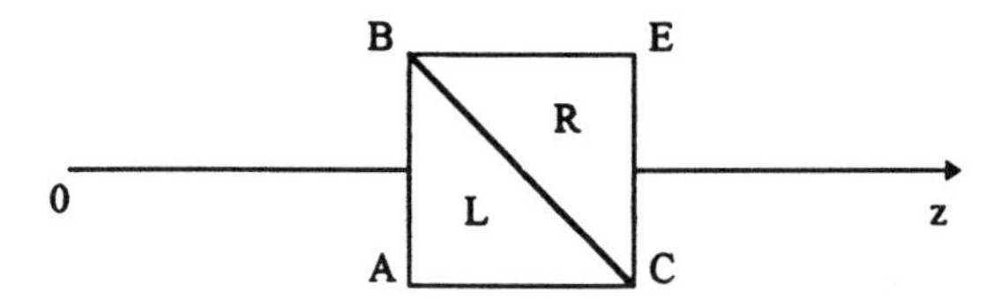

ABC is a left quartz prism (levorotatory); BCE is a right quartz prism (dextrorotatory). The optical axes of these two quartz prisms are perpendicular to the entrance and exit faces of the resulting plate, and lie along the Oz axis.

The plate is placed between parallel polarizer P and analyzer A and the whole entrance face of the plate is illuminated by a monochromatic plane wave. The emergent wave falls on a screen.

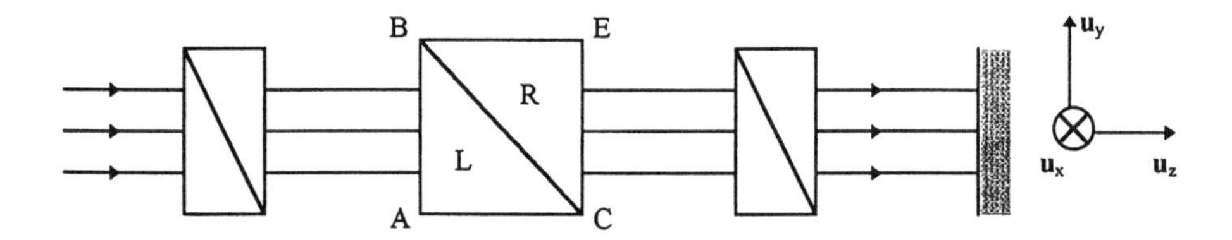

a. Show that parallel fringes appear on the screen. Specify their direction. What is the nature (dark or bright) of the central fringe?

b. Analyzer A is rotated towards the left around Oz. In what direction do the fringes move? How is the system of fringes modified when P and A are crossed?

c. Determine the distance between two consecutive bright fringes on the screen.

Numerical data: ABC = BCE = 45°; the rotatory power of quartz is $|\rho| = 21.7° \text{ mm}^{-1}$ for the wavelength used.

4. Rotational dispersion.

4.1. Biot's approximate law.

The rotatory power $\rho = \alpha/e$ varies very rapidly with λ_0. In quartz, in a first approximation, the variation follows the relation $\rho = A/\lambda_0^2$, where A is an algebraic constant (Biot's law). $|\rho| = 21.7°\text{mm}^{-1}$ for $\lambda_0 = 0.589$ µm (yellow sodium D line).

a. How does the difference between the refractive indices $\Delta n = n_R - n_L$ vary as a function of λ_0 in the case of quartz?

b. What must the thickness of a quartz plate be in order to give rise to a rotation $\alpha = \pi/2$ at the wavelength $\lambda_0 = 0.468$ µm (blue cadmium line)?

c. The plate of 4.1.b is placed between parallel polarizer and analyzer and is illuminated by white light. The transmitted light appears orange. Explain.

4.2. Spectrum with striations (or "banded spectrum")

A quartz plate C, of thickness e = 3.00 cm and which has been cut perpendicularly to the optical axis, is placed between parallel polarizer P and analyzer A. The whole is illuminated by a parallel beam of white light consisting of all the radiations lying between 0.4 and 0.8 µm. The emergent beam is focused on the slit F of a spectroscope.

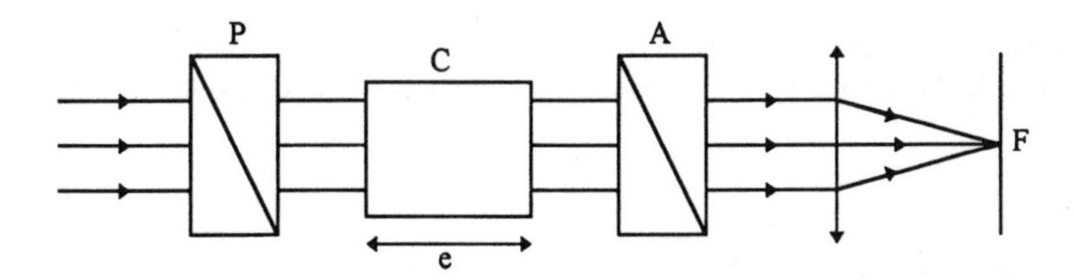

a. The observed spectrum shows striations. Explain why. Determine the wavelengths of the extinguished radiations and those of the radiations corresponding to the intensity maximum.

b. The polarizer P and the plate being fixed, the analyzer is turned towards the left about the Oz axis. The striations change from purple to red. Deduce whether the quartz sample is dextrorotatory or levorotatory. What does one observe when the plate is turned towards the left about Oz, the polarizer and the analyzer staying fixed? What is observed when the polarizer is turned towards the left about the Oz axis, the plate and the analyzer staying fixed?

1.a. We use Fresnel's kinetic theory. At the entrance of the plate, the linear vibration of amplitude $E_{x0} = 2a$ is decomposed into a levorotatory and a dextrorotatory circular vibration written as $(x_L = a\cos\omega t \; y_L = a\sin\omega t)(x_R = a\cos\omega t \, y_R = -a\sin\omega t)$.

At the exit, phase changes φ_L and φ_R are applied, respectively, to the two circular vibrations and we obtain

$$x' = x'_L + x'_R = E_{x0}\cos\left[\omega t + (\varphi_L + \varphi_R)/2\right]\cos\left[(\varphi_L - \varphi_R)/2\right]$$
$$y' = y'_L + y'_R = E_{x0}\cos\left[\omega t + (\varphi_L + \varphi_R)/2\right]\sin\left[(\varphi_L - \varphi_R)/2\right]$$

These equations describe a linear vibration which has rotated (it no longer lies along the Ox axis).

b. We write $\mathrm{tg}\alpha = y'/x' = \mathrm{tg}\left[(\varphi_L - \varphi_R)/2\right]$ and the rotation is

$$\rho = \alpha / e = \pi\,(n_L - n_R)/\lambda_0.$$

c. In a dextrorotatory quartz, the right-turning vibration travels with the greatest velocity. We find $n'_R = n''_L$ and $n'_L = n''_R$. As to the rotatory powers, they are opposite to each other.

2. We find $\alpha = 130.3°$; $\Delta n = 7.1 \times 10^{-5}$.

3.a. The fringes are straight and parallel to the direction of the vibration allowed by both the polarizer and the analyzer. The central fringe corresponds to an exact compensation of the rotations and it is bright.

b. If we turn the analyzer towards the left, the central bright fringe corresponds to a global left rotation, i.e. to a longer optical path in the left quartz. It moves in the negative y direction (towards the lower part of the figure). All fringes move in the same direction. When P and A are crossed, we observe the complementary fringe system, so the central fringe is dark.

c. Let us call i the distance between fringes. The path difference between the L and the R quartz plates is 2i. The interfringe distance corresponds to a phase shift of π so we have $2i\rho = \pi$. Therefore the interfringe distance is $i = 4.15$ mm.

4.1.a. Δn varies as $\lambda_0\rho$, i.e. as A/λ_0.

b. We have $\rho_B = \rho_Y(\lambda_Y/\lambda_B)^2$ $e = (\pi/2\rho_Y)(\lambda_B/\lambda_Y)^2 \approx 2.62$ mm. (B: Blue; Y: Yellow).

c. We observe an orange coloration because orange is the complementary color of blue which was extinguished by the plate.

4.2.a. Those radiations whose wavelengths λ satisfy the relation $\rho(\lambda)e = (2p+1)(\pi/2)$ are extinguished. Those satisfying the relation $\rho(\lambda)e = p\pi$ reach the spectroscope. p is an integer and the rotatory power is written as $\left[\rho(\lambda) = 21.7(0.589/\lambda)^2\right]°.\text{mm}^{-1}$. In this relation, the wavelength is expressed in microns. *Numerical application:* The 6 wavelengths corresponding to the values 2, 3, 4, 5, 6 and 7 of parameter p (i.e. the 6 wavelengths 0.408, 0.43, 0.477, 0.52, 0.598 and 0.7 micrometer) are extinguished. The 6 bright lines, corresponding to the same values of p, are those of wavelengths 0.79, 0.645, 0.56, 0.5, 0.456 and 0.42 micrometer. We observe a rainbow with six dark striations or stripes.

b. The change of the color from purple to red indicates that the rotation introduced by quartz is smaller, i. e. that quartz is dextrorotatory. If the quartz plate is turned towards the left, we observe no change (isotropic property around the Oz axis). If the polarizer is turned towards the left, the rotation must be increased in order to extinguish the light and the striations move towards the purple.

V.5.2. Circular dichroism (from "licence"; Paris; 1965)

By using the eye as a receiver, we want to study the polarization states of a plane monochromatic light wave produced by sodium (λ = 0.589 μs) which crosses a vessel containing an absorbing liquid with a natural rotatory power.

1.a. In a first experiment, light is linearly polarized by a polarizer P and the azimuth of the vibration is determined before and after crossing a cell C of thickness 0.5 cm. Knowing that the vibration turned clockwise (for the observer) by an angle α = 1.2°, and remembering that the rotatory power is due to circular birefringence, calculate the algebraic value of the difference between the refractive indices of the right and the left circular vibrations for the above defined wavelength.

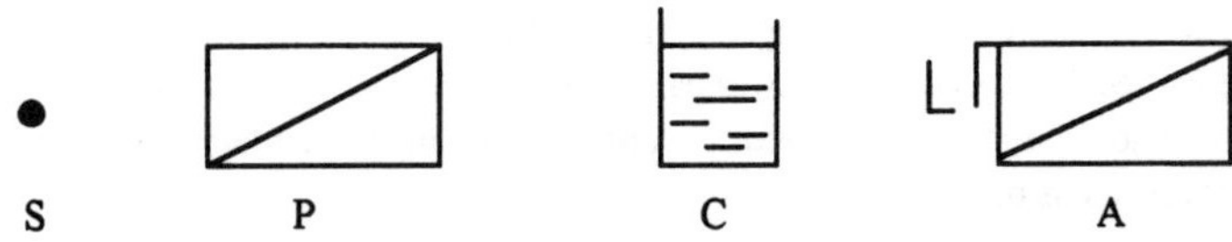

b. The orientation of the vibration is measured by means of a half-shadow analyzer consisting of a half-wave plate L, rigidly bound to a Glazebrook prism A and covering half the light beam. The transmission direction of this prism makes a small angle ε with one of the neutral lines of L. The whole device can be turned by a set of known angles. Explain how the device works.

2.a. In a second set of experiments, we want to obtain circularly polarized light. For this, a linear polarizer (for example a Glazebrook prism) and a quarter-wave plate with known neutral line directions can be used. Indicate briefly how left-polarized light, then right-polarized light, can be produced.

b. What is the thickness of a quarter-wave plate made of mica (for this crystal, the principal refractive indices in the cleavage plane are n = 1.5977 and n' = 1.5936 for sodium light)?

c. By using a receptor sensitive to the intensity of light, it is observed that a 1 mm thick cell, filled with the above described liquid, transmits 0.520 of left circularly-polarized light intensity while a similar but 2 mm thick cell transmits only 0.320. Calculate the absorption coefficient of the liquid, expressed in cm^{-1}. Why should measurements be taken for two cells of different thickness?

Same question for right circularly-polarized light. The fractions transmitted by the two cells are now equal to 0.503 and 0.301. Remember that in a homogeneous absorbing substance, the relative loss undergone by a monochromatic light flux F when crossing a layer of thickness dx is $-\,dF/F = K\,dx$, where K is the absorption coefficient.

3. Calculate the amplitude reduction factors of left and of right circular vibrations which cross a liquid of thickness 0.5 cm. Show that the emergent vibration is an ellipse if the liquid is illuminated by linearly polarized light, and determine the ratio of the axes of this ellipse.

4. The measurement on the rotatory power of the liquid described in question 1 is repeated. Then, without modifying the half-shadow analyzer, a quarter-wave plate whose neutral lines are parallel to those of the half-wave plate is placed between the cell and the analyzer. By what

angle should the half-shadow analyzer be turned in order to re-establish the equality between the brightness of the two parts of the field? Explain the result.

5. The absolute uncertainty on the positioning of a vibration, performed with the eye and a half-shadow analyzer, varies as the inverse of the square root of the light flux received by the analyzer. What thickness of the active and absorbing liquid minimizes the relative uncertainty on the measurement?

1.a. We write (*cf.* Problem V.4.1) $\alpha / e = \pi(n_R - n_L)/\lambda$. The rotation is dextrorotatory and θ is positive. n_R is smaller than n_L. We find $n_L - n_R = 7.85 \times 10^{-7}$.

b. The part of the incident linear vibration which travels through the half-wave plate comes out as a vibration which is symmetrical to the neutral lines of the plate. For the brightness of the two halves of the field to be equal, the vibration must be parallel to one of the neutral lines of the plate. The intensity varies as ε^2, so it is very weak. This means that the eye works in relative darkness, which optimizes its sensitivity to intensity differences between the two parts of the field.

2.a. A levorotatory (dextrorotatory) vibration is obtained by placing the fast neutral line, corresponding to the smallest refractive index, at + (–) 45 degrees from the incident polarization direction.

b. The relation $e\Delta n = (2p+1)(\lambda/4)$ must be satisfied, meaning that the thickness of the plate must be equal to an uneven multiple of 35.91 μm.

c. We write

$$(F_2/F_1) = \exp[-K(x_2 - x_1)]$$
$$K = [1/(x_2 - x_1)] \mathrm{Ln}[(F_1/F_0)/(F_2/F_0)].$$

We find K = 4.85 (5.13) cm^{-1} for a levorotatory (dextrorotatory) vibration. Two measurements are needed so as to eliminate the integration constant of the differential equation.

3. The amplitude reduction coefficient is K' = K/2. We find

for the left circular vibration: $K'_L = 0.297$

for the right circular vibration: $K'_R = 0.277$

Using relation 5.2.23, we obtain :

$$\frac{E_x}{E_y} = \frac{E'_L + E'_R}{E'_L - E'_R} = \frac{K'_L + K'_R}{K'_L - K'_R} = \frac{0.297 + 0.277}{0.297 - 0.277} = 28.7$$

The fact that the left and right circular vibrations are absorbed differently only modifies the nature of the emergent vibration, without affecting the rotation which depends only on the difference between the phase velocities of left and right circular vibrations in the medium.

4. Before insertion of the ¼ wave plate, the half-shadow analyzer points towards the small axis of the ellipse. The neutral lines are therefore parallel to the axes of the ellipse. So at the exit of the ¼ wave plate we obtain a linear vibration which is parallel to the diagonal of the rectangle containing the ellipse. The rotation of the analyzer is equal to

$$\theta \approx 1/28.7 = 0.035\text{rd} = 2° .$$

The absolute uncertainty $\Delta\theta$ varies as the inverse of the square root of the intensity, i.e. as exp (Kl/2) where l is the length of the sample. The angle θ varies as l. The relative uncertainty therefore varies as [exp (Kl/2)]/l. Its minimal value corresponds to $l = 2/K \approx 2/5 = 0.4$ cm.

V.5.3. Inversion kinetics of saccharose measured by polarimetry (from CAPES; P; 1991)

We want to study the reaction kinetics of saccharose inversion at room temperature.

$$\underset{\text{saccharose (S)}}{C_{12}H_{22}O_{11}} + H_2O \rightarrow \underset{\text{D-glucose (G)}}{C_6H_{12}O_6} + \underset{\text{D-fructose (F)}}{C_6H_{12}O_6}$$

1. Since the three substances S, G and F are optically active, the reaction kinetics will be studied using the technique of polarimetry.

a. Define the rotatory power of an optically active substance.

b. Give an example of such a substances in both organic and inorganic chemistry.

2. We shall need Biot's law which gives the algebraic rotatory power ρ of such a substance $\rho = [\alpha]\ lc$ (this law is additive for a mixture). $[\alpha]$ is the specific rotatory power of the substance, c is the mass concentration of the substance in solution (expressed in $g.cm^{-3}$) and l is the length cf substance crossed by the light (in dm).

a. What do we call a substance for which $[\alpha] > 0$? $[\alpha] < 0$?

b. What factors can influence $[\alpha]$?

c. Parameter α is measured with the help of a Laurent polarimeter. Using just a simple diagram, explain how this device works.

The reaction is an irreversible hydrolysis. It is also a slow reaction, catalyzed by H_3O^+ ions.

3.a. At time t = 0 a volume of 25.0 cm^3 of an aqueous solution of substance S at the concentration of 200 $g.l^{-1}$ is mixed with 25.0 cm^3 of a hydrochloric acid solution with a concentration of 4 $mol.l^{-1}$. Knowing that M_S = 342 $g.mol^{-1}$ and $M_G = M_F$ =180 $g.mol^{-1}$, calculate the initial concentrations $[S]_0$ and $[H_3O^+]_0$ in the mixture.

b. The following numerical data are provided: $[\alpha]_S = +66.6°$ $[\alpha]_F = -92.0°$ $[\alpha]_G = +52.5°$. The length of the polarimetric tube is l = 2.0 dm.

1. Express the value of the rotatory power α_0 of the solution at time t = 0 literally, and calculate it numerically.

2. Express literally the value of the rotational power α_∞ of the reacting mixture after a very long (i.e. quasi infinite) reaction time, and calculate it numerically.

c. Define the volumic velocity v of the reaction with respect to substance S.

d. We want to check that the reaction is of order 1 with respect to substance S. The velocity can be written as $v = k\ [S]_t$, where k denotes the velocity constant of the reaction in the given experimental conditions and $[S]_t$ the concentration of S at time t.

1. Derive the expression of $[S]_t$ as a function of k, t and $[S]_0$.

2. Expressed the rotatory power α_t of the solution at time t, then show that

$$\ln [(\alpha_0 - \alpha_\infty) / (\alpha_t - \alpha_\infty)] = kt$$

3. The following set of measurements were obtained:

α_t (0)	10.0	7.5	5.7	4.1	2.8	1.7	0.8
t (min)	5	10	15	20	25	30	35

Check that the experimental results are compatible with a velocity of order 1. Deduce the value of k from these results. What is the half-reaction time of this reaction?

e. In fact, k is an just an apparent constant and should be written as $k = k'[H_3O^+]_t$. Why do the experimental conditions justify the use of such an apparent constant?

1.a. *cf.* Section V.2.1.

b. Example in organic chemistry: Oses (such as saccharose, glucose etc.).
Example in inorganic chemistry: Some kinds of complex ions.

2.a., b. and c. *cf.* Section V.2.1.

3.a. $[S]_0 = 100 / 342 = 0.292\,\text{mol.l}^{-1}$ $[H_3O^+]_0 = 2\,\text{mol.l}^{-1}$

b.1. $\alpha_0 = 2 \times 100 \times 10^{-3} \times 66.6 = 13.3°$.

2. After an infinite time, the mixture contains only substances G and F (molar mass 180 g) at the concentration of $0.292\ \text{mol.l}^{-1}$.

$$\alpha_\infty = 2 \times 0.292 \times 180 \times 10^{-3}(52.5 - 92.0) = -4.15°.$$

c. $v = -d[S]/dt$

d.1. $-d[S]/dt = k[S]$. We obtain $[S] = [S_0]\exp(-kt)$ after integration.

2. The rotatory power at time t is expressed as

$$\alpha_t = l\{\alpha_S[S] + (\alpha_G + \alpha_F)([S_0] - [S])\}$$

so

$$\alpha_t = l[S_0]\{\alpha_S \exp(-kt) + (\alpha_G + \alpha_F)[1 - \exp(-kt)]\}$$

and, with the expressions

$$\alpha_0 = l[S_0]\alpha_S \text{ and } \alpha_\infty = l[S_0](\alpha_G + \alpha_F)$$
$$\alpha_t = \alpha_0 \exp(-kt) + \alpha_\infty[1 - \exp(-kt)]$$

the rotatory power varies with time as

$$\ln[(\alpha_0 - \alpha_\infty)/(\alpha_t - \alpha_\infty)] = kt$$

3. A linear adjustment using the least squares method and performed with a scientific pocket calculator confirms the order 1 and provides the value 0.035 mn^{-1} for constant k. The corresponding half-reaction time is $\ln 2/k \approx 19.9$ mn.

e. The relative variation of $[H_3O^+]$ is at most 7 %. One can assume that $[H_3O^+]$ remains constant during the reaction (degeneracy of the order). The use of an apparent constant is therefore justified.

V.6. Appendices

V.6.1. True or pseudotensor?

Let us consider the matrix $(M) = \begin{pmatrix} -1 & 0 & 0 \\ 0 & -1 & 0 \\ 0 & 0 & -1 \end{pmatrix}$ representing a space inversion in the orthonormed basis Oxyz. If the spatial inversion changes the sign of each component of a vector V as follows

$$\{V'\} = (M)\{V\}$$

the vector is called a *true vector*.

Now, let us consider for example the particular vector arising from a vectorial product and defined by $V = V_1 \times V_2$. If (x_1, y_1, z_1) and (x_2, y_2, z_2) represent the components of V_1 and V_2, respectively, the components $(y_1z_2 - y_2z_1,\ z_1x_2 - z_2x_1,\ x_1y_2 - x_2y_1)$ of the vectorial product remain unaltered under space inversion, and the transformation of vector V can be written as

$$\{V'\} = -1(M)\{V\} = J_{(M)}(M)\{V\}$$

where $J_{(M)} = -1$ is the Jacobian of the transformation, i.e. the determinant constructed with the elements of the matrix representing the transformation in the chosen basis.

Vectors which change under space inversion according to the above expression using the Jacobian of the transformation, are called *pseudovectors*. This definition is generalized to tensors of all ranks and, in particular, to scalars which are zero-rank tensors.

Under the effect of a space inversion, *true* entities changes as $(-1)^N$, where N is the rank of the tensor, whereas *pseudo* entities transform as $(-1)^{N+1}$.

V.6.2. Levi-Civita tensor

The Levi-Civita tensor is a tensor of rank three exclusively defined by the value of its components which are independent of the chosen coordinates:

$\varepsilon_{ijk} = \pm 1$ if indices i, j and k are all different. If, starting from the initial ordering of the subscripts i, j, k, an even number of permutations must be carried out to obtain ijk, then $\varepsilon = +1$; if the number of permutations is uneven, $\varepsilon = -1$.

$\varepsilon_{ijk} = 0$ if at least two of the indices are equal.

Let us perform a space inversion which transforms frame (α, β, γ) into frame (i, j, k). Let us calculate the component ε_{123} in this last frame. If $c_{i\alpha}$, $c_{j\beta}$ and $c_{k\gamma}$ are the directional cosines associated to the transformation we can write $\varepsilon_{ijk} = c_{i\alpha} c_{j\beta} c_{k\gamma} \varepsilon_{\alpha\beta\gamma}$ or $\varepsilon_{123} = c_{1\alpha} c_{2\beta} c_{3\gamma} \varepsilon_{\alpha\beta\gamma}$. The only nonzero directional cosines are $c_{11} = c_{22} = c_{33} = -1$ and we have $\varepsilon_{123} = -\varepsilon_{123} = -1$. But ε_{123} must by definition be equal to -1 in all frames. Consequently, we need to use the Jacobian (-1) of the transformation (*cf.* Appendix V.6.1) and write $\varepsilon_{ijk} = (-1) c_{i\alpha} c_{j\beta} c_{k\gamma} \varepsilon_{\alpha\beta\gamma}$. The Levi-Civita tensor (rank 3) transforms under space inversion as $(-1)^{3+1}$. It is uneven, therefore it is an antisymmetrical pseudotensor.

The Levi-Civita tensor has some properties which are very often used.

The vectorial product of two Levi-Civita tensors (obtained by multiplying each component of one tensor by each component of the other) is a true tensor of rank 6 which can be written as

$$\varepsilon_{ijk} \cdot \varepsilon_{\alpha\beta\gamma} = \begin{vmatrix} \delta_{i\alpha} & \delta_{j\alpha} & \delta_{k\alpha} \\ \delta_{i\beta} & \delta_{j\beta} & \delta_{k\beta} \\ \delta_{i\gamma} & \delta_{j\gamma} & \delta_{k\gamma} \end{vmatrix}$$

Contractions — carried out by successively multiplying the vectorial product by the adequate Kronecker symbols — lead to

1st contraction: $(\varepsilon_{ijk} \cdot \varepsilon_{\alpha\beta\gamma})\delta_{i\alpha} = \delta_{j\beta}\delta_{k\gamma} - \delta_{j\gamma}\delta_{k\beta}$ tensor of rank 4

2nd contraction: $(\varepsilon_{ijk} \cdot \varepsilon_{\alpha\beta\gamma})\delta_{i\alpha}\delta_{j\beta} = 2\delta_{k\gamma}$ tensor of rank 2

3rd contraction: $(\varepsilon_{ijk} \cdot \varepsilon_{\alpha\beta\gamma})\delta_{i\alpha}\delta_{j\beta}\delta_{k\gamma} = 6$ tensor of rank 0 or scalar

V.6.3. Optical activity tensor (antisymmetrical tensor relative to the two first indices)

1. **Triclinic**
   Class 1 (C_1): 18 nonzero components, 9 of which are independent.

   xyy = – yxy, yzz = – zyz, xyz = – yxz
   yxx = – xyx, zxx = – xzx, xzy = – zxy
   xzz = – zxz , zyy = – yzy, yzx = – zyx

   Class $\bar{1}$ (C_i): All components vanish.

2. **Monoclinic**
   Class 2 (C_2): 10 nonzero components, 5 of which are independent.

   zxx = – xzx, zyy = – yzy
   xyz = – yxz , xzy = – zxy, yzx = – zyx

   Class m (C_S): 8 nonzero components, 4 of which are independent.

   xyy = – yxy, yxx = – xyx
   xzz = – zxz, yzz = – zyz

   Class 2/m (C_{2h}): All components vanish.

3. **Orthorhombic**
   Class 222 (D_2): 6 nonzero components, 3 of which are independent.

   xyz = – yxz , xzy = – zxy, yzx = – zyx

   Class 2mm (C_{2v}): 4 nonzero components, 2 of which are independent.

   zxx = – xzx, zyy = – yzy

   Class mmm (D_{2h}): All components vanish.

4. **Quadratic**
   Class 4 (C_4): 10 nonzero components, 3 of which are independent.

   zxx = – xzx = zyy = – yzy
   xyz = – yxz
   xzy = – zxy = zyx = – yzx

   Class $\bar{4}$ (S_4): 8 nonzero components, 2 of which are independent.

   zxx = – xzx = yzy = – zyy
   xzy = – zxy = yzx = – zyx

   Class 4/m (C_{4h}): All components vanish.

   Class 422 (D_4): 6 nonzero components, 2 of which are independent.

   xyz = – yxz
   xzy = – zxy = zyx = – yzx

   Class 4mm (C_{4v}): 4 nonzero components, 1 of which is independent.

   zxx = – xzx = zyy = – yzy

   Class $\bar{4}$ 2m (D_{2d}): 4 nonzero components, 1 of which is independent.

   xzy = – zxy = yzx = – zyx

   Class 4/mmm (D_{4h}): All components vanish.

5. **Rhombohedral**
   Class 3 (C_3): 10 nonzero components, 3 of which are independent.

   zxx = – xzx = zyy = – yzy
   xyz = – yxz
   xzy = – zxy = zyx = – yzx

   Class $\bar{3}$ (S_6): All components vanish.

Class 32 (D_3): 6 nonzero components, 2 of which are independent.

xyz = – yxz
xzy = – zxy = zyx = – yzx

Class 3m (C_{3v}): 4 nonzero components, 1 of which is independent.

zxx = – xzx = zyy = – yzy

Class $\bar{3}$ m (D_{3d}): All components vanish.

6. **Hexagonal**
Class 6 (C_6): 10 nonzero components, 3 of which are independent.

zxx = – xzx = zyy = – yzy
xyz = – yxz
xzy = – zxy = zyx = – yzx

Class $\bar{6}$ (C_{3h}): All components vanish.

Class 6/m (C_{6h}): All components vanish.

Class 622 (D_6): 6 nonzero components, 2 of which are independent.

xyz = – yxz
xzy = – zxy = zyx = – yzx

Class 6mm (C_{6v}): 4 nonzero components, 1 of which is independent.

zxx = – xzx = zyy = – yzy

Class $\bar{6}$ 2m (D_{3h}): All components vanish.

Class 6/mmm (D_{6h}): All components vanish.

7. **Cubic**
Classes 23 (T) and 432 (0) : 6 nonzero components, 1 of which is independent.

xyz = – yxz = yzx = – zyx = zxy = – xzy

Classes m3 (T_h), m3m (O_h) and $\bar{4}$ 3m: All components vanish.

8. **Isotropic medium**
Classes I and K: 6 nonzero components, 1 of which is independent.

xyz = – yxz = yzx – zyx = zxy = – xzy

Classes I_h and K_h : All components vanish.

9. **Axially symmetric body**
Class ∞ (C_∞): 10 nonzero components, 3 of which are independent.

zxx = – xzx = zyy = – yzy
xyz = – yxz
xzy = – zxy = zyx = – yzx

Class ∞m ($C_{\infty v}$): 4 nonzero components, 1 of which is independent.

zxx = – xzx = zyy = – yzy

Classes ∞/m ($C_{\infty h}$) and ∞/mm ($D_{\infty h}$): All components vanish.

V.7. Bibliography

5.7.1. An introduction can be found in:

HENDERSON G. L. — Optical Rotary Dispersion, *J. of Chem. Educ.*, 45, 515 (1968).

JONES L.L. and EYRING H. — A Model for Optical Rotation, *J. of Chem. Educ.*, 38, 601 (1961).

5.7.2. Optical activity is described in:

BORN M. — *Optik*, Springer Verlag, Berlin (1933).

KAUZMANN W. — *Quantum Chemistry*, Academic Press, New York (1957).

KUHN W. — *Stereochemistry*, part 1, FREUDENBERG K. Editor, Deuticke, Leipzig (1932).

5.7.3. The role played by symmetry in chemistry can be found in:

JAFFE H.H. and ORCHIN M. — *Symmetry in Chemistry*, J.Wiley (1965).

5.7.4. Fumi's method can be studied, for example, in:

NYE J.F. — *Propriétés physiques des cristaux*, Dunod (1961).

5.7.5. Fundamentals of optical properties of liquid crystal can be found in:

CHANDRASEKHAR S. — *Liquid Crystals*, 2sd edition, Cambridge University Press (1992).

DE VRIES H. — Rotatory Power and Other Optical Properties of Certain Liquid Crystals, *Acta Cryst.* 4, 219-226 (1951).

5.7.6. A demonstration of the antisymmetry of the optical activity tensor relative to the first two indices can be found in:

LANDAU L. and LIFCHITZ E.— *Statistical Physics*, Mir (1967).

Chapter VI

Photoelectron absorption spectroscopy

VI.1. Principles

Photoelectron spectroscopy is the equivalent of the photoelectric effect in the case of a molecule. It seems therefore quite natural to adopt the phenomenological approach developed by Einstein (*cf.* Sections II.1 and IV.3.1.2) for its description. Under the impact of a photon whose energy hν lies in the UV-visible domain (about twenty electron-volts), the molecule, assumed to be in its fundamental vibrational energy state before the interaction, releases a photoelectron from one of its outer energy states, which electron then has a velocity v. Its kinetic energy $(1/2)m_e v^2$ (within the non relativistic approximation) is of the order of 1 to 10 eV. The molecule forms a cation which generally lies in an excited vibrational energy state

As we will see in the subsequent sections, the measure of the kinetic energy of the ejected photoelectron (which is not subject to any auxiliary potential at the time of its ejection) will give information about the energy of the electron as it was within the molecule.

From an historical point of view, we see that, paradoxically, more than half a century was needed before this technique saw the day. Indeed, this spectroscopy became popular only in 1960, mostly through the work of Turner. Also, if the energy of the incident photons is chosen to match the core energy levels, simultaneous ejection of *two photoelectrons* may be observed (the so-called *Auger spectroscopy*). Furthermore, it is also quite possible to use *X photons*, and this spectroscopy can then be used for chemical analysis purposes (ESCA: *Electron Spectroscopy for Chemical Analysis*). In the following, we shall limit ourselves to the description of photoelectron spectroscopy of which the starting process is described on Figure 6.1.1.

The energy of the electrons is purely kinetic, and since their velocities are

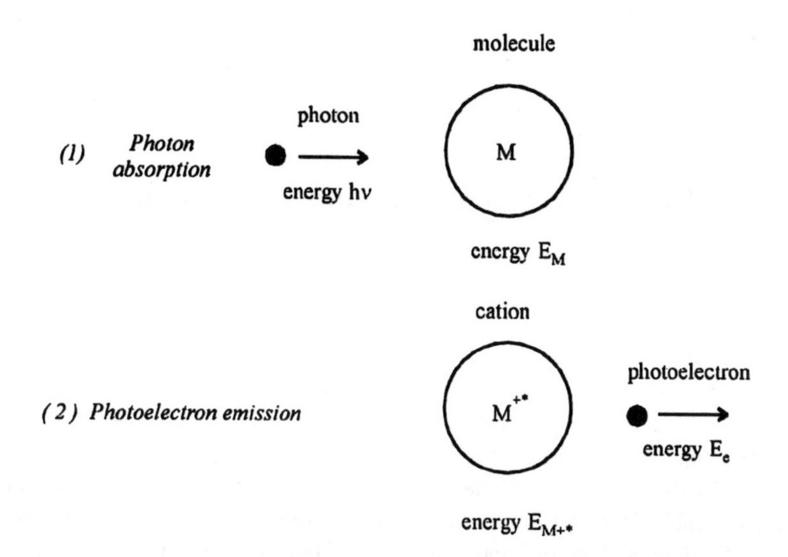

Figure 6.1.1: Schematic illustration of the photoelectron spectroscopy principle. M: molecule in its ground energy state; $M^{+\bullet}$: ion in an excited vibrational state; hν and E_e, respectively stand for the energy of the incident photon and of the ejected photoelectron; E_M and $E_{M+\bullet}$ are the total energies of the molecule and of the ion.

small compared to the speed of light, we will use the non relativistic approximation. Assuming an elastic shock, the energy conservation will be expressed as

$$h\nu + E_M = E_e + E_{M+\bullet} \tag{6.1.1}$$

The energy difference $E_{M+\bullet} - E_M$ is equal to the difference between the potential energies of the ionized and the non ionized species in their fundamental vibrational energy state, which we will call the *first ionization potential* (P_i), increased by the vibrational excitation energy of the ion ($E_{v'}$) and by the kinetic energy difference (ΔT) between the ion and the molecule:

$$E_{M+\bullet} - E_M = P_i + E_{v'} + \Delta T \tag{6.1.2}$$

and Equation 6.1.1 becomes

$$P_i = h\nu - E_e - E_{v'} - \Delta T \tag{6.1.3}$$

ΔT is usually small compared to E_e and $E_{v'}$. Indeed, the conservation of momentum during the elastic shock is expressed as

$$(h\nu / c)\mathbf{u} \approx m_e \mathbf{v}_e + M\Delta \mathbf{v}_M \tag{6.1.4}$$

M being the average mass of the molecule and of the ion, $\mathbf{v_e}$ the velocity vector of the ejected electron and **u** the unit vector defining the direction of the incident photon. Typically, $h\nu \approx 20$ eV and $E_e \approx 5$ eV, $h\nu/c \approx 10^{-26}$ Kg.m.s^{-1} and $m_e v_e \approx 1{,}6\times10^{-25}$ Kg.m.s^{-1}. The momentum of the incident photon being very small compared to that of the ejected photoelectron, we can write

$$\Delta T / E_e \approx \Delta v_M / v_e \approx m_e / M \qquad (6.1.5)$$

This ratio being at most equal to 1/1850 in the case of the hydrogen atom and ΔT remaining smaller than 2.5×10^{-3} eV, $\Delta T/E_e$ can be neglected before E_e. In addition, we will see later that for polyatomic ions, the order of magnitude of the vibrational energies $E_{v'}$ is about one tenth of an electron-volt and ΔT, whose value is inversely proportional to M, will also be small compared to $E_{v'}$. Therefore, Equation 6.1.3 takes the simplified form

$$P_i \approx h\nu - E_e - E_{v'} \qquad (6.1.6)$$

This equation is the basic equation of photoelectron spectroscopy. Indeed, the *knowledge* of the energy of the incident photon and the *measurement* of the kinetic energy of the ejected photoelectrons provide a *determination* of parameters P_i and $E_{v'}$. This determination being the major contribution of this type of spectroscopy.

Before delving more precisely into the determination of P_i et $E_{v'}$, let us quickly sketch the experimental techniques that are used.

VI.2. Experimental techniques

Figure 6.2.1 describes the principle of the spectrometer most commonly used.

The spectrometer is mainly composed of a source of very monochromatic radiation and of a device measuring the velocity of the ejected photoelectrons.

VI.2.1. Radiation sources

We have already indicated that the energy of the incident photons used in such experiments usually measures about twenty electron-volts. Such radiation is generally produced either by discharges between the anode and cathode of a lamp filled with ultra pure gas, or by using the carefully filtered radiation produced by particle accelerators. In this last case, the available intensities are considerable, allowing the detection of a very weak signal and therefore suitable for the investigation of a great number of molecules.

Table 6.1.1 specifies the nature of some frequently used sources together with some properties of the radiation that they produce.

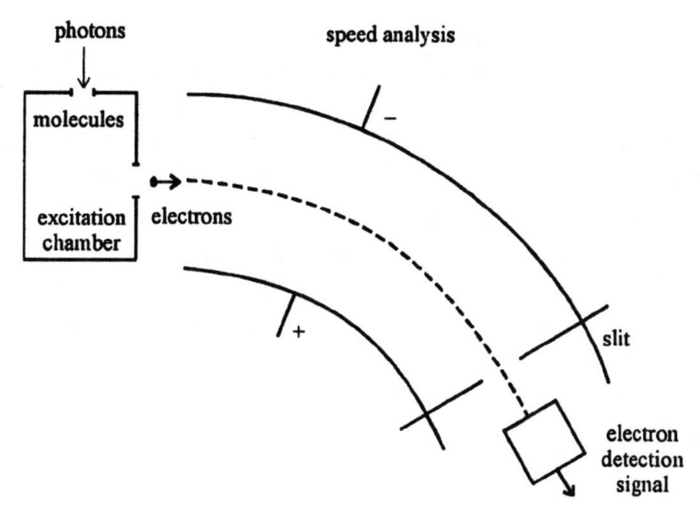

Figure 6.2.1: Principle of a photoelectron absorption spectrometer.

Table 6.1.1 — **Main source types used in photoelectron absorption spectroscopy.**

Source nature	Source type	Wavelength (nm)	Photon energy (eV)
He discharge	continuous	58.4	21.22
He^+ discharge	continuous	30.4	40.81
Ne discharge	continuous	74.4 and 73.6	16.7 and 16.8
synchrotron radiation	pulsed in the ns time domain	tunable with a monochromator	tunable with a monochromator

VI.2.2. Measurement of the velocity of photoelectrons

This is usually done by using either (i) a decelerating electric voltage V applied between a retarding grid and an entrance grid, or (ii) an electron filter formed by a sector of a conducting bicylinder (R being its average radius, r the inter-electrode distance and V the applied voltage V). In each case, the device is completed by an electron detector. In the first case (i), the selected velocity is such that $(1/2)m_ev^2 = eV$. By slowly varying V, minima (V_m) are recorded corresponding to velocity values $v = \sqrt{2eV_m / m_e}$ of the ejected photoelectrons. In the second case (ii), the only electrons leaving the cylindrical selector are those having a circular trajectory whose radius is R, i.e. those for whom the inertial centrifugal force mv^2/R is compensated by the radial electric force eV/r. In this last case, when slowly varying V, maxima of the photoelectric current are detected at values V_M of the voltage corresponding to ejected photoelectrons having velocity $v = \sqrt{eV_M R / m_e r}$. Of course, numerous technical improvements (for instance, use of multiple grids and hemispherical filters) have been designed for these two schematic set ups.

VI.2.3. Detection of photoelectrons

This is done with the help of electron multipliers, devices which are similar to photomultipliers and which, like these, include many multiplying dynodes allowing the detection of very weak signal intensities. Of course, multiple acquisition devices are used to ensure the simultaneous detection of the whole of the velocity distribution of the ejected electrons. The complete set up is controlled by a microcomputer.

The average resolution of these techniques is of about one hundredth electron-volt (the record currently stands at 4 or 5 meV).

VI.3. Results and contributions of photoelectron spectroscopy

VI.3.1. Spectra and their interpretation

The appearance of a typical photoelectron spectrum of an atom or a many-electron molecule is presented on Figure 6.3.1. Traditionally, the energy of photoelectrons is displayed in the abscissa.

We distinguish a few *bands*, each of them made up of an undefined number of *peaks*. Our knowledge of the electronic structure (*cf.* Section III.3) and equation 6.1.6, which can also be written as $E_e = E_i - (P_i + E_{v'})$, show us that each band corresponds to the ejection of a photoelectron from the microscopic system (the shell-like energy distribution leads to several values of the ionization potential P_i). Moreover, the peak structure of each band corresponds to the vibrational structure of the electronic state of the created ion (several energy levels $E_{v'}$ being associated to the same electronic level).

On the other hand, we observe that the bands do not all have the same number of peaks. What causes this difference? Let us explain this point by considering the

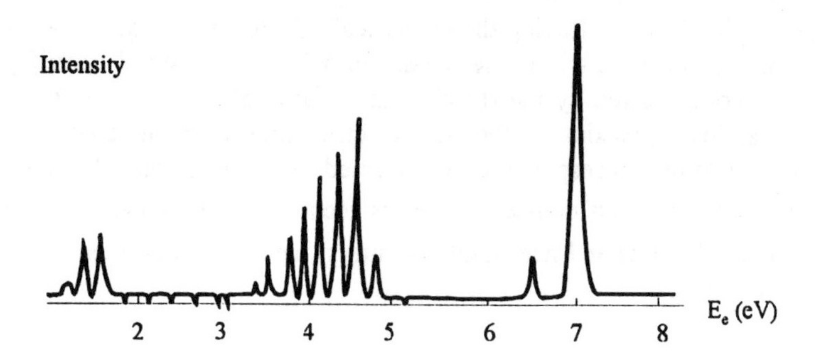

Figure 6.3.1: Schematic representation of a photoelectron spectroscopy spectrum.

simple example of a diatomic molecule AB. Figure 6.3.2 shows the two electronic ground states both for the AB molecule and for the AB^+ ion, together with the absorption photoelectron spectra likely to be recorded.

The curves represent the electronic energies of the ground state of the AB molecule (m) and of the AB^+ ion (i) as a function of the internuclear distance R_{AB}. These are the Morse potential energy curves that we have already used in Section IV.3.2.3 (see, for example, Figure 4.2.9). The energy of the ion is greater than that of the molecule. These curves have a minimum at the equilibrium distance R^e_{AB} of the molecule. We distinguish three cases, called a, b and c, respectively, corresponding to the cases $(R^e_{AB})_m = (R^e_{AB})_i$, $(R^e_{AB})_m > (R^e_{AB})_i$, $(R^e_{AB})_m < (R^e_{AB})_i$. Chemists term the corresponding ejected electrons as nonbonding, bonding, and anti-bonding, respectively. In the first case, the ejection of the nonbonding electron — that is, the one which does not take part in the bond — does not alter the equilibrium distance.

In the second case, a bonding electron is ejected and since the resulting ion is less stable, the equilibrium distance increases. Finally, in the third case describing the ejection of the anti-bonding electron, a disruptive source is suppressed so that the equilibrium distance decreases since the stability of the resulting ion is greater than that of the molecule. The fundamental electronic state of the AB molecule consists only of the vibrational level of lower energy, corresponding to the value v = 0 of the vibrational quantum number. Indeed, at the usual temperatures that we consider here, only state v = 0 is significantly populated (*cf.* Problem IV.4.13).

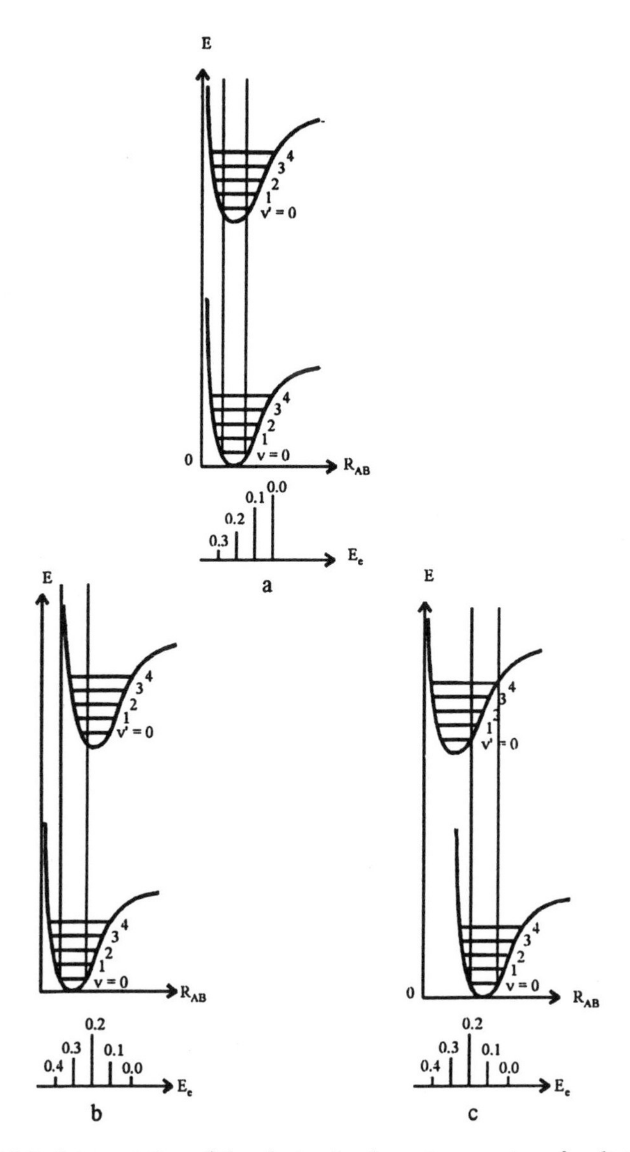

Figure 6.3.2: Interpretation of the electronic absorption spectra of a diatomic molecule AB. a: Nonbonding electron; b: Nonbonding electron; c: Anti-bonding electron.

On the other hand, the ejection of the photoelectron leads to the population of several vibrational states, corresponding to values v'= 0, 1, 2, ..., of the fundamental electronic energy state of the AB^+ ion.

How does the absorption of the incident photon and the transition $AB \rightarrow AB^+$ take place? We already mentioned (*cf.* Section IV.3.2.3) that the Franck-Condon principle imposes the so-called *vertical transitions*, shown on Figure 6.3.2, as we start from the lower energy state v = 0. Thus, several states v' will be populated. With what probability? Equation 6.3.47 provides the answer within the framework of the electric dipole approximation. This probability is proportional to

$$S_{v,v'} = \left|\left\langle \psi'_{v'}(R_{AB}) \middle| \psi_v(R_{AB}) \right\rangle\right|^2 \tag{6.3.1}$$

In this expression we recognize the nuclear parts, which depend on the nuclear coordinate R, of the state vectors representing the two vibration-modulated electronic states. $S_{v,v'}$ is the *overlapping factor* of the vibrational states v and v'. Problem VI.4.2 gives the details of the calculation of this term in the case of a nonbonding electron and shows that the probability transition is maximum for the $v = 0 \rightarrow v' = 0$ transition. Basically, we have to evaluate an integral over the internuclear R_{AB} coordinate. We see on Figure 6.3.2 that for the two cases of bonding and anti-bonding electrons, the overlapping factor is small for small and for large values of v'. A maximum must therefore exist for a certain intermediate value of v' (for v'= 2 in the present case). Finally, a simple glance at the band structure of the photoelectronic spectra provides an immediate way to solve the case corresponding to a nonbonding electron. This is the only one where the maximum falls on the first peak on the right hand side (the $\nu = 0 \rightarrow \nu' = 0$ transition) when the photoelectron energy is plotted in abscissa. This energy is maximum and corresponds to a minimum of the sum $P_i + E_{v'}$. For the two other cases, the band structures appear more symmetrical. A confirmation of this assignment can be found in the measurement of the energy difference between the peaks of the same band pattern. In the case of a nonbonding electron, the value of this energy difference is very close to that of the molecule before the ejection of the electron, this last value being traditionally measured by IR spectroscopy. In the cases of bonding and anti-bonding electrons, this energy difference is clearly distinct from that measured for the molecule before ionization, but it is more difficult to distinguish between the bonding and the anti bonding electrons, since their spectra are quite similar. However, in order to suggest a correct assignment, the experimenter can use the fact that the pattern corresponding to the ejection of an electron from a degenerate state (*cf.* Section III.1) has an intensity which is proportional to the degeneracy of the involved state. All these aspects will be emphasized in Problem VI.4.1 and Problems VI.4.3-VI.4.6, which introduce some results concerning simple molecules and which discuss their interpretations.

Selection rules for photoelectron spectroscopy are quite straightforward. In the case of an atom which absorbs a photon, gets ionized, and then ejects a photoelectron, the whole process obeys the rules established in Section IV.3.2.3 (*cf.* Equation 6.3.45). However, as the linear momentum of the ejected photoelectron can take on any value, the resulting ion is not specified and *selection rules do not exist* for the process as a whole. Of course this results also holds in the case of a molecule. As a consequence, all the energy states of a molecule can be excited and no selection rules can be derived.

VI.3.2. Contributions of photoelectron spectroscopy

Its contribution is twofold. On the one hand, photoelectron spectroscopy can be used for the verification of predictions suggested by certain quantum treatments of the atomic and molecular electronic structure — here we refer to the self-consistent Hartree-Fock approach (*cf.* Section III.3.2) — and, on the other hand, it can be of use for a critical study of certain traditional pictures of the molecular electronic structure such as, for example, the one by Lewis dealing with covalent bonding. Let us examine these two points successively.

- *Ascertainment of the theoretical predictions of the HF-LCAO formalism*

We just showed that photoelectron spectroscopy provides a way to measure the ionization potential P_i of the investigated system. However, Koopmans demonstrated that in the case of a closed shell molecule, the ionization potential should be approximately equal to the opposite of the MO energy from which the ejected electron originates, providing that this energy is calculated by a *self consistent field* (SCF) method, such as the one by Hartree-Fock which is described in Section III.3.3.2. This results reads as

$$P_i \cong -E^{CAC} \tag{6.3.2}$$

It cannot be a strict equality. The value of P_i is experimental in nature whereas E^{CAC} is basically a theoretical value. In fact, at least two phenomena have been neglected in this approximation:

— The *electronic correlation* (*cf.* Bibliographical reference 1.2.2), which is usually neglected in SCF calculations

— And the *electronic reorganization* occurring after the ejection of the electron which may lead to a value for $E^{CAC}(M)$ which is significantly different from $E^{CAC}(M^+)$.

However, for most of the cases, Equation 6.3.2 is a reasonable approximation and photoelectron spectroscopy turns out to be *the* technique to test the validity of the so-called self-consistent field for AO and MO energy calculations. The fundamental result is that in most cases, the measured first ionization energies do coincide with SCF calculations within 0.1 eV (let us recall that the precision of the experimental determination is at least equal to 0.01 eV). However, the measured energies span a range extending from one tenth of an eV to a few eV. The order of magnitude of the coincidence between theory and experiment seems to be about one or a few percent. This value is quite satisfying and largely contributed to the popularity of this type of calculation method.

- *Critical inspection of some traditional physical pictures*

We noted that photoelectron spectroscopy can distinguish between the nonbonding, bonding and anti-bonding nature of a MO. It thus becomes possible to test the validity of some traditional physical pictures, especially those used in teaching. Let us examine, for example, the case of the nitrogen molecule N_2, discussed in Problem V.4.3, and to which we refer the reader for an inspection of the corresponding MO diagram. An important result is that the photoelectron spectrum does not support the bonding and anti-bonding nature of the σ_z and σ_s^* MO's although this is largely proclaimed in numerous university text books! Although this molecule is rather simple, the Lewis triple bond sketch seems quite different from reality in this case, and may lead to an erroneous picture of the electronic molecular structure.

Similarly, the photoelectron spectrum of the benzene molecule cannot be understood without taking into account the σ orbitals, which are, however, very rarely considered, and are usually neglected in simplified approaches such as the Hückel approach.

These facts demonstrate that photoelectron spectroscopy is a powerful technique to investigate electronic properties of atoms and molecules. Complementary to UV-visible absorption spectroscopy, it leads to an accurate determination of the first ionization energies and appears essential to establish correct molecular orbital diagrams.

VI.4. Exercises and problems

VI.4.1. Vibronic transitions in photoelectron spectroscopy

1. Explain photoelectron spectroscopy with UV excitation. Which kind of information can you expected to obtain?

2. The spectrum of a homonuclear diatomic molecule is never made up of a small number of lines. Most of the time, one sees a fine structure superimposed on some of the peaks. What is its origin? Why does this fine structure not always exist? Can one infer important information from it ?

1. *Cf.* Section VI.1.

2. *Cf.* Section VI.3.

VI.4.2. Search for the intensity maximum in a vibronic band

The goal of this problem is to show that the line of maximum amplitude of the vibronic fine structure of a photoelectron band of a nonbonding electron is due to an excited ion in the vibrational energy state v' = 0. Parts A and B are independent.

A. Preliminary calculation of the transition probability $|a\rangle \rightarrow |b\rangle$

We consider the interaction of a sinusoidal wave of a tunable laser whose electric field has an amplitude E_0 and a pulsation ω_L, with a collection of identical molecules, assumed to be isolated, (pure state), and considered to be two-level ($|a\rangle$ and $|b\rangle$) electronic systems. $|a\rangle$ and $|b\rangle$ are orthonormal. Pulsation ω_L is very close to pulsation $\omega = \omega_b - \omega_a$ ($\omega_b > \omega_a$). Let H_0 be the molecular Hamiltonian of the isolated molecule and **p** the electric dipole moment. It is assumed that only the ground state $|a\rangle$ is populated at time t = 0 when the laser is switched on. $|b\rangle$ is the first electronic excited state.

Let us consider the nonstationary state characterizing the system at time t and described by

$$|\phi(t)\rangle = a(t)\exp(-i\omega_a t)|a\rangle + b(t)\exp(-i\omega_b t)|b\rangle$$

1. Setting $p = -\langle a|\sum_i e r_i|b\rangle = p^*$, write the evolution equations of a(t) and b(t) in presence of the laser wave, using the electric dipole and the rotating wave approximations.

2. Frequency ω_L is varied in order to reach resonance ($\omega_L = \omega$). Show that the population of state $|b\rangle$ (proportional to the transition probability $|a\rangle \rightarrow |b\rangle$) varies as p^2 as soon as the optical wave is applied.

B. Calculation of the maximum intensity of the photoelectron emission

We assume that the result obtained in A.2 still holds.

1. Let $\psi(r, R)$ be the state vector representing a diatomic molecule in the electronic ground state (r and R stand for the modulus of the electronic and the modulus of the nuclear coordinates, respectively). Let $\psi'(r, R)$ be the state vector representing the first electronic excited state.

Express the transition moment p of the electronic transition $\psi(r, R) \rightarrow \psi'(r, R)$.

2. Within the frame of the BO approximation, show that these vectors can be rewritten as $\psi(r, R) = \psi_e(r)\ \psi_N(R)$. In fact, the nuclear state vector $\psi_N(R)$ depends on the quantum vibration number v, and in the following we shall write $\psi_v(R)$.

3. Show that p is proportional to $S_{v',v} = |\langle\psi'_{v'}(R)|\psi_v(R)\rangle|^2$. $S_{v',v}$ being the overlapping factor of vibrational levels v and v'.

4. Show that the photoelectron emission is proportional to $|S_{v',v}|^2$.

5. Calculate $S_{v',v}$ for v = v' = 0 (fundamental vibrational states). We shall use the Gaussian functions defined underneath:

— Fundamental electronic and vibrational states, internuclear equilibrium distance R_e:

$$\psi_0(R) = \left(\alpha / \sqrt{\pi}\right)^{1/2} \exp\left[-\alpha^2 (R - R_e)^2 / 2\right]$$

— Excited electronic and vibrational states, internuclear equilibrium distance R_e^*:

$$\psi_0^*(R) = \left(\alpha / \sqrt{\pi}\right)^{1/2} \exp\left[-\alpha^2 (R - R_e^*)^2 / 2\right]$$

We recall that $\int_{-\infty}^{+\infty} \exp\left(-x^2\right) dx = \sqrt{\pi}$.

6. Show that the maximum photoelectron intensity emission occurs when $R_e = R_e^*$, i.e. for a nonbonding electron.

A.1. Within the frame of the electric dipole approximation, we can write $H = H_0 - p\, E_0 \cos \omega_L t$. Using postulate P_6 of quantum mechanics (*cf.* Section III.1.2) and taking into account that states $|a\rangle$ and $|b\rangle$ are orthogonal, we obtain the following differential system:

$$\dot{a} = \left(i\pi p E_0 / h\right) \cos \omega_L t \exp\left(-i\omega t\right) b$$

$$\dot{b} = \left(i\pi p E_0 / h\right) \cos \omega_L t \exp\left(+i\omega t\right) a$$

Let us write the cosine as a sum of complex exponentials and let us use the rotating wave approximation. The system can then be written as

$$\dot{a} = \left(i2\pi p E_0 / h\right) \exp\left[-i\left(\omega - \omega_L\right) t\right] b$$

$$\dot{b} = \left(i2\pi p E_0 / h\right) \exp\left[+i\left(\omega - \omega_L\right) t\right] a$$

2. Defining the Rabi pulsation as $\omega_R = 2\pi p E_0 / h$, at resonance we obtain

$$\dot{a} = \left(i\omega_R / 2\right) b \qquad \dot{b} = \left(i\omega_R / 2\right) a$$

This differential system can be easily solved, and, considering that at $t = 0$ only state $|a\rangle$ is populated, i.e. $|a(0)|^2 = 1$ and $|b(0)|^2 = 0$, we obtain

$$|a(t)|^2 = \cos^2\left(\omega_R t / 2\right) \quad \text{et} \quad |b(t)|^2 = \sin^2\left(\omega_R t / 2\right)$$

If $(\omega_R t/2) << (\pi/2)$, the population of the excited state $|b(t)|^2$ varies as ω_R^2 i.e. as p^2 (*cf.* Postulate P4, Section III.1.2.).

B.1. We have

$$p = -e \left| \psi'(r,R) \left(\sum_i r_i \right) \psi(r,R) \right|^2$$

2. We can write [*cf.* Section III.3.3.1]

$$\begin{aligned} H\psi_e(r)\psi_N(R) &= (H_e + T_N)\psi_e(r)\psi_N(R) = E_e\,\psi_e(r)\psi_N(R) + T_N\,\psi_e(r)\psi_N(R) \\ &= \left[(T_N + E_e)\psi_N(R)\right]\psi_e(r) = E\psi_e(r)\psi_N(R) \end{aligned}$$

The total energy E appears as the sum of both the kinetic energy of the nuclei and the electronic energy (represented through its potential term). The eigenvalue equation of the total Hamiltonian is then split in the two following eigenvalue equations:

$$\begin{aligned} (H_e + V_{NN})\psi_e(r) &= E_e\,\psi_e(r) \\ (T_N + E_e)\psi_N(R) &= E\psi_N(R) \end{aligned}$$

Let us recall that $\psi_e(r)$ in fact depends *parametrically* on R (it is the electronic state vector for a given nuclear configuration). But H_e "ignores" this *implicit* dependence.

3. Using the Born-Oppenheimer approximation, we obtain

$$p = -e\left|\psi'_{v'}(R)\psi_v(R)\right|^2 \left|\psi'_e(r)\left(\sum_i r_i\right)\psi_e(r)\right|^2 = S_{v'v}\left|\psi'_e(r)\left(\sum_i r_i\right)\psi_e(r)\right|^2$$

4. Photoelectron emission is proportional to p^2, that is, to $|S_{v'v}|^2$.

5. The integration over R is performed by setting $x = \alpha\left[R - (R_e + R_e^*)/2\right]$. We find $S_{00} = \exp\left[-\alpha^2 (R_e - R_e^*)^2 / 4\right]$.

6. The maximum photoelectron emission is obtained for $R_e = R_e^*$, that is to say for a nonbonding electron whose ejection takes place without any change of the equilibrium internuclear distance.

VI.4.3. Photoelectron spectroscopy of the nitrogen molecule

The two following figures qualitatively reproduce:

a. An experimental result concerning the N_2 molecule,

b. An extract from a photoelectron handbook concerning N_2.

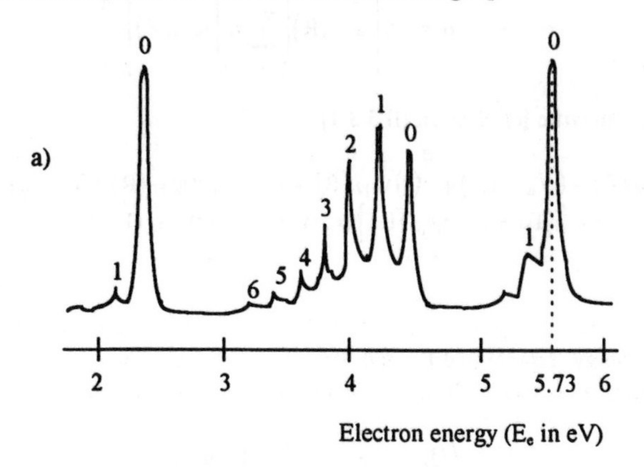

1. Calculate the approximate wavelength of the UV incident beam.

2. In the case of the N_2 molecule, suggest a molecular orbital diagram which is compatible with the spectrum. Please keep in mind that when degenerate orbitals are occupied, the total band intensity is greatly enhanced. We also remind you that the vibrational wavenumber of N_2 is 2 350 cm^{-1}.

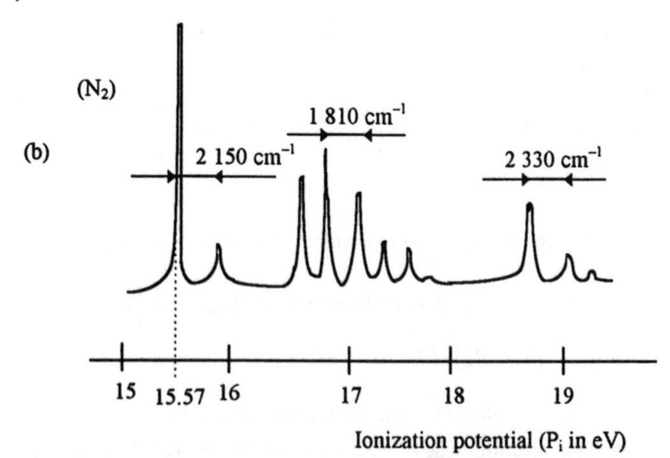

3. In fact, the observed spectrum is rather different from the one that should be observed in the case of the traditional so-called triple bond N ≡ N. Give a qualitative sketch of such a spectrum. Which conclusion can be drawn from this discrepancy?

1. We have $h\nu = P_i + E_e \approx 15.57 + 5.73 \approx 21.3$ eV. $\lambda = hc/(h\nu) \approx 58.4$ nm (helium ultraviolet line).

2. The suggested diagram is sketched below (the nitrogen molecule is invariant by rotation around the Oz axis).

The band situated at 15.57 eV (weaker ionization potential) corresponds to the ejection of an electron from the σ_z MO. The small number of vibrational peaks shows the nonbonding character of the electron, which is confirmed by the value of their energy difference of 2 150 cm^{-1}, a value close to that of the N_2 molecule. The intermediate band observed at 17 eV is very well structured, and the energy difference between its vibrational peaks (1 810 cm^{-1}) is very different from the case of the N_2 molecule. This points to the ejection of a bonding electron from the π_x or the π_y MO. The band situated at about 48.7 eV corresponds to a nonbonding electron σ_s^* (the frequency of the N^+_2 ion is almost identical to that of the N_2 molecule).

3. The traditional picture of the so-called triple bond is not confirmed and the Lewis model should be used with care.

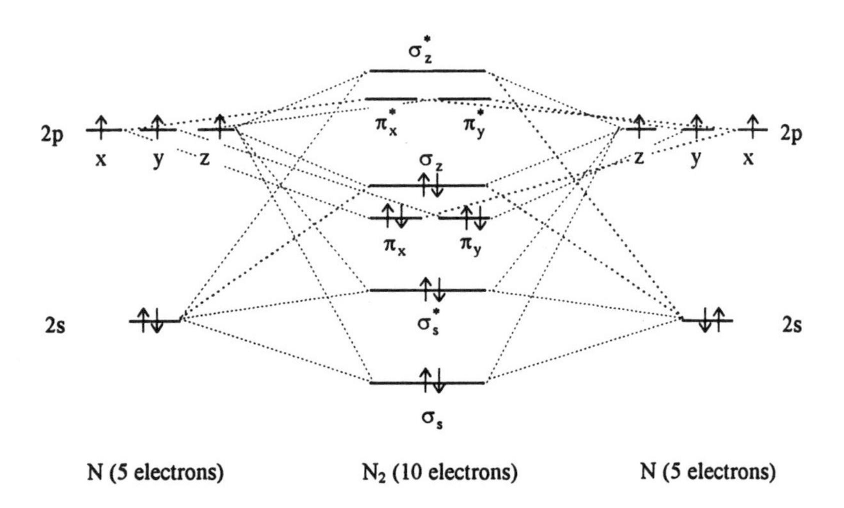

VI.4.4. Photoelectron spectroscopy of the carbon monoxide molecule

The figure below shows the photoelectron spectrum of the CO molecule (which is isoelectronic with the N_2 molecule).

1. Give a qualitative representation of the HF- LCAO diagram of this molecule.

2. Give the attribution of the three observed bands.

3. Calculate the excitation wavelength.

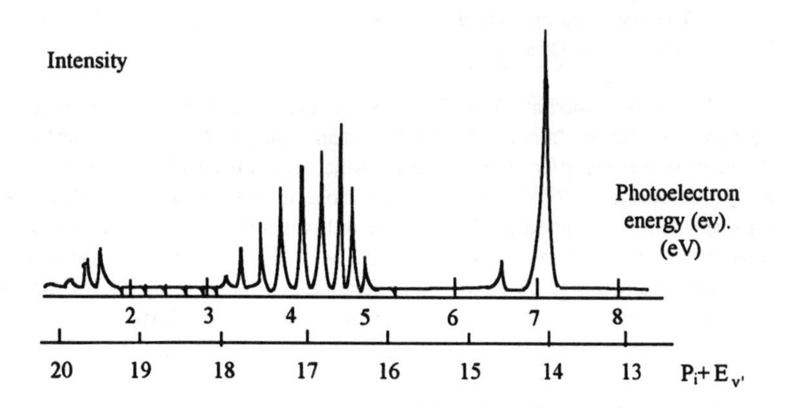

1. A qualitative sketch of the MO diagram is given below (the Oz axis is the molecular symmetry axis).

2. The attribution is similar to the one of the isoelectronic N_2 molecule (*cf.* Problem VI.4.3).

3. In order to calculate the wavelength of the incident radiation, we add up the kinetic energy of the photoelectron and the corresponding ionization potential energy. For example $7.15 + 14 \approx 21.15$ eV. The value of the corresponding wavelength is about 58.8 nm (it lies in the ultraviolet line of the helium atom, frequently used in photoelectron spectroscopy).

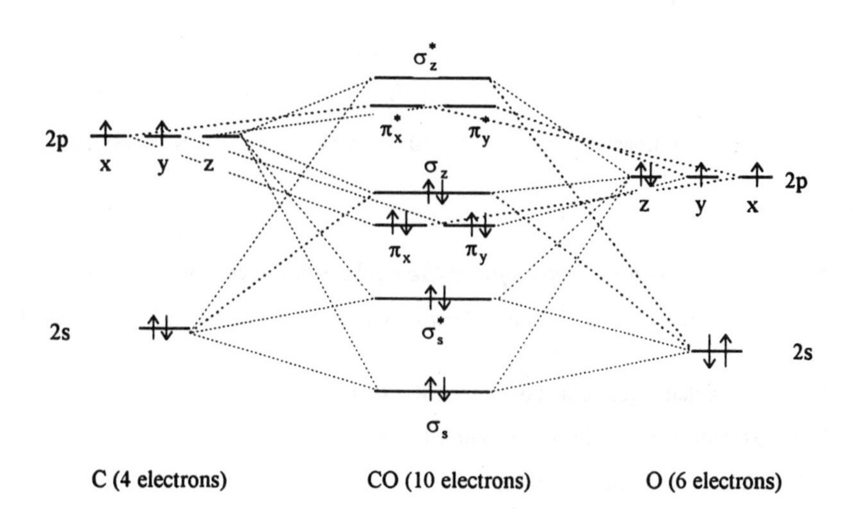

VI.4.5. Photoelectron spectroscopy of the water molecule

The most energetic photoelectrons of the photoelectron spectrum of the water molecule (excited by the 21. 21 eV radiation of He) correspond to about 10 eV. In these conditions, the resulting excited H_2O^+ ion exhibits a weak vibrational structure with a periodicity of about 0.45 eV. Let us recall that the symmetric stretching mode of the unionized water molecule has a frequency of 3 652 cm^{-1}. The figure below gives the results of a minimal basis HF-LCAO calculation performed on a water molecule.

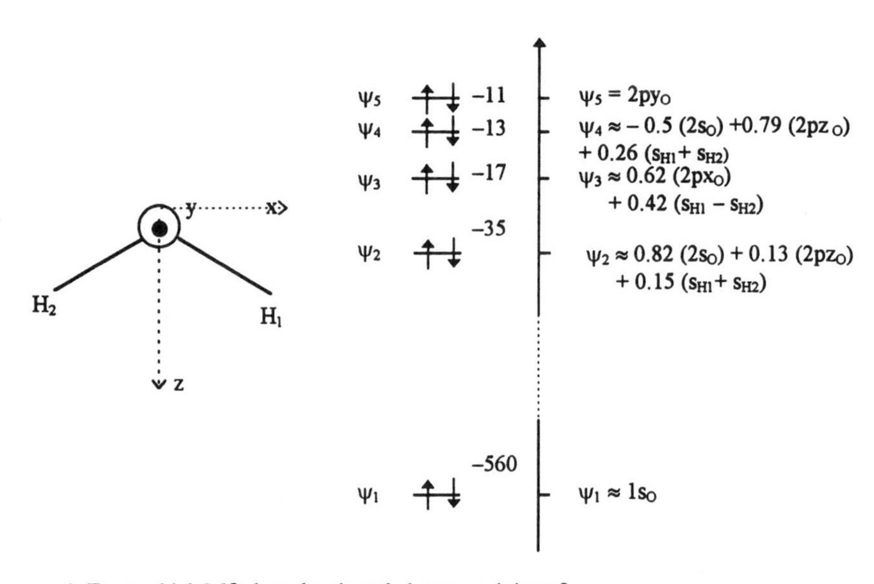

1. From which MO does the ejected electron originate?
2. What does the weakness of the vibrational structure suggest?
3. How does the measured periodicity support the previous comment?
4. What is the nature of the MO mentioned in question 1?

1. The ionization potential is about 21.21 – 10 = 11.21 eV. The ejected electron originates from the orbital ψ_5.

2. It is a nonbonding electron.

3. The energy difference 0.45 eV corresponds to a wavenumber of 3 620 cm^{-1}. The similarity between the observed vibronic signature of the H_2O^+ ion and that of the H_2O molecule support the nonbonding character of the ψ_5 MO.

4. It is the $2p_y$ AO of the oxygen atom (so-called "free pair").

VI.4.6. Photoelectron spectroscopy of the benzene molecule

The photoelectron spectrum of the benzene molecule is obtained by using an incoming radiation with a wavelength of 30.4 nm. The maximum kinetic energy of the photoelectrons is 31.5 eV. What is the first ionization energy of the HOMO of the benzene molecule? What would the maximum kinetic energy of the photoelectrons have been if an incoming radiation with a 58.4 nm wavelength had been used?

The first ionization energy of the HOMO of the benzene molecule is $40.8 - 31.5 = 9.3$ eV. The maximum kinetic energy of the photoelectron would have been $21.2 - 9.3 = 11.9$ eV.

VI.5. Bibliography

6.5.1. Survey articles may be found in:

ALLAN M. — Electron Spectroscopy Methods in Teaching, *J. of Chem. Educ.*, 64, 419 (1987).

DUKE B.J. — Non-Koopmans' Molecules, *J. of Chem. Educ.*, 72, 501 (1995).

6.5.2. Photoelectron spectroscopy is described in detail in:

BAKER A.D. and BETTERIDGE D. — *Photoelectron Spectroscopy*, Pergamon Press (1977).

BRUNDLE C.R. and BAKER A.D. — *Electron Spectroscopy: Theory, Techniques, and Applications*, vol.1 et 2, Academic Press (1977).

ELAND J.H.D. — *Photoelectron Spectroscopy*, Butterworths (1974).

RABALAIS J.W. — *Principles of Ultraviolet Photoelectron Spectroscopy*, Wiley (1979).

TURNER D.W., BAKER C., BAKER A.D. and BRUNELLE C.R. — *Molecular Photoelectron Spectroscopy*, Wiley-Interscience (1970).

Chapter VII

Laser absorption spectroscopy

VII.1. About laser sources

VII.1.1. What is a laser?

A laser is an optical oscillator. In general and whatever the domain of physics considered, an oscillator consists of the association of an *amplifier* and a *feedback* procedure, also called *resonator*, which sends all or part of the output back to the input of the device and in phase with the input signal. Figure 7.1.1 shows an example of this kind of optical device.

Mirror M_1 is only partially reflecting and ensures that the optical wave (frequency ν_L) of the resonator (also called optical cavity) can get out and be put to use. At every successive round trip inside the cavity, the energy of the wave both decreases (the losses $L_{\nu L}$ are mainly due to the exit mirror) and increases (the gain $G_{\nu L}$ is due to the amplifying material). A stationary regime is obtained when the gain compensates the losses at the output frequency of the device. In the lower part of Figure 7.1.1, the wave is folded up on itself, and so it only needs two of the four mirrors of the ring cavity. This configuration corresponds to the so-called *Fabry-Pérot linear cavity*. The energy which must of course be brought to the system is injected into the amplifying material in various forms. The first fundamental work concerning the laser is attributed to Basov, Prokhorov, Townes and Schawlov (1958), while the first actual lasers were built by Maiman (Hughes, 1959) and Javan (MIT, 1960).

– *Amplifying material:* Let us consider a microscopic system with two energy levels (calling the lower level 1, and the upper level 2), similar to the system described in Chapter IV for the analysis of the processes of absorption, of stimulated emission and of spontaneous emission using Einstein's phenomenological theory. If we also take into account spontaneous and stimulated emission, Equation 4.3.14 takes the form

$$\dot{N}_1 = -N_1 B_{12}\bar{\rho} + N_2 B_{21}\bar{\rho} + N_2 A_{21}$$

$$\dot{N}_2 = N_1 B_{12}\bar{\rho} - N_2 B_{21}\bar{\rho} + N_2 A_{21}$$

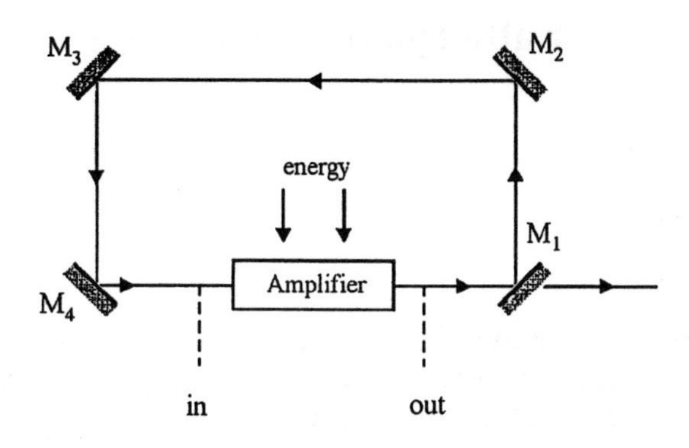

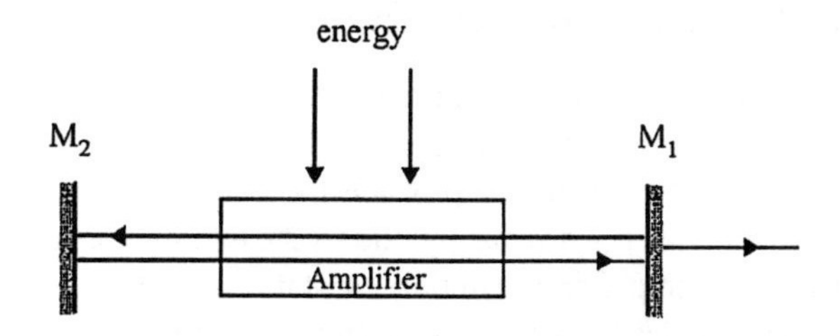

Figure 7.1.1: The two types of laser cavities. Top: Ring cavity (a dextrorotatory wave can also propagate in the system); Bottom: Linear cavity. Phase conditions are not illustrated here.

This differential system constitutes the *rate equations* and it expresses the change of the populations with time. B_{12} and B_{21} are the Einstein coefficients for absorption and stimulated emission respectively. A_{21} is the coefficient of spontaneous emission and $\bar{\rho}$ represents the average energy density of the wave. The stationary state (labeled by the subscript s) is obtained by setting the time derivatives of the populations equal to zero. With these equations we can express the amplification condition. When a wave (with average intensity $\bar{I}$) crosses a material sample of thickness dz (of section S and with refractive index n at the frequency of the wave), simply expressing the conservation of the energy during the interaction yields

$$-d\bar{I}(z) = (h/2\pi)(N_{1s} - N_{2s})B\bar{\rho}\,\omega_0 S dz \qquad (7.1.1)$$

ω_0 is the difference between the resonance angular frequencies associated to the energy levels of the system. We assumed that $B_{12} = B_{21} = B$ (same degeneracy for levels 1 and 2) and neglected spontaneous emission. We can see immediately that if there is to be an amplification of the energy of the light wave, we must have $N_{2s} > N_{1s}$. This condition is called the *population inversion condition*. But we must note that:

— In absence of any applied radiation and at thermal equilibrium, the populations of the stationary state obey to the Maxwell-Boltzmann statistics and always satisfy the condition $N_{1s} > N_{2s}$.

— In presence of radiation, the rate equations show that the stationary-state populations are at best equal when the energy density of the wave is infinite, and this means that population inversion can never be achieved for a two level system.

Amplification cannot arise from a two-level system, so the medium used in the laser must necessarily consist of at least three level systems. In fact, in the examples used thereafter four-level systems will be used to describe the amplifying substances (dye molecule for example). Under certain conditions, applying an auxiliary radiation between two levels (optical pumping, Kastler, 1940-1950) leads to a population inversion between two other levels (one at least being different from the two first levels) and thereby induces amplification. This amplification can only be obtained for a frequency range which depends on the width of the levels concerned (linewidth).

— *Optical cavity:* The properties of optical cavities are illustrated in Problems VII.5.1 and VII.5.2. An optical cavity is a resonator, i.e. a structure which generates *modes* under certain conditions. Modes are space-time energy distributions characteristic of the resonator. The optical cavity is a 3D structure, with its longest dimension lying along the propagation axis of the optical wave, in which a transverse spatial filter of small diameter is inserted. In this kind of cavity, the cavity modes are called *transverse electromagnetic modes* and noted TEM_{00q}, the letter q referring to the z coordinate of the longitudinal propagation. These modes correspond to transverse distributions of the electric field of the wave satisfying the *self-consistency condition* inside the resonator, i.e. they are identical to themselves after traveling back-and-forth once through the cavity (with a possible change of phase of course). This condition is needed for the gradual construction of a stable stationary optical wave by successive back-and-forth trips of the wave inside the resonator. This condition is the cause of a fundamental property of the TEM_{00q} or *fundamental mode* of the laser. Indeed, the propagation of a wave in a material structure is governed by the phenomenon of diffraction. In a first approximation, it is possible to show that diffraction is described mathematically by a Fourier transform. An energy distribution is preserved during its propagation if it remains unchanged by Fourier transform. Now, the Fourier transform of a Gaussian remains a Gaussian. Therefore, the fundamental mode of a laser will have a Gaussian distribution of its energy in the plane normal to the wave vector. The properties of Gaussian waves have already been described in Section II.2.2.3 and the complete mathematical treatment of this structure can be found in Bibliographical references 7.6.2 and 7.6.3. Here, we only report the two main results concerning linear cavities:

— A cavity consisting of two mirrors (with radii of curvature R_1 and R_2) separated by length L is characterized by the two parameters $h_1 = 1 - L/R_1$ and $h_2 = 1 - L/R_2$. In order to progressively generate a stationary laser wave, it must be stable, i.e. it must satisfy the relation $0 \le h_1 h_2 \le 1$.

— In these conditions, the spatial part of the expression describing the electric field can be written as

$$\begin{aligned} E(x,y,z) &= E_0\left[\omega(0)/\omega(z)\right]\exp\left[-(x^2+y^2)/\omega^2(z)\right] \\ &\times \exp - i\left\{k_q\left[z+(x^2+x^2)/2R(z)\right] - \mathrm{arctg}\left[\lambda z/\pi\omega^2(0)\right]\right\} \end{aligned} \tag{7.1.3}$$

with

$$\begin{aligned} \omega^2(0) &= (\lambda L/\pi)\sqrt{h_1h_2(1-h_1h_2)/(h_1+h_2-2h_1h_2)^2} \\ \omega^2(z)/\omega^2(0) &= 1+(z^2/L^2)(h_1+h_2-2h_1h_2)^2/h_1h_2(1-h_1h_2) \\ R(z) &= z\left[1+(L^2/z^2)h_1h_2(1-h_1h_2)/(h_1+h_2-2h_1h_2)^2\right] \end{aligned} \tag{7.1.4}$$

We indeed recognize a Gaussian radial distribution of the electric field, with characteristic diameter $\omega(z)$, affecting a spherical wave (*cf.* Section II.2.2.3). As said earlier, such a wave is called a *quasi-spherical wave*. It becomes planar for $z = 0$ [$R(0)$ is infinite]. At the origin of the propagation coordinate, $\omega(z)$ takes on its minimal value $\omega(0)$ called the *beam waist.* It corresponds to the smallest spatial extension of the beam (usually a few hundred microns). Its position is easily determined by writing $R(z_1) = R_1$ or $R(z_2) = R_2$ (z_1 and z_2 are the coordinates of the two mirrors). Of course, a propagation term depending on time and including the frequency of the mode is associated to this spatial term.

Now we want to determine the phase, which is very important since it controls the correct behavior of the amplifier. After traveling back-and-forth once in a linear cavity of length L, the phase condition can be written as

$$\begin{aligned} \varphi(2L) = (q+1)2\pi &= k_q 2L - 2\mathrm{arctg}(\lambda z_2/\pi\omega_0^2) \\ &+ 2\mathrm{arctg}(\lambda z_1/\pi\omega_0^2) \end{aligned} \tag{7.1.5}$$

where q is an integer. We can thus determine the allowed values of k_q from which we can calculate the frequencies of the longitudinal modes of the laser

$$\nu_q = (c/2L)(q + 1 + \arccos\sqrt{h_1 h_2}/\pi) \tag{7.1.6}$$

There is an infinite number of frequencies separated by c/2L, so there is also an infinite number of longitudinal TEM_{00q} modes. Problem VII.5.2 shows that these modes have a frequency with a width of

$$\Delta\nu \approx c(1-r)/2\pi L\sqrt{r} \tag{7.1.7}$$

where r is the reflection coefficient of the exit mirror. When r = 1 modes are infinitely thin. Losses are responsible for the undesirable width affecting the resonance.

What happens if we insert an appropriately pumped amplifier inside the resonator? Figure 7.1.2 illustrates the answer and indicates the main features of laser emission.

For laser emission to occur the non saturated gain of the amplifier must be at least equal to the losses of the resonator. These losses are assumed to be independent of the frequency. In a *longitudinal multimode operation*, the laser oscillates on all the frequencies of its cavity satisfying this condition. As a result, the emission of this kind of laser is not monochromatic. We shall discuss this point later and will show how to remedy to this inconvenience. Let us also observe that (i) the wave inside a resonator is by nature a *stationary wave* (resulting from the superposition of the two wave propagating in opposite directions in either type of cavity) and (ii) the properties of the modes not only depend on the cavity but also on the amplifier which is inserted inside it (Problem VII.5.2 shows that the spectral width of the resonance narrows down very fast as the gain of the amplifier grows, and vanishes completely at the laser threshold). These points will be discussed later.

But first, let us recall briefly the properties of lasers.

VII.1.2. Main properties of laser oscillators

The properties of lasers are linked to a very important characteristic of emission: *Space-time coherence*. While in a conventional source, the light wave consists of a superposition of spherical waves emitted by different points and with phases which (i) are defined individually for short periods of time and (ii) vary randomly with respect to each other, we showed that the fundamental TEM_{00q} wave of a laser was a *spherical Gaussian wave* generated by the material amplifier as a whole.

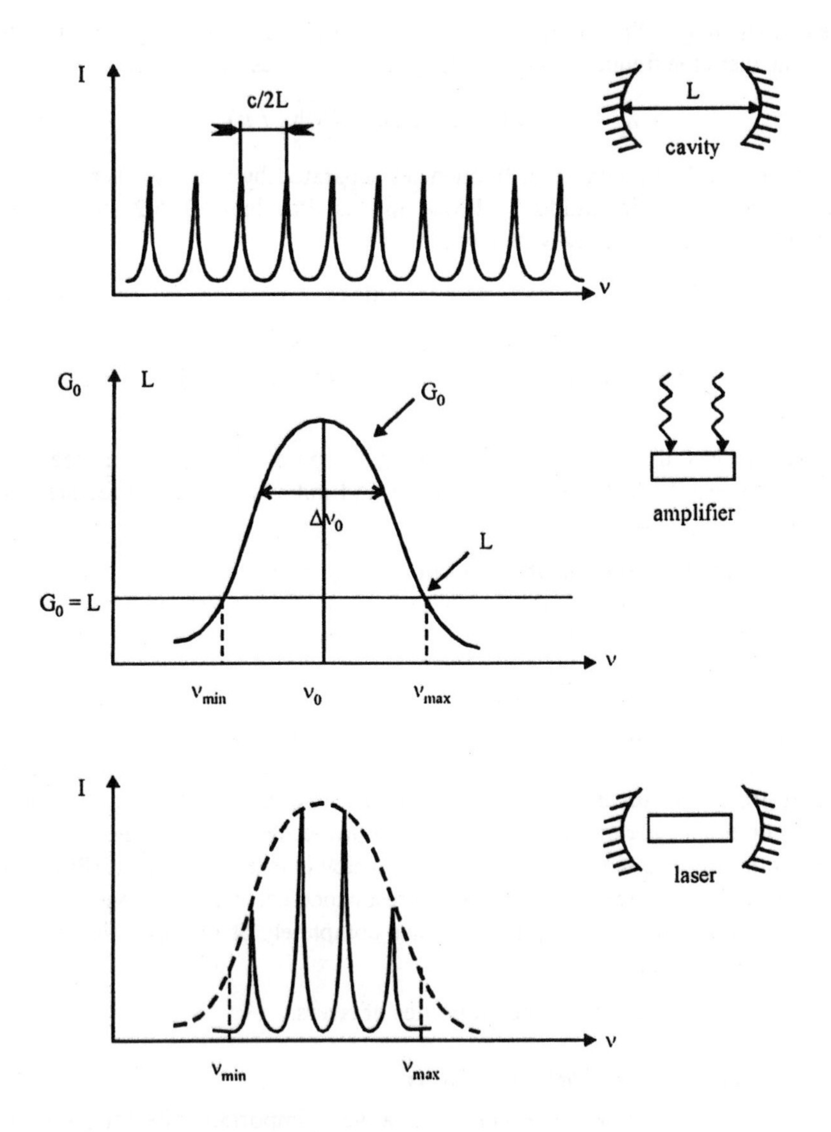

Figure 7.1.2: Laser emission illustrated in frequency space. Top: TEM_{00q} modes of the cavity. Middle: Non-saturated gain of the amplifier and losses (assumed to be independent of the frequency) of the resonator. Bottom: Multimode emission of the laser.

This remarkable difference is due both to the cavity which imposes a gradual construction of the steady state structure by its back-and-forth trips in the cavity and, of course, to the very nature of stimulated emission, namely that inside the amplifying medium every photon generates a second stimulated photon which is emitted in the same direction and in phase with the first. Thus, the wave laser results from a *cooperative emission* of the microscopic systems *monitored* both by stimulated emission and by the resonator.

This also leads to an effect of phase correlation for:

— The phase between two different points of the wave surface at the same instant of time, and this at any moment (*space coherence*)

— The phase between two different moments of time at the same point of the wave surface, and this at any point of space (*time coherence*).

A few properties of *concentration* ensue, which are stated briefly in the following four paragraphs.

The two first concern more directly space coherence.

— *Angular concentration*: All the energy of the laser is emitted within a very small solid angle, of the order of the milliradian, i.e. of the minute of arc. This divergence can be reduced further to a few microradians by using adequate technical improvements. The main applications concern wave guidance, for long and intermediate distance information transmission (Earth-moon and Earth-satellite telemetric measurements, etc.) and guided propagation through optical fibers. This last point concerns us directly because recent "single molecule" experiments (*cf.* Section VII.4.3) often use wave injections operated with the help of optical fibers.

— *Surface concentration:* All the energy of the laser can be concentrated in a surface element of the order of the square micron, reaching intensities as large as the gigawatt per square centimeter in the c.w. regime and considerably larger values in the pulsed regime. The most important applications concern mechanical actions (soldering, cutting, surface treating, etc.) and medical applications (laser surgical knife). But, of most direct concern to us, it allows to reduce the interaction volume between wave and matter, and therefore the number of molecules which can be studied at the same time (*cf.* Section VII.4.3).

The two last paragraphs are more directly concerned with *time coherence*.

— *Time concentration:* All the energy of the laser can be concentrated in a very small time interval (a few femtoseconde). This has some applications in applied research (fast commutation in integrated optics, for example) and in fundamental research with the development of a time-resolved spectroscopy to which we shall come back later.

— *Frequency concentration:* All the energy of the laser can be concentrated in a very narrow frequency band (about ten Hertz with sophisticated technologies).

There are a great number of applications in holography, metrology, telecommunications, video and compact disks, selective excitation of molecules and its application in isotope separation and, of course, frequency-resolved absorption spectroscopy which we shall study in more detail below.

VII.1.3. What can laser technology contribute to absorption spectroscopy?

As we just saw, the electromagnetic wave delivered by the laser is characterized by outstanding properties of energy concentration. The frequency and time concentrations brought about a total revolution in several important fields of fundamental research, especially those in which matter-radiation interaction is used as a tool for the investigation of matter, like spectroscopy, for instance. Let us look at this more closely.

— *Time concentration:* In liquid and solid materials, molecular interactions and collective processes play a fundamental part. This results in very low values of the time relaxation because of the great intensity of the interactions. Before the discovery of lasers, pulsed sources (flash lamps for example) were used along with detection systems such as photomultipliers and photodiodes, which limited the experiments to studies with characteristic time constants greater than the nanosecond. Hence, some domains like picosecond studies of vibrational relaxation in liquids remained totally inaccessible with these classical methods. With the ultrafast pulses delivered by mode-locked lasers it became possible not only to obtain a well adapted excitation source for this kind of study, but also to realize a time-resolved method to analyze the evolution of the material after excitation. Moreover, the extremely high density of the available energy delivered by pulsed lasers makes it possible to study the mechanisms of relaxation with very low probabilities which cannot be observed with the excitation provided by a classical light source.

Thanks to lasers, a real-time spectroscopy has been born which gives rise to new experimental studies which in turn have already yielded countless new results in physics (induced transparency, spin echo, to mention just a few examples), and in physical and biophysical chemistries (the study of fast fundamental mechanisms for instance). The reader can discover these in Bibliographic references 7.6.3 and 7.6.4.

— *Frequency concentration:* By analyzing both theoretically and experimentally some typical fundamental research experiments, we intend to show how accurate frequency analyses of the optical properties of atoms and molecules in gases, liquids, and solid states (*cf.* Problem VII.5.4) are made possible by using lasers with a high frequency stability. Before the discovery of lasers, the positions and the widths of the characteristic spectral lines of an atom or a molecule were measured with spectrographs. Spectrographs analyze the light absorbed or emitted by the sample. The incoherent light source emits a wave whose spectrum is wide compared to the width of the lines to be analyzed, and the quality of the experiment (especially the resolution of the system) is tied to the quality of the spectrograph. It is very natural therefore that spectrograph technology evolved very rapidly between the nineteen forties and the sixties.

On the contrary, in a laser spectroscopy experiment, the classical source (a lamp with wide spectrum) is replaced by a laser which has a tunable central frequency and whose spectrum can be very narrow with respect to the spectral width of the resonance line. Since the atomic material is excited selectively, there is no longer any need to analyze the light absorbed by the material with a spectrograph. The experiment therefore simply consists in recording the variation of the signal emitted by an appropriate radiation detector (photodiode, photomultiplier or photon-counting device) as the frequency of the quasi-monochromatic laser sweeps a given range. The spectrum obtained in this way shows all the possible absorption frequencies of the material as well as their corresponding line shapes. The resolution of the experiment (that is, the ratio $R = \nu/\Delta\nu$ of the average frequency ν used to the minimal distance $\Delta\nu$ between two separable lines), is no longer determined by the detection system as in the previous case, but by the properties of the source. By using lasers, we currently reach resolutions R of about 10^9, while the best spectrographs have their resolution limited to about 10^6.

In the following, we shall only speak about frequency concentrated laser spectroscopy.

However, to be exhaustive, we want to call attention to a possibility which is presently little exploited in electronic absorption technologies, but which could lead to interesting experiments in the future, namely *Intracavity Laser Absorption Spectroscopy* (ICLAS ; *cf.* Bibliographic reference 7.6.5). The basic idea is very simple: A very weakly absorbing sample of usual length (about a centimeter) placed inside a laser cavity is swept by the wave at every back-and-forth trip of the light in a linear cavity, or at every round trip in a ring cavity. In a first approximation, the wave acts as if the optical length of the sample were extraordinarily long. In a recent experiment, an equivalent length of 900 Km was obtained! Because a multimode laser is highly sensitive to very weak absorptions inside the cavity, extremely weak absorptions (on the order of 10^{-9} cm^{-1}) can be detected with very good spectral resolution (on the order of 10^{-6}). Various application in domains such as the environment (detection of trace molecules), microelectronics (detection of trace poisons), biology and medicine are currently exploited.

VII.2. Laser sources most frequently used in absorption spectroscopy and their characteristics

In order to record the frequencies and widths of the transition lines of atoms and molecules, we need to know the specific properties of the source used to study them. Here we recall the properties of the laser:

— *The source frequency can be adjusted*: To analyze a transition frequency ω_0, the source frequency ω should be adjusted so as to lie in the vicinity of the characteristic frequency ω_0. It should therefore be possible to use the source correctly in a frequency domain as wide as possible so that a large number of different transitions can be studied. Such an adjustable source is said to be *tunable*.

— The linewidth of the laser is very much smaller than the linewidth of most studied transitions.

— *Possibility of sweeping the source frequency:* Once tuning is achieved, it should be possible to vary the frequency ω continuously in a controlled way in a restricted domain around ω_0, whose width is larger than the width of the studied transition.

— *Possibility of locking the frequency ω:* The above described sweeping is efficient only if the time fluctuations of the frequency are very small during one sweep. It is therefore essential to be able to lock the frequency of the source.

— *Intensity of the source:* The recorded signals are usually very weak. Moreover, as we shall see later, some spectroscopy techniques require that the studied transitions be saturated. In this case, the laser source should therefore be rather powerful, even though the careful experimenter will always keep its power as small as possible in order to avoid undesirable secondary phenomena.

— *Possibility of intensity stabilization of the source:* The to be detected signals are often so weak that experiments in laser spectroscopy sometimes last a long time (up to several days). It is therefore of the utmost importance to stabilize the intensity of the sources used in the experiment as well as possible.

These six properties define the ideal laser source in frequency-resolved spectroscopy. Almost all laser sources satisfy the last five criteria: They are powerful and their intensity can be stabilized, their linewidth can be reduced by selecting a single longitudinal mode, they can be frequency stabilized and frequency swept. In fact, it is the first criterium which imposes the choice of the laser. To have a tunable laser imposes the choice of a liquid or solid amplifying material with a large bandwidth, thus eliminating gas lasers, also called fixed-frequency lasers, which can sweep only a very limited spectral range. The two amplifying materials most used at present are liquid dye solutions (*dye laser*) and the sapphire monocrystal doped by Ti^{3+} titanium ions (*titanium-sapphire laser*). In the case of the Ti-sapphire laser, it is often essential to generate, using nonlinear optics, one of the first harmonic waves (usually the first or the second one) of the fundamental wave which is unfortunately situated at the upper limit of the visible part of the spectrum. This handicap explains why at present the ideal laser source remains the dye laser, in spite of the drawbacks due to the liquid state of its amplifier and the complicated technology resulting from this. To be exhaustive, let us add that in certain cases, gas lasers can be used for a high-frequency resolution analysis of transitions in atoms and molecules whose resonance frequency ω_0 happens to coincide with one of the emission frequencies of the gas laser. But this possibility remains very limited.

We shall restrict ourselves to the description of the dye laser and try to show how all the six requisite properties can be conferred to this laser.

VII.2.1. Choice of the frequency range and tuning

In a dye laser, the liquid amplifier consists of dye molecules (rhodamines, coumarins, xanthens) in a solution (10^{-3} to 10^{-4} mol.l^{-1}) of water, methanol or ethylene glycol. Every electronic state has a large number of vibrational-rotational states and each of these states presents a strong homogeneous broadening due to the environment. This results in an overlapping of the energy of the states so that a continuous energy band is formed (*cf.* Figure 7.2.1).

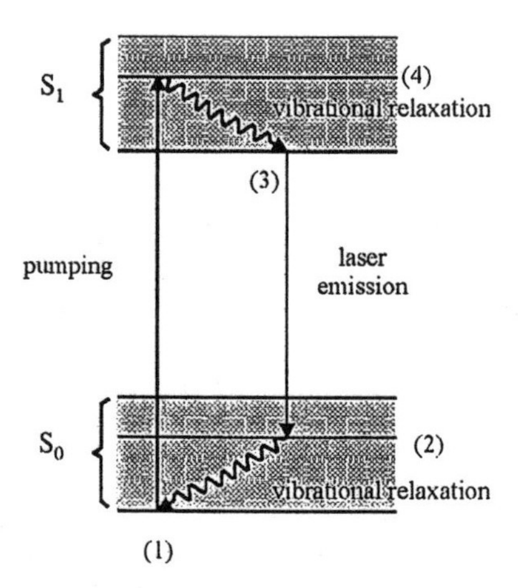

Figure 7.2.1: Diagram of a dye amplifier. The vibrational structures of the electronic bands S_0 and S_1 are not specified.

The 1 → 4 excitation is achieved by optical pumping using a traditional lamp or an auxiliary laser (with intensities around a few hundred kilowatt per squared centimeter). Vibrational state 4 relaxes rapidly (in a few picoseconds) to state 3, whose population is greater than the population of each of the vibrational states of the thermally weakly populated band S_0. A population inversion occurs between level 3 and a continuous energy band of levels 2. S_0 and S_1 represent the fundamental (S_0) and the excited (S_1) electronic singlet states. The laser functions according to the simplified model of the four-level laser. Actually, this ideal behavior is somewhat perturbed by the possible intersystem conversion of the singlet S_1 towards a triplet state of lower energy T_1, followed by a possible absorption of the wave pump by a second triplet T_2. To suppress these disastrous phenomena which reduce the laser intensity, the molecules which reach the triplet state are eliminated by using a circulating dye solution in the form of a few mm thick jet flowing with a velocity approximately equal to the ratio between the diameter of the pumped dye (about ten microns) and the population time of the triplet (which depends on the solvent and can reach the microsecond in a viscous material like ethylene glycol). These velocities are of the order of ten meters per second. The laser is tunable over an approximately 100 nm wide band, which corresponds to almost all of the fluorescence band with the exception of the spectral zone where absorption and fluorescence overlap.

Tuning is traditionally achieved by inserting a Lyot filter in the optical cavity. It consists of a set of several birefringent plates of different thickness. The laser beam, linearly polarized because the dye jet satisfies the Brewster condition, falls at Brewster incident angle on the quartz plates cut parallel to the optical axis. These plates are placed in the Brewster condition so as to minimize the losses of the cavity. The incident vibration is split into an ordinary vibration (refractive index n_o) and an extraordinary vibration (with a refractive index depending on the position of the optical axis with respect to the incidence plane). At the exit of the Lyot filter, a vibration with the same linear polarization as the incident vibration appears only if the relation $\Delta = k\lambda$ is satisfied [Δ is the optical path difference $(n_e-n_o)l$ induced by the filter crossed at a thickness l]. The wavelengths which are thus selected and whose polarization does not change during the crossing of the plates, undergo the least losses by reflection at the various dioptres positioned at Brewster angle inside the cavity. These wavelengths are therefore dominant and prevent other wavelengths from oscillating. To obtain a narrow bandwidth (of the order of a hundred nanometers) around one single wavelength, several plates (usually three) of different thickness are used. The central wavelength selected by the Lyot filter can be changed by rotating the plates around their normal direction. Varying the angle between the optical axis and the incidence plane changes n_e and therefore Δ and λ.

VII.2.2. Mode selection and width of the selected line

Many longitudinal modes oscillate in the tuning band. Of course, in order to benefit from the very narrow spectral width of the laser, just one single longitudinal mode must be selected within the band.

Since the longitudinal modes are spaced approximately 150 MHz apart for a cavity length of 1 m, about one hundred modes exist in the tuning band. The selection of a single mode is operated by inserting an auxiliary Fabry-Pérot filter inside the cavity. This filter is a transparent plate of thickness e and of refractive index n, slightly tilted on the optical axis by an angle θ, and it introduces minimal losses for those wavelengths λ which satisfy the relation $k\lambda = 2ne\cos\theta$, where k is an integer. Since it is impossible to calculate the relative losses needed to obtain a monomode behavior because of the complexity of the intermode competition phenomena, the choice of these filters is relatively empirical : The parameters n, e and q are chosen in such a way that the resonance spectral width of the filter is smaller than the intermode interval inside the laser cavity. Under these conditions, only one mode oscillates in the cavity.

VII.2.3. Frequency sweep of the selected mode

The usual method to obtain a continuous sweep of the central frequency of the longitudinal mode selected in the tuning band, is to vary the length of the cavity continuously because the frequencies of the modes giving rise to laser emission depend very closely on this length. To obtain this change in length, one of the mirrors is fixed on a piezoelectric device with which small displacements of the mirror can be achieved along the axis of the cavity. A displacement of about 2µm of the mirror corresponds to a change of laser frequency of the order of 1 GHz in the visible range. However, so as not to lose the selected longitudinal mode during this change of wavelength, the incidence angle θ of the mode selector must be modified simultaneously by rotating the selector. This is also operated under piezoelectric control.

VII.2.4. Frequency locking of the selected mode

The linewidth of the longitudinal monomode laser is determined by the spectral width of its mode. But the fluctuations of the optical length of the cavity induce fluctuations in the central frequency. Both the mechanical variations of the length and the variations of the refractive index (mainly due to acoustic modes in the cavity, to the uneven flow of the dye, and to thermal variation of the optical length of the Fabry-Pérot selector) cause both long term drifts and fast fluctuations of the optical length of the cavity. It is therefore indispensable to lock the frequency of the longitudinal mode selected for the experiment. The optical length of the laser must be servo-controlled to a reference length which is more stable than the cavity length to avoid frequency fluctuations of the mode. Usually, the reference length consists of an auxiliary Fabry-Pérot cavity, maintained at sufficiently constant temperature and pressure so that its optical length can be considered as stable and taken as a reference. To achieve servo-control, a piezoelectric control of the position of one of mirrors of the laser can be inserted so that the optical length can be varied in such a way that the frequency of the laser mode remains locked on the edge of the transmission curve of the reference resonator. The principle of a recent realization is shown on Figure 7.2.2.

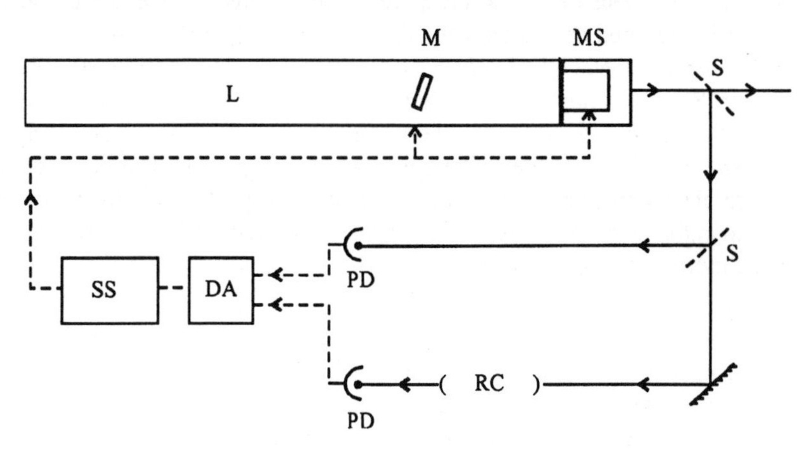

Figure 7.2.2: Principle of a frequency servo-control of a longitudinal monomode laser. L: laser; MS: exit mirror on a piezoelectric mount; S: beam splitter; RC: reference cavity; MS: mode selector; PD: photodiode; DA: differential amplifier; SS: servo system.

A small part of the exit intensity of the laser is sent onto the servo system by a beam splitter. It is divided into two beams by a separator. The first beam falls directly on a photodiode, the second one crosses the reference cavity before falling on a second photodiode. The electric currents delivered by the two photodiodes are sent onto the entrance of a controlled differential amplifier so that the error signal has zero value when the frequency of the laser mode corresponds to a predetermined transmission of the reference resonator. Any frequency fluctuation around the reference value gives an error signal whose sign depends on the sign of the fluctuation. The servo system then sends a corrective voltage to the piezoelectric holder carrying the exit mirror of the laser, thereby stabilizing the frequency of the mode. To sweep this frequency-locked laser, one can just vary the optical length of the resonator by using a mirror fixed on another piezoelectric holder or by varying the pressure, and thereby the refractive index, inside the reference cavity. The resonance curve of the resonator is thus frequency displaced when the length varies, and the frequency of the laser mode, locked on the edge of the transmission curve, follows this transfer thanks to the corrective voltage delivered by the servo system. However, as we mentioned before, the resonance frequency of the longitudinal mode selector must be displaced simultaneously during the sweep.

This type of servo system presents some essential drawbacks. The main one is that the high-frequency fluctuations are difficult to correct. Indeed, to get a sweep on an extended range, the laser cavity length must be submitted to large piezoelectric displacements which need low resonance frequencies. This requires a servo system bandwidth of the order of the kilohertz which does not correct the rapid frequency fluctuations very well. Various improving processes have been suggested and experimented (using two different holders simultaneously, one with a low resonance frequency and the other with a high resonance frequency, fixed on two different mirrors of the laser ; modulation of the refractive index of an element inserted in the cavity of the laser to avoid piezoelectric devices, etc.). Servo-system with bandwidths up to the megahertz were thus achieved.

Last, we want to point out that the stability of the laser is that of the reference cavity. However, it remains sensitive to the temperature and pressure fluctuations which always subsist, in spite of all precautions taken. In more sophisticated systems, the length of the reference cavity is locked on an atomic reference (He-Ne laser locked on a transition of molecular iodine), and a stability of the order of the kilohertz can be obtained by using high tech devices. We shall see that this is more than sufficient in most cases.

VII.2.5. Source power

Hitherto, we did not speak about the structure of the optical cavity of the dye laser. We know that there are two main types of cavities: The linear cavity and the ring cavity. The simplicity of the linear cavity is attractive. But it leads to powers which are too weak for our purpose since they are of the order of a few hundred milliwatt in the monomode regime. Indeed, the competition between longitudinal modes is not the same for all the modes because of the stationary wave structure inside the amplifying material. Inside the amplifying material, two modes do not interact very much if the antinodes of the electric field of a mode correspond to the nodes of the other. The large difference which exists between the available powers in the multimode and in the monomode regimes is due to this small competition. In order to get large powers in monomode regime, we are tempted to increase the power of the pump laser, but then intermode interaction disrupts the monomode regime.

The ring cavity therefore imposes itself. Indeed, if we encourage one of the two propagation directions for the electromagnetic wave by creating losses on the other direction with the help of an element called *optical diode* (*cf.* Bibliographic reference 7.6.3) Thus, only one progressive wave can oscillate since the progressive wave turning in the opposite direction is prevented from satisfying the oscillation condition. So, contrarily to the case of the linear cavity where a stationary wave exists, here the amplifying material is submitted to a progressive wave. All points of the amplifier interact identically with the wave. In the monomode regime, when the power of the pump laser is increased, the selected longitudinal mode prevents the others from oscillating, and this for all values of the pump power. Parasitic modes no longer exist so they cannot destroy the monomode regime. In this way, powers of the order of the watt can be obtained from a pump power of about 5 Watt. This is more than sufficient for most applications.

VII.2.6. Power stabilization

As in most c.w. lasers, an intensity stabilization on a predetermined value is obtained by using a servo system acting on the pump power.

VII.3. Doppler free spectroscopies

As indicated earlier, discrete spectroscopic structures exist in gases and liquids but they are completely masked by the Doppler effect. These structures are very important from a theoretical point of view. The use of a highly monochromatic light source, like a monomode and frequency-locked laser, does not solve this problem entirely: We need to imagine Doppler-free methods to benefit from the excellent resolution of the laser source. In fact, there are several types of Doppler-free spectroscopies, some of which existed before the discovery of lasers.

First of all, we can mention the so-called *optical-pumping methods* (Kastler, 1950), in which an excited state is populated by using a classical source and then the evolution of this state is analyzed as it is voluntarily submitted to an external constraint (electric or magnetic field, for example). In certain conditions, this evolution is not affected by the Doppler widening. We can mention, for example, the *Hanle effect* (fluorescence under magnetic constraint). The fluorescence depolarization is independent of the molecular velocity and therefore of the Doppler effect. In fact, the width of the curve plotting the variations of the depolarization factor as a function of the amplitude of the applied magnetic field yields the product of the Landé factor by the lifetime of the considered level. But the Hanle effect can not reveal hyperfine structure because the recorded signal is the sum of the responses of all the populated hyperfine levels.

A second example is given by the *double resonance* method. This method consists of inducing transitions between neighboring levels (hyperfine levels, for example) in an excited state by submitting the sample to the field of a hyperfrequency, radiofrequency or infrared wave. The resonance between the frequencies of this field and those of the induced transitions can then be detected. In fact, these resonances are not Doppler free. However, because the Doppler widening in the radio or the hyperfrequency domains is so small compared to the homogeneous widths of the transitions, this method can be considered as a Doppler-free method. Unfortunately, the amplitude of the low-frequency field must be sufficiently great, especially in the molecular domain, for any detection to be possible, so that this method remains difficult to operate in most cases.

The second type of Doppler-free methods is based on the principle of selecting a molecular axial-velocity class in the sample. The signal arising from this velocity class is of course not widened by the Doppler effect. The use of *atomic or molecular jets* constitutes a first and already old solution which has given ample proof of its efficiency. However, besides the inherent technological difficulties to obtain good molecular jets for certain categories of molecules, the requirement of a correctly collimated molecular jet is incompatible with that of a large density of molecules in the jet. This means that the larger the required resolution, the weaker the signal which must be detected. This constitutes a major drawback for the study of optical transitions with small probabilities.

A second way to select an axial velocity class takes advantage of the nonlinear interaction which exists between matter and the electric field carried by an intense optical wave. This method is possible only with lasers. It was practically born with them and is called *laser-saturated spectroscopy*. When a gas or a liquid interacts with a monochromatic wave (of wavevector $\mathbf{k}$) whose frequency ω is very close to the molecular resonance frequency ω_0, only those molecules whose axial velocity satisfies the equation $kv_z = \omega - \omega_0$, or $v_z = c(\omega - \omega_0)/\omega$ (if the homogeneous width is assumed to be infinitely narrow) are optically excited. The detection of this perturbation, which concerns only one velocity class, provides a Doppler-free signal. In particular, in the so-called *saturated absorption spectroscopy* method, the perturbation created in the material by an intense wave (pump wave) is tested by a second wave with the same frequency propagating in opposite direction (probe wave). The perturbation created by the pump wave can also be detected by using a second transition which has a common level with the saturated transition. This is the case of optical-optical *double-resonance methods* which present the drawback that they require two different laser frequencies, one for the pump wave and the other for the probe wave.

Finally, a third type of Doppler-free methods exists. These also depend on the development of the laser and are called *multi-photon absorption methods* (*cf.* Problems VII.5.3 and VII.5.5). Let us consider a transition involving only one photon and two levels with the same parity. This transition is forbidden within the electric dipole approximation. However, it becomes possible by simultaneous absorption of two photons during which the atom passes through an intermediate non resonant state. If the two absorbed photons belong to two waves with opposite wavevectors, the resonance condition can be written as

$$\omega_0 = (\omega - kv_z) + (\omega + kv_z) = 2\omega$$

and this condition does not depend on the speed of the molecules.

When the atom is submitted to two waves of the same frequency ω and with opposite directions, one obtains a resonant Doppler-free signal for $\omega = \omega_0/2$. The advantage of this two photon Doppler-free method is that the forbidden one-photon transitions can be studied with very high resolution. Moreover, it is a method in which all the molecular speed classes contribute to the signal. However, the transition probabilities can be extremely weak in this type of spectroscopy.

These last two examples will now be presented in more detail.

VII.3.1. A first example: Saturated absorption spectroscopy

VII.3.1.1. Principle of the method

Let us consider a cell containing the substance to study. It is illuminated by an intense laser beam with a frequency ω in the vicinity of ω_0, the resonance frequency of the system (*cf.* Figure 7.3.1). This wave, which is a high intensity wave and is called the pump wave, modifies the population equilibrium of levels a and b of the molecules susceptible to absorb its photons, as was explained earlier. The absorption condition is satisfied if the molecule "sees" the photon at the transition frequency ω_0 in its own referential frame. This occurs when the equation $\omega - k_v z = \omega_0$ is satisfied. The distribution curve of the population difference D_0 in the absence of a saturating wave (*cf.* Section IV.3.2.2), plotted as a function of v_z, is a Gaussian curve, in agreement with the Maxwell-Boltzmann distribution. The pump wave populates level b and digs a hole in the distribution at the abscissa $(\omega - \omega_0)/k$. When ω varies, the hole moves in the distribution curve. The absorption signal of the saturating wave has a Gaussian shape which reflects this distribution. Let us now send a second wave onto the cell, with the same frequency (it is very often provided by the same laser), but propagating in opposite direction. This wave, which tests the perturbation created by the pump wave, is called the probe wave and it has a considerably lower intensity than the pump wave.

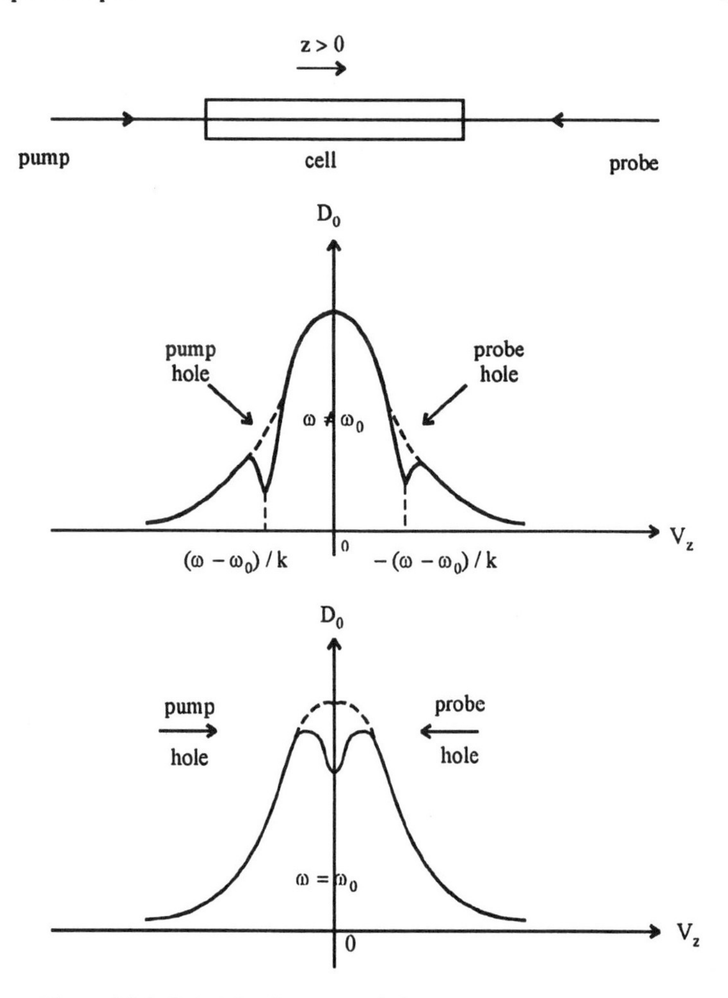

Figure 7.3.1: Principle of a saturated absorption experiment.

It interacts with molecules whose velocity is given by the equation $-kv_z = \omega - \omega_0$, that is, whose velocity is $v_z = -(\omega - \omega_0)/k$, and it digs a second hole in the velocity distribution. These molecules are different from those which interacted with the pump wave so that the probe wave is completely "unaware" of the presence of the pump wave in the cell. The two holes in the velocity distribution are symmetrical. Their width is given by the relation (*cf.* Section IV.3.2.2)

$$\Delta v_z = \gamma_{ab}\sqrt{1+S}\,/\,k \tag{7.3.1}$$

The absorption coefficient of the probe wave remains independent of the pump wave when ω varies *as long as these holes do not overlap*. However, when the frequency ω of the laser reaches the value ω_0, that is, when $v_z = 0$, the holes overlap. In this case, the probe wave can only interact with molecules which also interaction with the pump wave — those whose velocity is perpendicular to the wave vectors of both waves — and it analyzes the perturbation created in the material by the pump wave. If the studied material is initially absorbing, the probe wave "sees" a material whose population difference is smaller than that corresponding to thermal equilibrium. The transmitted intensity of the probe wave then presents the shape shown on Figure 7.3.2.

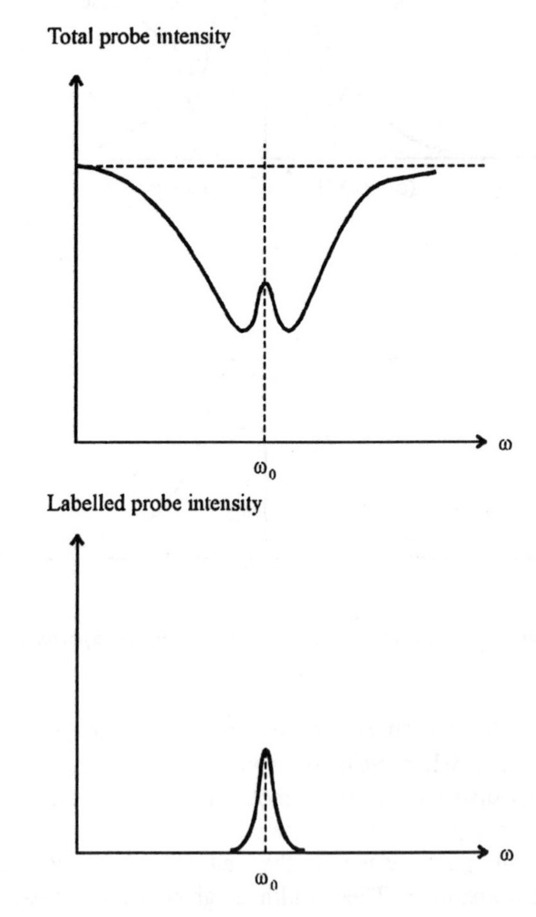

Figure 7.3.2: Variations of the intensity of the probe wave as a function of its frequency. Top: Total intensity of the probe wave. Bottom: Intensity of the part of the probe wave labeled at the modulation frequency of the pump wave.

By isolating in the intensity of the probe wave only that part which depends on the intensity of the pump wave (the so-called *labeled* intensity), we get a signal whose width is roughly equal to the homogeneous width of the transition. This method has been suggested independently and almost simultaneously by the research groups of C. Bordé (1970) and that of T. Hänsch (1971).

It has been widely exploited by the research group of B. Couillaud and A. Ducasse. It is interesting to note that this spectroscopy had a forerunner, the so-called " Lamb dip " spectroscopy. In the latter, an absorbing cell is placed inside a linear laser cavity and is submitted to two progressive waves propagating in opposite directions. The study of the exit power of such a laser as a function of the frequency reveals a hole which arises for the same reason as that of the above described signal and it is also exempted of Doppler effect.

VII.3.1.2. Theoretical analysis

To determine the probe signal, the propagation of the probe field in the material must be known. However, because of the probe-pump interaction it is incorrect to treat the propagation of the two fields independently. We must therefore write the wave equation for the total field in the form:

$$\partial^2\mathbf{E} / \partial z^2 - (1/c^2)(\partial^2\mathbf{E} / \partial t^2) = \mu_0(\partial^2\mathbf{P} / \partial t^2) \qquad (7.3.2)$$

with

$$\mathbf{E} = \mathrm{Re}\{\mathbf{A}_+(z)\exp[+i(\omega t + kz)] + \mathbf{A}_-(z)\exp[+i(\omega t - kz)]\} \qquad (7.3.3)$$

A_+ and A_- designate, respectively, the slowly variable complex functions of the pump and probe fields propagating along the +Oz and the – Oz axes. For simplicity, it is assumed that the fields are collinear and linearly polarized along the Ox axis. In this case, we only need to solve a few scalar equations. **P** is the polarization of the material submitted to the influence of the two fields. We shall write **P** in the form given by Equation 4.3.10.

By assuming that $\partial^2 A / \partial z^2 << k(\partial A / \partial z)$ (*cf.* Section II.2.2.3 and Problem II.3.2), we get the equation

$$(\partial A_+ / \partial z)\exp(ikz) + (\partial A_- / \partial z)\exp(-ikz) = (ik / 2\varepsilon_0)\underline{P}(z) \qquad (7.3.4)$$

To continue the calculation, we must know the function P(z). This can be determined only by writing the evolution equation of the coefficient ρ_{ba} of the population rate matrix of the material. This equation has already been written (*cf.* Equation 4.3.30) for the case when the center of mass of the system was assumed to be fixed. Let us now assume that this center of mass is linearly and uniformly moving with a speed v_z, as given by the relation $z = z_0 + v_z t$. The total time derivative of the population rate operator then takes the form

$$(d/dt)[\rho(z,t)] = \partial\rho/\partial t + v_z(\partial\rho/\partial z) \tag{7.3.5}$$

and Equation 4.3.30 can be written as

$$\begin{aligned}[\partial/\partial t + v_z(\partial/\partial z)]\rho_{ba} &= -(i\omega_0 + \gamma_{ba})\rho_{ba} \\ &+ (2\pi i/h)P(\rho_{aa} - \rho_{bb})\{A_+ \exp[i(\omega t + kz)] + A_- \exp[i(\omega t - kz)]\}\end{aligned} \tag{7.3.6}$$

The integration of this equation, which we shall not detail here, leads to the following approached expression of the susceptibility

$$\begin{aligned}\chi = &-(2\pi i P^2 D_0/h\varepsilon_0)\int_{-\infty}^{+\infty}\exp\left(-v_z^2/v_0^2\right)dv_z \\ &\times\left\{1 - S_+/\left[1 + S_+ + (\Delta\omega_+/2)^2\right]\right\} \\ &\times\left\{1/\left[\gamma_{ba} + i(\Delta\omega_-/2)\right]\right\}\end{aligned} \tag{7.3.7}$$

D_0 and S have been defined in Section IV.3.2.2. S_+ concerns the pump wave. We define $\Delta\omega$ by the equation $\Delta\omega_\pm = 2(\omega - \omega_0 \pm kv_z)/\gamma_{ba}$ and v_0 is the characteristic velocity of the Maxwell-Boltzmann distribution given by $v_0^2 = 2RT/M$. The above expression consists of two terms. The first corresponds to the already formulated expression of the susceptibility in the absence of saturation (it is the term corresponding to the weak intensity probe wave). As an exercise, the reader can check that this term corresponds to the expression of $<\mathbf{P}>/\varepsilon_0 E_0$ given by Equation 4.3.35 in the case of $S = 0$. The integral of the second term is a product of two resonant denominators for the values $-kv_z$ and $+kv_z$ of the angular frequency ω. The integral will be different from zero only when these two resonance conditions are satisfied simultaneously. It yields the Doppler-free contribution.

It is easy to isolate the imaginary parts of the refractive index and of the intensity absorption coefficient of the probe wave from Equation 7.3.7. By defining $\Delta\omega = 2(\omega - \omega_0)/\gamma_{ba}$, we obtain the expression

$$\begin{aligned}\alpha = &-\left\{4\pi^2 P^2 D_0/\varepsilon_0 h\lambda\gamma_{ba}\left[1 + (\Delta\omega/2)^2\right]\right\} \\ &\times\left\{\left[\Delta\omega/\left(1 + \sqrt{1+S_+}\right)\right]^2 + 1/\sqrt{1+S_+}\right\} \\ &/\left\{1 + \left[\Delta\omega/\left(1 + \sqrt{1+S_+}\right)\right]^2\right\}\end{aligned} \tag{7.3.8}$$

When $S_+ = 0$, this expression gives the weak intensity limit ($S = 0$) of Equation 4.3.36, valid for only one wave propagating in the material. The saturated absorption signal is a Lorentzian with a width given by

$\gamma_{ba}\left(1+\sqrt{1+S_+}\right)/2$ and an amplitude proportional to $1/\sqrt{1+S_+}$ (total probe signal) or to $1-1/\sqrt{1+S_+}$ (probe signal labeled at the pump wave modulation).

VII.3.1.3. Experimental setup

A possible setup is described on Figure 7.3.3.

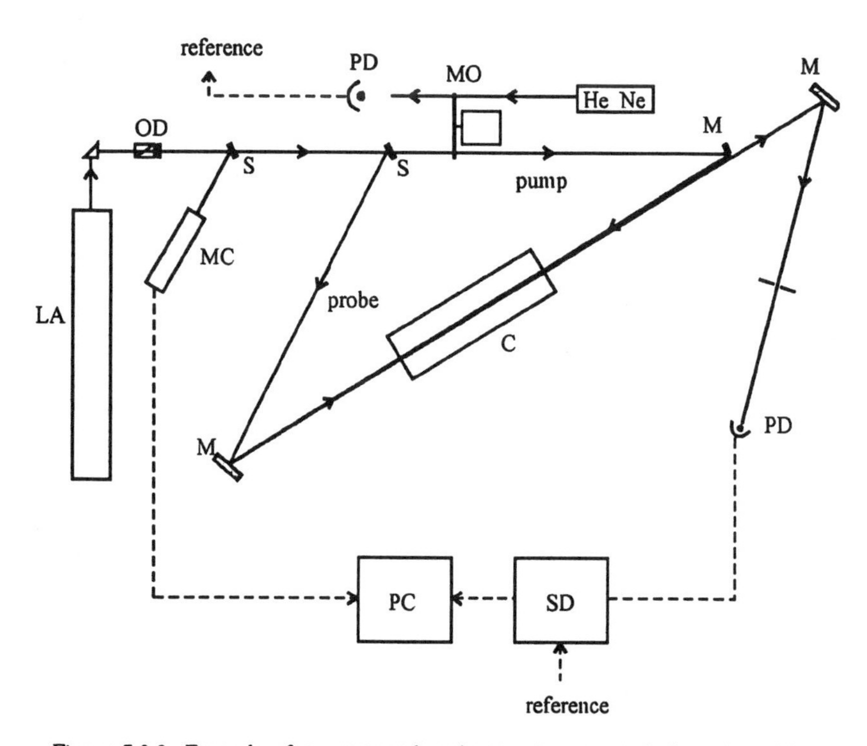

Figure 7.3.3: Example of experimental realization in saturated absorption. LA: tunable laser; MC: mode control unit; OD: optical diode; M: modulator; C: cell; M: mirror; PD: photodiode or photomultiplier; SD: synchronous detection; PC: microcomputer.

In this configuration, some care must be taken so as to avoid a few technical difficulties. In order to obtain an accurate superposition of the propagation directions of the two waves provided by the same laser inside the absorption cell, the ring cavity built at the exit of the laser should be of very high quality. Simple geometrical considerations show that, in these conditions, part of the intensity of each of the two waves will inevitably travel back inside the laser in a certain direction, and the better the probe and pump waves are superposed, the closer this direction will be to that of the stationary wave inside the laser. Such a return of the waves in the laser source is catastrophic for its characteristics (in particular for the purity and stability of its frequency). We use an optic insulator to minimize this problem, but it is difficult to get a very good insulation in a simple way. The second technical problem which arises is of the same type as the first one and concerns the separation of the useful signal from the probe wave which carries it. This useful signal is generally very small (1/10 or 1/100) with respect to the linear absorption signal on which it is superposed. To detect it, the classical technique of synchronous detection must be used. The pump wave is modulated in amplitude at a low-frequency, very often with the help of an electro-optical device. Only that component which is modulated synchronically with the pump modulation is recorded.

However, care should be taken to send exclusively the probe wave onto the detector, and to exclude all reflected or parasitic scattering coming from the pump wave. A last very important point concerns the constant control of the spectral quality (frequency and intensity) of the tunable monomode laser source. As mentioned before, a monomode laser (dye or Ti/sapphire laser) is locked onto one of the resonance frequencies of a Fabry-Pérot resonator which is very carefully frequency stabilized and which is tunable by modifying its length (with the help of a piezo-electric holder) or by varying the pressure of the gas that it contains. The whole of the servo-control is driven by a microcomputer which also controls the detection of the absorption signal. The power of the pump wave varies between about ten milliwatts and a watt. The probe wave is less intense (a few milliwatt). Generally, it is advisable to avoid — if possible — to focus the waves too much inside the cell (the laser beam waists very rarely measure less than about a few hundred microns). Recorded bandwidths are typically of the order of a few megahertz.

VII.3.1.4. Example of a result

We shall only recall one of the first very important results obtained in the seventies and concerning the *hyperfine structure* of the iodine molecule. Hyperfine structure of a vibrational-rotational level is induced by the nuclear spin which must be taken into account in the calculation of the total molecular kinetic moment. For example, the hydrogen atom has a nuclear spin with a value of ½ and the $^2S_{1/2}$ level in fact consists of two satellite lines separated by 0.0475 cm^{-1}. To this very small energy gap corresponds a wavelength of about 21 cm, and the detection of this radiofrequency emission has been extensively used in the past to determine the hydrogen density in interstellar space. The nucleus of the iodine atom I has a nuclear spin of 5/2, and every vibrational-rotational level of each electronic state consists of 15 or 21 hyperfine components according to the parity of the kinetic moment J. The distance between these components lies between 10 and 150 MHz whereas the Doppler width of the line associated to each of the hyperfine transitions is of the order of 450 MHz. The hyperfine structure of the iodine molecule has been observed correctly only by using atomic jets and, in more deep going investigations, by saturated absorption experiments. Figure 7.3.4 reports some partial results concerning the iodine molecule.

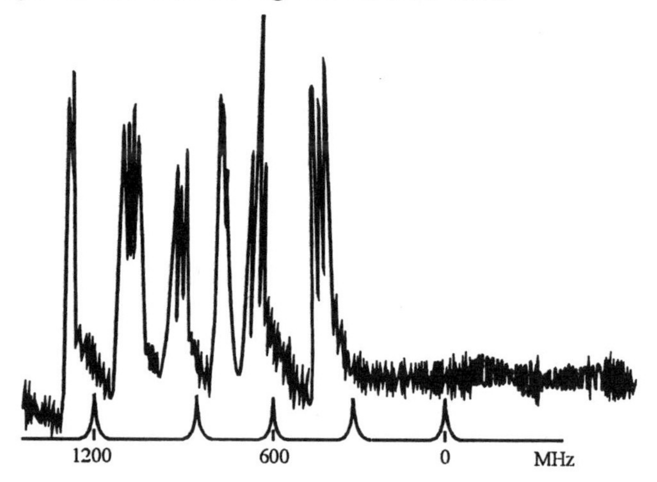

Figure 7.3.4: Hyperfine structure of the R(73) 13-2 line of $^{127}I_2$ (20 of the 21 expected lines are visible); saturation intensity: 2 W.cm^{-2}; iodine vapor pressure: 20 mtorr [from B. Couillaud and A. Ducasse, Optics Communications, 13, 4 (1975)]; reproduced with the appreciated authorization of the editor.

In Bibliography 7.6.6, the reader will find many examples of results, as well as a more detailed description of saturated absorption and of the many derived techniques developed about twenty years ago.

VII.3.1.5. Limitations

They are of many types:

— *Limitations due to the structure of the pump and probe waves*

The probe and pump waves are not perfectly collinear and planar. The non-collinearity of the waves gives rise to a signal which contains a residual Doppler effect. It is then more difficult to calculate the detected signal but it can still be done. It is much more difficult to reduce the limitation due to fact that the waves are not plane and that they have finite dimensions (Gaussian waves). The molecules which give rise to the useful signal in saturated absorption spectroscopy are those corresponding to velocity class $v_z = 0$, that is, those whose velocity is perpendicular to the propagation axis of the wave. If we call ω the radius of both the probe and the pump laser beams, the molecule with velocity v_0 perpendicular to the Oz axis needs an average time to cross the beam equal to $\tau_t = \omega/v_0$ and called the *transit time* of the molecule in the beam. If the relation $\tau_t >> \gamma_{ba} - 1$ is satisfied, the system is in interaction with a field whose amplitude and phase vary little, so that it can be correctly assumed to be driven by a plane wave. On the other hand, if $\tau_t << \gamma_{ba} - 1$, the molecule has considerably moved during the interaction and the results obtained for a plane wave no longer apply. The complete calculation shows that the signal has widened (with a widening of the order of $1/\tau_t$) and reveals a displacement of the resonance frequency which must necessarily be taken into account.

— *Limitations due to the non homogeneity and to the modulation of the saturation parameter*

A molecule does not move much in the beam during its interaction time with the wave. It is submitted to the effect of Gaussian waves whose intensity depends on the distance to the common axis. The absorption coefficient $\alpha(r, z)$ therefore depends not only on coordinate z but also on the distance r of the point in consideration to the axis. This means that the center of the probe beam crosses a more saturated material and is less absorbed than its wings. The detected signal will be affected by this radial dependence of the intensity of the laser wave to the distance from the propagation axis.

In the same way, the low frequency modulation of the pump wave imposed by the synchronous detection complicates the exploitation of the experimental results (width and amplitude of the signal).

— *Ultimate fundamental limitations*

In fact, line broadening can be due to other reasons than just to the first order Doppler effect, the only one we took into account in the foregoing analysis. Let us mention:

– The recoil effect of a molecule when it absorbs a photon (a few kHz in the I_2 molecule)

– The second order Doppler effect, proportional to v^2/c^2 (a broadening of about 100 Hz in the I_2 molecule)

– The effect of the gravity on the molecular trajectories

– The Zeeman effect due to the magnetic field of the earth.

In conclusion, let us recall that here we only want to give one important example of the so-called saturation methods. Several other methods exist, all based on the detection of a saturation induced in the atomic material by a saturating laser beam. They differ by the method used to detect the induced signal. Let us mention, for example, *intermodulated fluorescence* and *saturated polarization spectroscopy* which obtained important results these last few years in molecular spectroscopy.

VII.3.2. A second example: Two-photon absorption spectroscopy

VII.3.2.1. Principle of the method

We showed (*cf.* Section IV.3.2.3) that one-photon transitions between two states with the *same spatial parity* are forbidden within EDA. This is, for example, the case of the simplest transition ls ($^2S_{1/2}$) $\rightarrow$ 2s ($^2S_{1/2}$) in hydrogen. The study of this transition is obviously of high interest: The hydrogen atom is the system for which theoretical predictions have been pushed the farthest and it is indispensable to obtain experimental results which are as precise as possible to control the theoretical validity of the predictions. However, because of the radiative decoupling between the two levels, due to the fact itself that the electric dipole transition is forbidden, the 2s level has a long lifetime (of the order of the second) and, consequently, the homogeneous width of the 1s - 2s transition is small. This means that here there exists a potential for obtaining very precise measurements of the energy gap between the two levels. And it is obvious that studies on other "simple" atoms of the many transitions existing between states with the same parity is definitely of great interest.

But how can this "ban on" be circumvented? The idea to use several photons in order to escape the rule is not recent. However, the laser had to be invented for many-photon absorption techniques to see the light. Figure 7.3.5 presents an illustration of various kinds of spectroscopy which use these techniques. We can see that here there is no resonance phenomenon as in a one-photon absorption. State v is a *virtual* state. Traditionally, it is expressed by using an LCAO expansion developed on the eigenvector basis of the Hamiltonian operator of the microscopic system. The two-level system approximation is not used in this treatment. Here, we shall develop only the case of the two-photon absorption, although the procedure can easily be generalized to many-photon absorption. Two photon absorption was first put in evidence experimentally by Kaiser and Garrett in 1961 (absorption of the red emission of the ruby laser by Eu^{2+} ions included in a fluorine crystalline matrix).

VII.3.2.2. Theoretical analysis

First of all, we have to keep in mind that in the absence of direct resonance, two-photon transition probability is probably very small. Since the population of state b, induced by the very little absorbed optical wave, always remains very small with respect to the population of the ground level a, we need not take into consideration the saturation phenomenon we spoke about earlier.

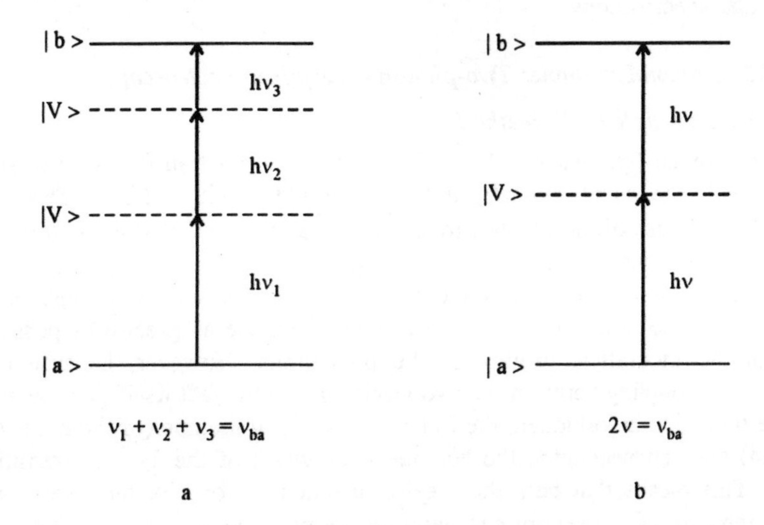

Figure 7.3.5: Illustration of the many-photon absorption; (a) absorption of three different photons; (b) absorption of two identical photons. ν represents a virtual energy state of the system.

To calculate the probability of the transition of state "a" towards state "b", we must analyze the perturbation induced by the electric field of the wave (assumed to be linearly polarized along Ox) on the system initially in state "a". Let us first assume that the systems are at rest and that they interact with only one optical wave with amplitude E_0 and frequency ω. We can develop $\psi(t)$, the state vector of the system on basis (a, b, n) where n indicates any state of the system other than a or b:

$$\left|\psi(t)\right\rangle = a(t)\left|a\right\rangle + b(t)\left|b\right\rangle + \sum_n n(t)\left|n\right\rangle \tag{7.3.9}$$

By using postulate P6 of quantum mechanics (*cf.* Section III.2.1.3) and writing the Hamiltonian within EDA in the form given by Equation 4.3.39, we obtain the evolution equations. The following change of variables

$$\begin{aligned} A(t) &= a(t)\exp(i\omega_a t) \\ B(t) &= b(t)\exp(i\omega_b t) \\ N(t) &= n(t)\exp(i\omega_n t) \end{aligned} \tag{7.3.10}$$

leads to the differential system

$$\dot{A} = (2\pi i E_0 / h)\cos\omega t \sum_n N \exp(-i\omega_n t)\langle a|\boldsymbol{p}|n\rangle \exp(i\omega_a t)$$

$$\dot{B} = (2\pi i E_0 / h)\cos\omega t$$
$$\times \sum_n N \exp(-i\omega_n t)\langle b|\boldsymbol{p}|n\rangle \exp(i\omega_b t) - \gamma_b B / 2 \qquad (7.3.11)$$

$$\dot{N} = (2\pi i E_0 / h)\cos\omega t$$
$$\times \left[A \exp(-i\omega_a t)\langle n|\boldsymbol{p}|a\rangle + B \exp(-i\omega_b t)\langle n|\boldsymbol{p}|b\rangle\right]\exp(i\omega_n t)$$

To integrate the third equation, we assume that $A^{(1)} \approx 1$ and $B^{(1)} \approx 0$ (1st order of the perturbation). We find the following expression for $N^{(1)}$, at the first perturbation order:

$$N^{(1)} = (\pi E_0 / h)\langle n|\boldsymbol{p}|a\rangle\left\{\exp\left[i\left(\omega_{na} - \omega\right)t\right] - 1\right\} / \left(\omega_{na} - \omega\right) \qquad (7.3.12)$$

and if we insert this result into the second equation of 7.3.11, only keeping the main terms depending on $(\omega_{ba} - 2\omega)$ and using AOT, we can calculate the value of B up to the 2nd order of the perturbation:

$$B^{(2)} = -i(\pi E_0 / h)^2 \sum_n \langle n|\boldsymbol{p}|a\rangle\langle b|\boldsymbol{p}|n\rangle / (\omega_{na} - \omega)$$
$$\times \left\{\exp\left(-\gamma_b t / 2\right) - \exp\left[i\left(\omega_{ba} - 2\omega\right)t\right]\right\} / \left[\left(\gamma_b / 2\right) + i\left(\omega_{ba} - 2\omega\right)\right]$$

and the expression of the stationary-state population of level b is expressed by

$$\left|B^{(2)}\right|^2 = (16\pi^2 E_0^4 / h^4 \gamma_b^4)\left[1 / (1 + \Delta\omega^2)\right]$$
$$\times \left|\sum_n \langle n|\boldsymbol{p}|a\rangle\langle b|\boldsymbol{p}|n\rangle / \delta\omega_n\right|^2 \qquad (7.3.13)$$

In analogy with the procedure used for the one-photon absorption in Chapter IV, we write $\Delta\omega = 2(2\omega - \omega_{ba}) / \gamma_b$ and $\delta\omega_n = 2(\omega - \omega_{na}) / \gamma_b$.

If we assume that the population of level b (probability Pb) is given by the two-photon absorption less the relaxation, we write $\dot{\rho}_{bb} = Pb - \gamma_b \rho_{bb}$, or $Pb = \gamma_b \rho_{bb} = \gamma_b \left|B^{(2)}\right|^2$ at steady state. So, the probability of the two-photon absorption can be written as

$$Pb = (16\pi^2 E_0^4 / h^4 \gamma_b^3)\left[1/(1+\Delta\omega^2)\right] \times \left|\sum_n \langle n|p|a\rangle\langle b|p|n\rangle / \delta\omega_n\right|^2 \qquad (7.3.14)$$

Problem VII.5.5 can be considered as an application of the above procedure.

This expression calls for several physical comments:

— The denominator has a resonant form. The transition probability is greatest when $2\omega \approx \omega_{ba}$. In the seventies, Q-switched lasers with high intensities in the red part of the spectrum (ruby laser emitting at 694.3 nm, for instance) were very often used to induce two-photon transitions in certain aromatic polycondensated and heterocyclic molecules which have absorption bands in the vicinity of 350 nm. Indeed, high light intensities, on the order of some MW/cm^2, are needed because of the above mentioned small transition probability. In biophysics, the YAG/Nd^{3+} laser which emits at 1 060 nm was used successfully after frequency quadrupling, because a weak absorption exists at 265 nm (1 060 nm/4) in the aromatic amino acids contained in biological substances.

— The transition probability varies as the fourth power of the electric field, that is, as the square of the intensity of the wave. Two-photon absorption is therefore nonlinear. That fact explains among others why it is so weak. It can be expressed by the imaginary part of a *susceptibility of order n = 3*. This susceptibility induces a *nonlinear polarization of order 3* which, according to the convention defined in Section V.2.1.2, can be written as

$$P_i^{NL}(\omega) = i\chi''_{ijkl}(-\omega,-\omega,\omega,\omega)E_j^*(-\omega)E_k(\omega)E_l(\omega) \qquad (7.3.15)$$

The ensuing tensor of rank n + 1 = 4 is symmetrical with respect to the two first and to the two last indices (similar to the elasticity tensor, for example). It is proportional to the power n + 1 = 4 of the transition moment, in agreement with Equation 7.3.14.

Hitherto, we assumed the presence of only one optical wave in systems at rest. In fact, and as indicated in the introduction, this spectroscopy is used very often in Doppler-free technology. As in saturated absorption, this technology requires two laser waves with the same amplitude E_0, propagating in opposite directions and in interaction with systems whose velocities are distributed according to a Maxwell distribution. Problem VII.5.3 studies such a geometry. In this case equation 7.3.14 becomes

$$\mathrm{Pb} = (16\pi^2 E_0^4 / h^4 \gamma_b^3)\left|\sum_n \langle n|\boldsymbol{p}|a\rangle\langle b|\boldsymbol{p}|n\rangle / \delta\omega_n\right|^2 \times\left[1/(1+\Delta\omega_-^2)+1/(1+\Delta\omega_+^2)+4/(1+\Delta\omega^2)\right] \tag{7.3.16}$$

with $\Delta\omega_- = 2(2\omega - \omega_{ba} - kv_z)/\gamma_b$; $\Delta\omega_+ = 2(2\omega - \omega_{ba} + kv_z)/\gamma_b$.

Three essential contributions to the transition probability now appear. Each of the first two terms corresponds to a two-photon absorption of the same beam and they are affected by the Doppler effect. The last term corresponds to the absorption of one photon of one beam and one photon of the other beam. Taking into account all the speed classes, we obtain for this last term a Doppler-free line, whose amplitude is four times broader than that of the first two terms. The Doppler-free signal is therefore twice as broad as the sum of the Doppler contributions which constitute the background on which the effect appears. We can understand this phenomenon intuitively. Two photons absorbed on the same beam can not be distinguished by the system. On the other hand, when the system absorbs one photon of each beam, the photons can be distinguished in the reference frame tied to the atom since the waves travel in opposite directions. For this reason, the probability amplitude for a system to absorb two photons traveling in opposite direction is equal to the sum of the amplitudes of two processes (first photon on wave 1 and second photon on the wave 2; or first photon on wave 2 and second photon on wave 1). Each of these two processes has an amplitude equal to that of the absorption of two photons on the same wave, resulting in a factor 4 on the total probability.

VII.3.2.3. Experimental realization

A possible experimental setup is described on Figure 7.3.6.

In this experiment, a tunable laser is required, and a pulsed tunable laser is very often used because very high incident intensities are needed. The laser must be very carefully frequency stabilized (by using an auxiliary Fabry-Pérot interferometer) and stabilized in intensity (by using a servo-control). It can be frequency doubled, tripled or quadrupled with the help of nonlinear crystals. The very small population of the excited level is usually deduced from the fluorescence output. Of course, the weakness of both the incident and the transmitted light practically always imposes the use of photomultipliers. As said earlier, this kind of manipulation, usually fully controlled by microcomputers, frequently makes use of the many components provided by modern optoelectronics.

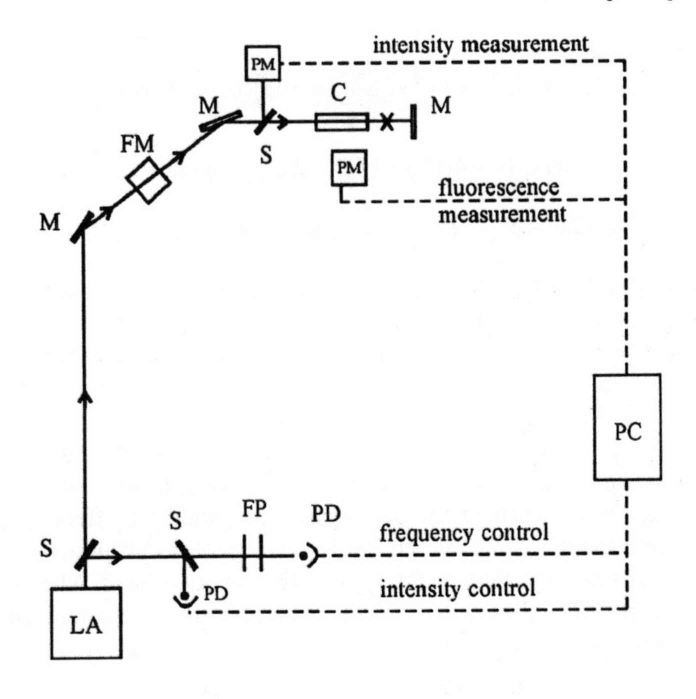

Figure 7.3.6: Principle of the Doppler-free two-photon absorption spectroscopy. LA: Tunable laser; C: cell; PM : photomultiplier; PD: photodiode; FM: frequency multiplier; FP: Fabry-Pérot; M: mirror; S: splitter; PC: microcomputer.

VII.3.2.4. Example of a result

One of the most important results obtained is most certainly the measurement of the *Lamb shift* of the hydrogen atom in state 1s. The shift of the energy of an atomic level is due to the coupling of the atom with the electromagnetic field of vacuum, which in turn induces spontaneous emission. Figure 7.3.7 indicates the energy structure of the singlet levels n = 1 and n = 2 of the hydrogen atom.

The coupling with the electromagnetic field of vacuum has the following effects:

– It shifts the $1^2S_{1/2}$ doublet.

– It breaks the $2^2S_{1/2}$ - $2^2P_{1/2}$ degeneracy.

– It shifts the $2^2S_{1/2}$ doublet.

Hyperfine coupling with the nuclear spin splits the $1^2S_{1/2}$ doublet.

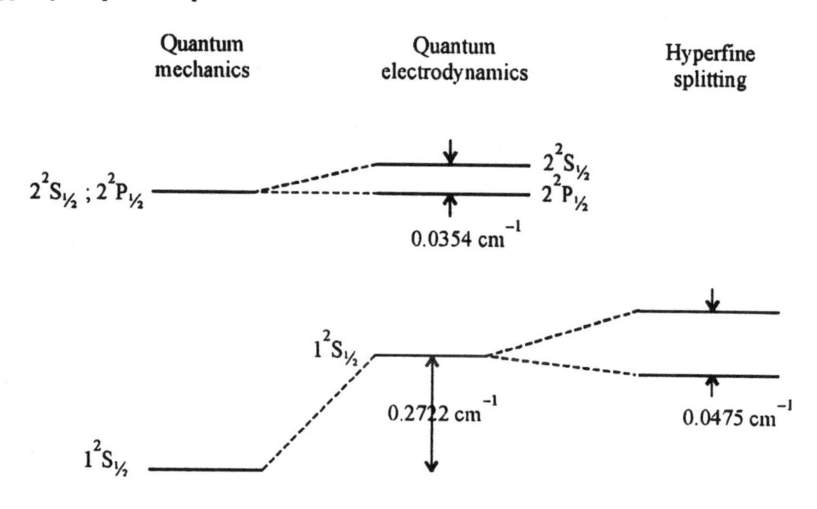

Figure 7.3.7: Energy structure of the singlet states of levels n = 1 and n = 2 of the hydrogen atom.

In spite of its great theoretical importance, the Lamb shift of state 1s of the hydrogen atom could not be measured before the advent of the laser because the 1s - 2s transition is forbidden as we indicated earlier. Calculations gave a value of about 8 GHz. On the other hand, the Lamb shifts of the excited states had been measured with good accuracy thanks to radiofrequency measurements. A extra difficulty came from the fact that value of the Rydberg constant was not known with a very good relative accuracy and this constant is needed for an absolute determination. The difficulty was solved at Stanford by Couillaud and Hänsch who used a relative comparison between the frequencies of the 1s - 2s transition and the β Balmer line (transition $2^2S_{1/2}$ - $4^2P_{1/2}$),. Thus, they were able to measure the Lamb shifts of hydrogen (8.16 GHz) and deuterium (8.18 GHz) atoms and they were also able to determined the value of the Rydberg constant with a relative accuracy of 10^{-9}.

With two-photon absorption techniques in the visible range, highly excited atomic or molecular levels can be reached without the need of an electromagnetic field in the UV range. This is a great advantage because lasers in the UV domain are rare and much more delicate to handle than those in the visible range. Moreover, in two-photon spectroscopy all the molecular velocity classes contribute to the useful signal so that the method is not sensitive to collisions which change the velocities and which complicate the interpretation of the results as is the case in for saturated absorption. On top of this, in spite of the small probability of two-photon transitions, the fact that all molecular velocity classes contribute to the signal considerably reduces the gap between the intensities needed in two-photon transition spectroscopy and those needed in one-photon transitions in saturated absorption. But of course, these two techniques are complementary since they can be used to reach different excited levels starting from the same ground level. Another property of two-photon spectroscopy is that there is no molecular recoil effect because the total momentum of the molecule is preserved during the absorption of two photons of opposite momentum. Last, we would like to point out that the useful signal is no longer affected by saturation because the induced transitions towards intermediate levels are usually far from resonance so that they are cannot be saturated.

VII.3.2.5. Limitations

Some limitations have already been alluded to in the section about saturated absorption spectroscopy:

— The second order Doppler effect limits the detected linewidth in a fundamental way.

— The finite transit time of the microscopic systems in the beam induces a widening of the detected signal.

However, the most important limitation comes from the *light shift* effect. Waves of high intensity are needed to obtain a non negligible probability of two-photon transition and such large values of the intensity induce a considerable displacement of the levels because of the so-called light shift effect. The light shift is a *dynamic Stark effect*, i.e. a shift of the energy levels under the influence of an intense electromagnetic radiation. Its understanding implies a knowledge of the formalism of the *atom dressed by photons* (Cohen-Tannoundji, *cf.* Bibliographic reference 2.5.3) which we shall not develop here. However, the shifts of levels a and b fortunately have a tendency to compensate each other so that the global effect due to the light induced shift usually remains smaller than the natural width of the transition.

VII.4. Selective molecular spectroscopies

VII.4.1. Introductory remarks

Since the beginning of this century, physicists have been looking for direct experimental evidence of the "isolated" atom or molecule. This is, for example, the case of J. Perrin (1920) who tried his whole life long to detect evidence of monomolecular fluorescences in thin films. From the nineteen fifties to the nineteen seventies, the soviet school (Shpol'skii, Personov, *cf.* Bibliographic reference 7.6.7) reported on absorption and emission phenomena revealing a fine-line structure, using cooled solutions of organic compounds in paraffinic solvents. In fact, this structure points to the existence of a population of absorbers (with individual linewidths of the order of a few tens of cm $^{-1}$) distributed in a broad profile envelope. At this time, using the laser and its monochromaticity, it was possible to show that those individual absorptions which were in resonance with the optical wave could be "burned" out of the wide distribution. The so-called *hole burning spectroscopy* was born (*cf.* Bibliographic reference 7.6.8). We shall give a short description of this technique in the next section.

In the nineteen eighties, two decisive steps toward the study of isolated particles were taken in very different ways:

— The discovery of the scanning electron *tunneling microscope* (1983) and, generally, of *near-field technologies* with which showed that the observation of single atoms and molecules had now become an accessible objective.

— The spectacular experiments of *microscopic levitation* and of atomic and molecule *laser-induced trapping*, which have taken a considerable extension in the domain of *optical cooling* these last years, opens up a new possibility for the study of single microscopic systems.

Another decisive step was taken in absorption spectroscopy in 1989 by Moerner and Kador (*cf.* Bibliographic reference 7.6.8), who published the first absorption experiment on single pentacene molecules in an organic crystal.

In recent years, this new so-called *Selective* or *Single Molecule Spectroscopy* has given rise to very important research concerning the emission spectroscopy of single molecules (Orrit and Bernard, 1990, Bibliographic reference 7.6.9). We shall describe some applications of this spectroscopy in this chapter.

VII.4.2. Hole burning absorption spectroscopy

VII.4.2.1. Principle of the method

Electronic absorption spectra of organic molecules dissolved in a crystalline lattice, traditionally recorded at low temperature (often at 4.2 K, the temperature of liquid helium), generally present a few bands which are broader than the predictions obtained from homogeneous linewidth calculations. We mentioned for example a total homogeneous width of the order of 1 000 GHz for a doping molecule in a crystal at room temperature (*cf.* Figure 4.3.1). When the temperature decreases by a factor of about 100, this width, which is essentially due to the phase change induced by the incoherence of the oscillators as seen in Chapter I, decreases by about 4 orders of magnitude to a few hundreds of megahertz. However, even recordings obtained at very low temperature often reveal widths on the order of several hundreds of GHz. We already proposed an explanation (*cf.* Section IV.3.1.1) based on *inhomogeneous broadening* due in our example to a distribution of resonance frequencies induced by the diversity and heterogeneity of the sites within the crystalline lattice (defects). When a very intense and very monochromatic wave laser illuminates such a material, all the molecules in resonance with the wave absorb the radiation. The molecules are lifted up in an excited state which can present some reactivity. These absorbing sites are thus eliminated — we shall say "burned out" — by the wave and the residual absorption spectrum presents a permanent "hole" which can subsist several days in the particular conditions in which it is reversible, or indefinitely if it is irreversible. Its *width is reduced* because of the molecular selection operated by the incident monochromatic wave. But it would be deceptive to believe that the homogeneous width of a transition could be obtained so easily. In fact, the electrons of the dopant are affected by intramolecular vibrations and by the vibrations of the crystalline lattice in solids (intermolecular phonons). A weak phonon-type vibrational coupling ensues which induces a dissymmetry of the shape of the hole. It consists of a very narrow line without electron-phonon coupling (zero-phonon line ZPL) which is usually stretched out towards the shorter wavelengths by a phonon wing (PW). When a phonon wing shows up, that means that part of the energy of the absorbed photon is transferred inelastically to the lattice by the creation of vibrational phonons. In principle, the ZPL contains the total homogeneous width of the transition. Because the probability of creating a vibrational phonon decreases very rapidly with temperature, we can understand that in order to detect exclusively the ZPL requires experiments performed at very low temperatures, implying they take place in the solid state. Here, the reader will recognize a description very similar to that of Section IV.3.2.3, when we used the Born-Oppenheimer approximation in our discussion of the intramolecular vibrational structure of the electronic state. But here vibrations can couple to the lattice. In particular, the relative importance of the two contributions depends, as already noted in traditional absorption spectroscopy, on the overlap of the nuclear parts of the state vectors representing the ground state with those representing the excited electronic states. Figure 7.4.1 gives a very schematic summary of the complete process described above.

VII.4.2.2. Experimental aspects

The experiment usually consists in a comparison between absorption spectra before and after irradiation by the laser wave. Spectra are recorded at low light energy, typically two to three orders of magnitude lower than the burning intensity. The burning intensity can vary from a few microwatt to a few watt per cm^2. The laser used is a frequency stabilized and tunable monomode laser. In the earliest experiments, a dye laser was often used. It is now usually replaced by a

Ti/sapphire laser, whose implantation in laboratories increased very quickly these last few years in connection with the modern frequency-multiplication techniques which are now well established and which offer high energy conversion outputs (several tens of percent). Resolutions of the order of the MHz are obtained this way.

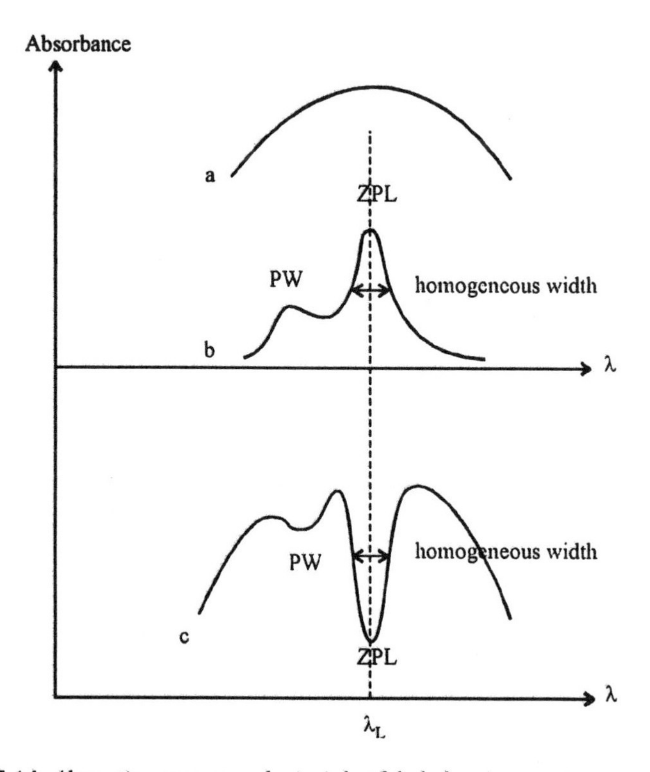

Figure 7.4.1: Absorption process and principle of hole-burning spectroscopy (we neglected the influence of the non resonant molecules in the processes described). a: Broadened spectrum observed in usual electronic absorption spectroscopy; b: Spectrum observed when using a laser source in resonance with the absorbers (very schematically drawn); c: Hole-burning spectroscopy.

The experiment also supposes a constant control of the burning intensity and all the lasers involved must have carefully stabilized intensities. The phenomenon of saturation would indeed be dangerous because it would induce an artificial widening of the measured width which could be erroneously attributed to the homogeneous width. As mentioned earlier, this danger exists in all saturation techniques. For this reason, the spectra are measured after irradiation for different incident intensities and, if necessary, extrapolated to zero burning intensity. The experiment takes place in a bath of pumped helium IV at very low temperature (1.7 K < T < 4.2 K).

VII.4.2.3. Some domains in which "Hole burning" can be used

Here we shall mention only three domains exploited these last years, referring the reader to Bibliographic reference 7.6.8 for a more exhaustive presentation which lies out of the frame of this work. Hole-burning spectroscopy can be used (i) for the study of the absorbing microscopic system itself or (ii) to study its environment, i.e. the solid host matrix.

(i) *Study of the absorbing microscopic system: Spectroscopy of the vibrational states in organic and inorganic molecules*

The vibrational levels of the ground and excited electronic states of many aromatic molecules (condensed and heterocyclic), ions, polymers, dyes, and biological molecules have been determined with a high accuracy over the last twenty years.

(ii) *Study of the host matrix*

— *Hole-burning in glasses and in disorganized solids*

This is one of the first domains explored by hole-burning spectroscopy. Glasses are especially interesting because of their dynamic properties. They are "out of equilibrium" systems which therefore evolve with time. Such a time relaxation induces spectral diffusion. When a glass is doped at very low concentration, the resonance frequency of the dopant presents a resonance frequency which varies with time, and hole burning spectroscopy turns out to be a particularly well adapted technique for the study of the evolution dynamics of this medium. It can be used to measure the characteristic times associated to spectral diffusion and, among others, hole-burning techniques revealed how very different these times can be (extending over up to six orders of magnitude!).

— *Measurement of the Zeeman and Stark displacements of the ZPL*

These measurements have provided original information on, respectively, the Jahn-Teller effect (molecular distortion which lifts the degeneracy due to symmetry, *cf.* Section VI.3.2) and the linear Stark effect presented by certain centro-symmetric systems in disorganized matrices.

— *"Hole burning" in thin films*

One of the most promising results obtained these last years (Orrit, Bibliographic reference 7.6.9) concerns thin Langmuir-Blodgett films. With their two-dimensional order, these systems can be used to simulate and to study surfaces and interfaces. They are also precious in the study of the organized phases of liquid crystals, because of their marked orientational effects. This is especially interesting to study anisotropic molecules which can be inserted in their matrices. Also, they are often of high optical quality and since they are so very thin, the dopant density in any studied volume is necessarily very small, and this is, as said earlier, a very important requisite for single molecule spectroscopy. A film (cadmium arachidate) is formed by stacking one by one thirty layers doped with an ionic dye (one dye molecule for 500 arachidate molecules) on a thin transparent plate. The "burning" is operated with a monomode dye laser, tunable between 570 and 640 nm, with an intensity between the microwatt and the milliwatt per cm^2. The intensity of the probe laser is about one hundred times less than the intensity of the burning laser and recordings are taken at liquid helium temperature. In traditional absorption spectroscopy, these doped films present a few broad and structureless absorption lines in a range of about ten THz.

This experiment shows that it should be possible to burn holes with widths of about 10 GHz in one-layer samples. The width decreases when the number of layers increases and reaches the GHz for a stacking of about ten layers. By doping just one particular layer within the stacking and studying the effect of the position of the doped layer within the stacking shows that (i) the broadening varies with the position of the doped layer within the stacking and (ii) at low temperatures, if we classify the layers by decreasing hole widths, we obtain the following order: Film/air interfacing layer > glass/film interfacing layer > inner layer of the stacking. This experiment shows there is a difference between the motion of the molecules inside the stacking and those which lie at its air interface, probably because the aliphatic tails of the amphiphilic molecules have more conformational possibilities at the air interface.

We would also like to mention the very interesting results obtained in the problem of the energy transfer between chromophores at strong dope concentrations, by studying the variations of the hole width with the burning wavelength. Moreover, a complete determination of the orientation of the dye molecule (here a cyanine) within the film can be obtained by studying the linear Stark shift of the maximum burning wavelength.

However, these last experiments are done at very high doping concentrations so that each measurement still concerns about ten million molecules ! We now want to give a very brief historic overview of single molecule spectroscopy and its development in recent years.

VII.4.3. Single molecule spectroscopy in dense media

VII.4.3.1. Introduction to the method

In last section we showed that *spectral selection* (made possible by the monochromaticity and the frequency stabilization of the laser source) results in a significant frequency narrowing of the absorption signals of organic molecules in cold solid state matrices. But how can we detect the signal emitted by only one single molecule ? The answer is one of simple common sense:

— By ***decreasing the number*** of molecules contributing to the signal, that is by using samples with a ***very small volume*** (we know how to focus laser beams and we can obtain beam waist of the order of a few microns) and containing a very *low concentration* of doping molecules, for example 10^{-9} mol.l^{-1}. At this concentration, a volume of about 1 mm^3 contains only one molecule.

— By exploiting as much as possible the *spatial dispersion* of the resonance frequency which is due to the inhomogeneous differences between the dope-matrix interactions in the ground and excited states, differences which arise from the specific configuration of the site occupied by each dope molecule in the solid matrix.

Figure 7.4.2, taken from Bibliographic reference 7.6.9, is a very telling numerical simulation of this method.

We can watch the birth of a spectrum consisting of single molecule lines, which progressively appears as a fluctuation on the wings of the wide spectrum characteristic of the large number of molecules of the initial sample, and finally spreads to the whole width of the spectrum as the number of molecules decreases.

VII.4.3.2. Experimental aspects

The experimental aspects of single molecule spectroscopy are similar to those of hole burning experiments, but spatial selection is improved by focusing the beam to the very limits of the possibilities of laser optics. Moreover, the effects of the laser absorption are no longer tested by the absorbance of an auxiliary probe wave but directly from the fluorescence emitted by the excited molecule (so that, from this point of view, this spectroscopy can also be considered as a fluorescence spectroscopy). To capture this fluorescence, the detection solid angle must be as wide as possible because the intensity to be detected is so small. Why do we want to detect the fluorescence?

This problem is well known by scientists who work with optics: It is easier to detecting an emission, even a very weak one, on a dark background than to recording small transmitted intensity variations on an intense background, even though modern "marking" techniques have considerably improved the sensitivity of the latter type of method. Furthermore, the choice of the doping molecule obeys to strict spectroscopic constraints (*cf.* Reference VII.6.9).

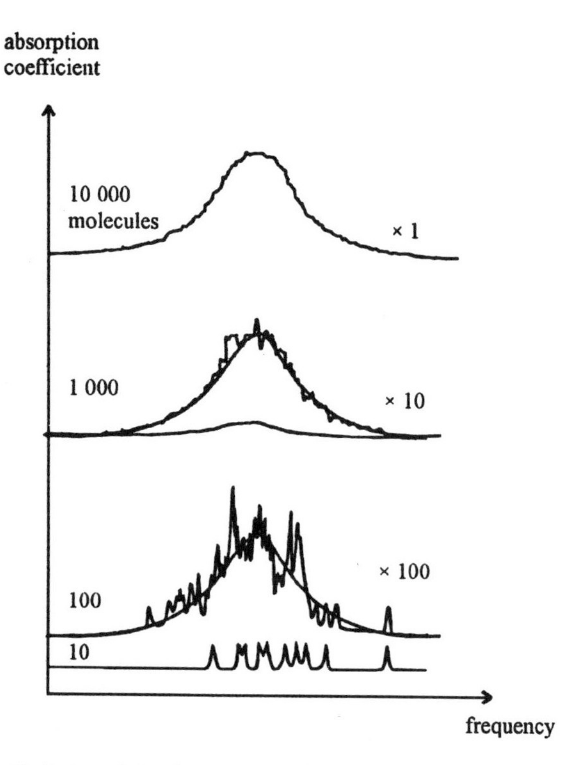

Figure 7.4.2: Variation of the absorption coefficient of an absorbing species in a solid matrix as a function of the frequency [from Orrit M., Bernard J. and Personov R.I., J. Chem. Phys., 97, 10256 (1993)]; reproduced with the authorization of the American Physical Society.

VII.4.3.3. A few results and some perspectives

The first spectra of single pentacene molecules in a p-terphenyl matrix were recorded between 1989 and 1990. The linewidth of about 8 MHz at 1.8 K corresponds to the natural width (time constant τ_2, *cf.* Section IV.3.1.1) of the electronic transition at 16 883.0 cm^{-1} (592.3 nm) for the insertion sites of the dopant in the matrix.

The past ten years saw a rapid extension of this new type of spectroscopy and many results were obtained both in traditional spectroscopy (measurements of the widths and profiles of the absorption lines, electric and magnetic field effects on the energy levels, etc.) and, more generally, in molecular physics where unknown effects, specific to this technique, such as the existence of isolated two-levels systems, were discovered and studied.

Clearly, in the coming years laser spectroscopy of the single molecule, which has now become a spectroscopy of both absorption and emission, will give new breathing space to traditional UV-visible absorption spectroscopy by providing a fertile research tool and choice educational examples in the study of the structure and the electronic properties of atoms and molecules.

VII.5. Exercises and problems

VII.5.1. Electromagnetic resonator and laser (from Agreg. P; A; 1995)

In this problem we adopt a scalar theory in which an electromagnetic wave is described at point M and at time t by the amplitude E(M, t) of an electric field, without specifying its direction. It will be admitted that this field vanishes at the surface of any perfect conductor.

1. Eigenmodes of a parallelepipedic cavity

We consider a parallelepipede such that $0 < x < a$, $0 < y < b$, $0 < z < d$ whose faces are assimilated to perfect conductors. Inside, it is empty. We look for solutions of the form $E = E_0 \sin(k_x x) \sin(k_y y) \sin(k_z z) \cos(t)$. The vector **k** is defined as $\mathbf{k} = k_x \mathbf{u}_x + k_y \mathbf{u}_y + k_z \mathbf{u}_z$ and we set $k = |\mathbf{k}|$. We admit that the propagation equation of E imposes the condition $k = \omega/c$.

a. Show that the limiting conditions imposed on E by the walls of the cavity imply the quantization of parameters k_x, k_y, k_z . Express these as functions of a, b, d and of three whole numbers n_x, n_y and n_z. Determine the frequency ω_0 of the fundamental mode. Calculate numerically the frequency corresponding to a cavity with the following dimensions: a = 2 cm, b = 4 cm, d = 3cm.

b. By using an analogy with a resonant R C L circuit, give a definition (involving energy) of the quality coefficient Q of a real cavity. What is the origin of the losses of such a cavity? The quality coefficient Q is equal to 25 000 for a cavity with the dimensions given in question 1.a. Calculate the bandpass of the fundamental mode.

2. Mode density

Now assume that $\omega >> \omega_0$ and do the calculations in the **k** vector space. We are looking for the number dN (assumed to be large) of modes with frequencies between ω and $\omega + d\omega$. Taking into account the relation between k and ω, what is the geometrical shape of the volume of **k** space where the modes whose frequencies are between ω and $\omega + d\omega$ are located? What is its volume ? Taking into account the quantization relations, what is the volume in **k** space occupied by one mode? Derive the expression of dN as a function of the volume of the cavity V = abd, of ω, $d\omega$ and c.

a. The walls of the cavity are infinitely conducting so the electric field should be normal to the surface of the walls. In scalar approximation used here, the field E vanishes on the walls of the parallelepipede and so we can write

$$k_x a = n_x \pi \qquad k_y b = n_y \pi \qquad k_z d = n_z \pi$$

and therefore $\quad k_x = (\pi/a)\, n_x \qquad k_y = (\pi/b)\, n_y \qquad k_z = (\pi/d)\, n_z$

For the fundamental mode, we have $n_x = n_y = n_z = 1$ and

$$\omega_0 = kc = \sqrt{k_x^2 + k_y^2 + k_z^2}\, c = \pi c \sqrt{a^{-2} + b^{-2} + d^{-2}}$$

b. Q = f/Δf. Δf is the resonance width of the eigenfrequencies ω given by: $\omega^2 / c^2 = \left(\pi^2 n_x^2 / a^2\right) + \left(\pi^2 n_y^2 / b^2\right) + \left(\pi^2 n_z^2 / d^2\right)$. Losses mainly arise from the nonzero absorption of the walls.

$$f_0 = \omega_0 / 2\pi = (c/2)\sqrt{a^{-2} + b^{-2} + d^{-2}} \approx 0{,}98\ \text{MHz} \quad \text{and} \quad \Delta f_0 = f_0 / Q \approx 39\ \text{Hz}$$

2. It is the volume included between two spheres with respective radii of ω/c and (ω + dω)/c. Its volume is equal to $4\ \pi\omega^2 d\omega / c^3$. The volume occupied by one mode is a cube with dimensions (π/a), (π/b) and (π/d).

$$dN = \left(4\pi^2\omega^2 d\omega / c^3\right) / \left(\pi^3 / abd\right) = 4\omega^2\, abd\, d\omega / \pi^2 c^3 .$$

VII.5.2. Fabry-Pérot interferometer and laser (from Agreg. P; A; 1995)

In this part, we adopt the scalar wave model in optics. The waves have a complex amplitude noted $\underline{a}$. The waves are emitted by a monochromatic point source of wavelength λ.

1. Fabry-Pérot interferometer.

The Fabry-Pérot interferometer consists of two plane and partially reflecting plates L_1 et L_2, assumed to be infinitely thin, parallel and separated by a distance e. A plane wave with amplitude $\underline{a}$ which falls on L_1 (respectively L_2) gives rise to a transmitted wave with amplitude $t_1\,\underline{a}$ (respectively $t_2\,\underline{a}$) and to a reflected wave with amplitude $r_1\,\underline{a}$ (respectively $r_2\,\underline{a}$) . The Fabry-Pérot interferometer is illuminated under normal incidence by a wave with amplitude a_0. The successive emerging waves have a complex amplitude $\underline{a}_k$ (the subscript k = 0 corresponds to the unreflected wave), and they interfere "at infinity", where they give rise to a wave with amplitude $\underline{a}$.

a. Determine the phase difference ϕ between two successive emerging waves as a function of e and λ, then as a function of e, c and ω.

b. Determine $\underline{a}_k$ as a function of a_0, r_1, r_2, t_1, t_2 and ϕ. Check that $\underline{a}_{k+1} / \underline{a}_k$ remains constant and can be expressed as a function of r_1, r_2 and ϕ.

c. Show that the intensity can be written as $I = I_{max} / [1 + m \sin^2(\phi/2)]$.

Give the expression of m.

d. The graph of I(ω) shows resonance frequencies ω_n whose full widths at half maximum (FWHM) are much smaller than the gaps between two successive resonance frequencies. What condition on r_1 and r_2 can lead to this result? Why is it interesting? Check that the resonance frequencies are equal to the frequencies of the eigenmodes of an imaginary cavity electromagnetic cavity corresponding to $r_1 = r_2 = 1$; $t_1 = t_2 = 0$.

2. Laser.

The cavity of the above Fabry-Pérot interferometer is filled with an amplifying material whose gain G varies with frequency. In these conditions, $r_1 r_2$ must be replaced by $g(\omega) r_1 r_2$ in the calculations. The graph of g(ω) shows a resonance frequency ω_M with a full width at half maximum equal to Γ. Its maximum value g_M can be tuned. We assume that $\Gamma >> \Delta\omega$. Show that an output wave with frequency ω_n ($\underline{a} \neq 0$) can be generated without incoming wave ($a_0 = 0$) if g (ω_n), r_1 and r_2 satisfy an inequality which we ask you to state (starting condition). With this theory, it is not possible to calculate the amplitude of the output wave if this starting condition is satisfied. What is the pertinent parameter here? We choose a dissymmetric Fabry-Pérot interferometer, with $r_1 \neq r_2$, to represent the laser cavity. Justify this choice in a few words.

3. Polarization of certain kinds of lasers

a. What is meant by the Brewster angle? Describe an experiment which shows its existence (we do not ask for theory).

b. In many lasers, a gas vessel of length L' whose extremities are placed at points A_1 and A_2 is placed between two planes M_1 and M_2 separated by a distance L > L' which form the laser cavity. In configuration, a Brewster window is installed at each of the extremities of the vessel A_1 and A_2 in such a way that the average light beam crosses A_1 and A_2 under the Brewster incident angle. What is the use of these Brewster windows? Why is the wave emitted by the laser polarized? In what direction? How can this property be verified experimentally?

4. Mode-locked laser

Let us consider a multimode laser with $\Gamma = 10^{12}$ rad.s^{-1} and e = 10 cm. The resonance frequencies of the cavity are $\omega_n = n\pi c/e$ (n is an integer). We assume, for simplicity, that the frequency ω_M is equal to the frequency ω_{n0} of the mode corresponding to $n = n_0$. Moreover, we also assume that the modes which satisfy the starting condition are those whose frequency ω_n lies between $\omega_{n0} - \Gamma/2$ and $\omega_{n0} + \Gamma/2$.

a. Calculate the number of modes whose frequency ω_n lies between $\omega_{n0} - \Gamma/2$ and $\omega_{n0} + \Gamma/2$. In the rest of this problem, we assume that this number is large enough that it can be replaced by the nearest uneven integer (2N + 1).

b. The following model is chosen to describe the mode-locked laser: In the laser cavity, we excite all the modes whose frequency is ω_n where n satisfies the double inequality $n_0 - N \leq n \leq n_0 + N$. All these modes are in phase and have the same amplitude. If we call u_x the unit vector normal to plates L_1 and L_2 of the cavity, each mode can be decomposed into two progressive waves propagating along the $+ u_x$ and the $- u_x$ directions. In these conditions, we assume that the complex amplitude of the output wave of the laser can be written as $\sum A \exp\{j\omega_n[t-(x/c)]\}$ where the sum is taken over the numbers n defined above. Show that $\underline{a}$ (t) can be written as $\underline{a}(t) = f\left[t-(x/c)\exp\{j\omega_{n0}[t-(x/c)]\}\right]$ where f(u) is a classical function with a width Δu of the order of $1/\Gamma$. Calculate Δu and give the physical interpretation of the mathematical expression of $\underline{a}$(t). What is the advantage of this kind of laser?

1.a. We find $\phi = 4\pi e/\lambda = 2e\omega/c$.

b. We have

$$\underline{a}_1 = a_0\, t_1\, t_2$$
$$\underline{a}_2 = a_0\, t_1\, (r_2\, r_1\, t_2)\exp(j\phi) = a_0\, r_1\, r_2\, t_1\, t_2\, \exp(j\phi)$$

$$\underline{a}_3 = a_0\ r_1 r_2\ t_1 \exp(j\phi)(r_2 r_1 t_2)\exp(j\phi) = a_0\ (r_1 r_2)^2\ t_1 t_2 \exp(j2\phi)$$

....

$$\underline{a}_k = \ldots = a_0\ (r_1 r_2)^{k-1} t_1\ t_2 \exp[j(k-1)\phi]$$

$$\underline{a}_{k+1} / \underline{a}_k = r_1\ r_2 \exp(j\phi)$$

c. The amplitude $\underline{a}$ of the field is the sum of a geometric progression of ratio $r_1 r_2\ \exp\ (j\phi)$ whose first term is $a_0 t_1 t_2$. After summation, we obtain $\underline{a} = a_0\ t_1\ t_2 / [1 - r_1\ r_2 \exp(j\phi)]$ and the intensity I is expressed as

$$I = \underline{a}\,\underline{a}^* = a_0^2\ t_1^2\ t_2^2 / (1 + r_1^2\ r_2^2 - 2 r_1\ r_2 \cos\phi)$$

For $\phi = 0$ $I = I_{max} = a_0^2\ t_1^2\ t_2^2 / (1 - r_1\ r_2)^2$ and I takes the form

$$I = I_{max}\left\{1/\left[1 + m \sin^2(\phi/2)\right]\right\} \quad \text{with } m = 4 r_1\ r_2 / (1 - r_1\ r_2)^2$$

1.4. Resonance is obtained for $\phi_n/2 = n\pi$ i.e. for $\omega_n = n\pi c/e$ with n = 1, 2, 3, ...

We have $I_{1/2} = I_{max}/2$ if $m \sin^2(\phi_{1/2}/2) = 1$ yielding

$$\phi_{1/2} = (2e/c)\omega_{1/2} \approx 2/\sqrt{m} \qquad \Delta\omega = 2\omega_{1/2} \approx 2c/\sqrt{m}\,e = c(1 - r_1\ r_2)/e\sqrt{r_1\ r_2}$$

and so we obtain $$\Delta\omega / \omega_n \approx (1 - r_1\ r_2)/ n\pi \sqrt{r_1\ r_2}$$

$\Delta\omega << \omega_n$ gives $$1 - r_1\ r_2 << n\pi\sqrt{r_1\ r_2}$$

This means that $r_1 r_2$ must have a value very close to 1, which implies that the reflectivity of plates L_1 and L_2 must be large. If this is true, the fineness $\omega_1/\Delta\omega$ is large so that we have an increase in the resolution of the interferometer. If $r_1 = r_2 = 1$ we have $t_1 = t_2 = 0$. The interferometer is equivalent to a one-dimensional closed cavity. Problem VII.5.1 (question 1.1) shows that, in this case, resonance is obtained when the relation $k_n e = n\pi$ is satisfied, i.e. when $\omega_n = n\pi c/e$ (look again at the equation which gives the eigenfrequencies and set $n_x = n_y = 0$, $n_z = n$).

2. By replacing $r_1 r_2$ by $g(\omega)\ r_1 r_2$, the intensity I can be expressed as

$$I = \frac{a_0^2\ t_1^2\ t_2^2}{[1 - g(\omega) r_1 r_2]\left\{1 + [4 g(\omega) r_1 r_2 \sin^2(\phi/2)]/[1 - g(\omega) r_1 r_2]^2\right\}}$$

From this equation, it is clear that if $g(\omega)\ r_1 r_2 = 1$, the intensity I can be different from zero at resonance even if $a_0 = 0$ (both the numerator and the denominator tend to zero). Moreover, $\Delta\omega/\omega$ also tends to zero: Light is emitted with an almost vanishing frequency width. Physically, the condition $g(\omega)\ r_1 r_2 > 1$ implies that the in-field gain for a back-and-forth trip in the cavity is greater than the losses due to the imperfect reflection of the two mirrors. This constitutes the self-oscillation condition of the laser.

In fact, the amplitude of the wave is determined by the equality $g(\omega)\ r_1\ r_2 = 1$. In a stationary regime, the gain in one back-and-forth trip (distance 2e) is equal to the transmission losses of the mirrors. We can write the following equations:

$$I_{(2e)} - I_{(0)} = I^*\ \ g(\omega)\ 2e = I_{(0)}\ (1 - r_1^2\ r_2^2) = I_{(s)}$$

I^* is a characteristic intensity of the laser and $I_{(s)}$ the transmitted laser intensity. In practice, we have $r_1=1$ $r_2 = r < 1$. The second mirror is the output mirror. We have $I_s = I_0\left(1 - r_2^2\right) = I^* g(\omega)\, 2e$.

3.a. The Brewster incidence angle i_B relative to the plane dioptre (1, 2) ($n_1 = 1$; $n_2 = n > 1$) is defined by tg $i_B = n$. For the incident angle i_B, the reflected light is totally linearly polarized and the polarization is normal to the incidence plane. This fact can be demonstrated by illuminating a plane dioptre by a non polarized wave and then using a linear polarizer to check that the reflected light can be extinguished completely only for a specific incidence angle, namely, the Brewster angle i_B.

b. In lasers, the use of exit dioptres placed at Brewster angle minimizes the reflection losses for the polarization within the incidence plane. This is the polarization of the laser emission. We can check this point by studying the exit polarization.

4.a. The gap between two successive modes is $\pi c/e = 9.42 \times 10^9$ rd.s^{-1}. Therefore, there are about $(2N + 1) = 107$ longitudinal modes which can oscillate ($N = 53$).

b. We have $$\underline{a} = \sum_{m=n_0-N}^{m=n_0+N} A \exp\left\{j\omega_n\left[t - (x/c)\right]\right\} \quad \text{with} \quad \omega_n = \omega_{n0} + m\ (\pi c/e)$$

And, by identification

$$f\left[t - (x/c)\right] = A \sum_{m=n_0-N}^{m=n_0+N} \exp\left\{jm(\pi c/e)\left[t - (x/c)\right]\right\}$$

Therefore f turns out to be the time-frequency Fourier transform of the frequency "comb" of width Γ consisting of the $(2N + 1)$ longitudinal in-phase modes [A is zero inside the intervals $(-\infty,\ n_0 - N)$ and $(n_0 + N, \infty)$]. The linewidth of each mode is assumed to be vanishingly small. So f is a function of time of infinite extension, of width $2\pi/\Gamma$, and consisting of a succession of peaks separated by 2e/c intervals. The numerical application gives $2\pi/\Gamma \approx 6$ pS. The laser emits the wave $\exp\left\{j\omega_{n0}\left[t - (x/c)\right]\right\}$ in pulses lasting a time equal to $2\pi/\Gamma$. The function $f\ [t - (x/c)]$ describes the time envelop of these oscillations of optical frequency ω_{n0}. This laser is used in many applications such as:

— Fundamental research: Time-resolved fast optical spectroscopy

— Applied research: Physics of plasma, research on nuclear fusion by inertial confining, etc.

VII.5.3. Two-photon absorption (from X; M'; 2^{sd} test; 1982)

1. Let us consider an electromagnetic wave consisting of the superposition of two waves of different frequencies, so that the x component at point O of the electric field can be expressed algebraically as:

$$E_x = E_1 \cos \omega_1 t + E_2 \cos \omega_2 t$$

The molecule susceptible to absorb this radiation has a fixed center of inertia placed at O, and the absorption is described by a classical model (*cf.* Problem IV.4.4.). The constant τ describes the average lifetime of the excited states populated by the absorption of radiations of frequencies ω_1 and ω_2 such that $|\omega_1 - \omega_2| >> 1/\tau$. Calculate the average absorbed power P, where the average is taken over a time longer than $2\pi/\omega_1$, $2\pi/\omega_2$ and $|\omega_1 - \omega_2|$. Write P as a function of the absorbed powers P_1 and P_2 corresponding, respectively, to the fields $E_1 \cos \omega_1 t$ and $E_2 \cos \omega_2 t$. For what values of the eigenfrequency ω_0 is the absorption by the molecule relatively large?

2. When the luminous flux is very intense, meaning practically inside a laser beam, it can happen that a molecule absorbs two photons simultaneously. To describe this phenomenon within our classical model, we just need to add to the previously considered forces a force with the same direction as the electric field vector (E_X, 0, 0) and whose norm is proportional to the square of the amplitude of this intense field. The coordinates of this force are therefore bE_X^2, 0, 0, where b is a real parameter depending on the molecule.

a. Determine the values of the eigenfrequency ω_0 which induce considerable absorption when the field E_X of question 1 is very intense.

b. Taking into account the fact that we are only interested in the absorption in the visible and the near infrared domains, study the case in which the two frequencies ω_1 et ω_2 lie in the near infrared, with a relative frequency difference $|\omega_1 - \omega_2| / \omega_1$ on the order of 10^{-2}.

3. Using a set of mirrors, two intense plane waves propagating in opposite directions in the Oz direction are superposed. The resulting electric field measured along Ox can be written as

$$E_x = E_0 \left\{\cos\omega\left[t - (z/c)\right] + \cos\omega\left[t + (z/c)\right]\right\}$$

E_0 is large. Let us consider a molecule moving with an uniform straight motion along the Oz direction and with a velocity v_z such that $v_z << c$. In the frame of the inertia center of the molecule, determine the frequencies ω_1 and ω_2 of the fields whose superposition constitutes E_X. Deduce the frequencies absorbed by a molecule with eigenfrequency ω_0 and velocity v_z. What is the advantage of this technique for the experimental determination of the lifetime of an excited state?

1. The differential equation which describes the motion due to the electric field E_x can be expressed as $\quad m\ddot{x} + (1/\sigma)\dot{x} + \omega_0^2 x = -(eE_1/m)\cos\omega_1 t - (eE_2/m)\cos\omega_2 t$

This linear equation implies a particular solution of the form

$$x = X_1 \cos(\omega_1 t + \varphi_1) + X_2 \cos(\omega_2 t + \varphi_2)$$

The average power is $P = \overline{-eE_x \dot{x}} = e\omega_1 E_1 \sin\varphi_1 / 2 + e\omega_2 E_2 \sin\varphi_2 / 2$. These two terms represent the powers P_1 and P_2 of fields E_1 and E_2 respectively (*cf.* Problem IV.4.4, question 2b; those terms which include both frequencies ω_1 and ω_2 have an average value of zero). The absorption is significant at resonance ($\omega_1 = \omega_0$ or $\omega_2 = \omega_0$).

2.a. Since the force induced by the electric field is proportional to the square of E_x, only terms in $\cos^2 \omega_1 t$ (frequency $2\omega_1$), $\cos^2 \omega_2 t$ (frequency $2\omega_2$) and $\cos \omega_1 t.\cos \omega_2 t$ [frequencies $(\omega_1 + \omega_2)$ and $(\omega_1 - \omega_2)$] contribute to the force. Significant absorption occurs when ω_0 is equal to one of the four frequencies $2\omega_1$, $2\omega_2$, $(\omega_1 + \omega_2)$, and $(\omega_1 - \omega_2)$,

b. If $\omega_2 \approx \omega_1$, only the three frequencies $2\omega_1$, $2\omega_2$ and $\omega_1 + \omega_2$ lie in the visible and the near infrared domains. The frequency difference $\omega_1 - \omega_2 \approx 10^{-2}\ \omega_1$ now lies in the far infrared.

3. Doppler effect must now be taken into account. We have

$$\omega_1 = [1 + (v_z/c)]\ \omega \quad ; \quad \omega_2 = [1 - (v_z/c)]\ \omega.$$

Resonance occurs for the three frequencies of 2.b. But the sum $(\omega_1 + \omega_2)$ does not depend on v_z because the two waves contributing to this sum frequency propagate in opposite directions. Molecules whose velocities are perpendicular to the propagation axis give rise to a Doppler-free signal, so that we obtain a correct determination of the homogeneous linewidth, i.e. of the lifetime of the excited state.

VII.5.4. One-photon atomic spectroscopy

We want to study a hydrogen jet in which a large number of atoms are in the metastable state 2s. A laser beam, linearly polarized along Ox, is sent onto the atoms of the jet. The light travels perpendicularly to the atomic velocities so that we can neglect the Doppler effect in the experiment. The frequency of the laser ω_L is close to the resonance frequency ω_0 of the transition $|2s\rangle_{2^2S_{1/2}} \rightarrow |3p\rangle_{3^2P_{1/2}}$; $E_{3p} - E_{2s} = h(\nu_{3p} - \nu_{2s}) = h\nu_0$. Upon interacting with the electromagnetic wave, the atoms which were initially in state $|2s\rangle$ are lifted into the nonstationary state $|\psi(t)\rangle = a(t)|2s\rangle + b(t)|3p\rangle$. The interaction is detected by measuring the intensity of the spontaneous emission from state $|3p\rangle$. This spontaneous emission is proportional to $|b(t)|^2$, and we assume that it does not perturb state $|\psi(t)\rangle$.

1. Show that the electric dipole transition between these states is allowed. Conclude that within EDA the Hamiltonian can be written as $V = -\boldsymbol{p}\,\mathbf{E}$, where $\mathbf{E} = \mathbf{E}_0 \cos(\omega_L t)$ is the electric field of the wave, and $\boldsymbol{p}$ the dipole moment operator of the atom. Show that $\boldsymbol{p}$ has no diagonal terms in basis $|2s\rangle$, $|3p\rangle$. For the calculations, please use the notation $\langle 2s|p_z|3p\rangle = \langle 3p|p_z|2s\rangle = p$.

2. Let us write $A(t) = a(t) \exp(i\omega_{2s} t)$; $B(t) = b(t) \exp(i\omega_{3p} t)$. Find the equations governing the evolution of A and B within RWA. Find particular solutions for B which are of the form $\exp(i\omega t)$. With the help of these solutions, show that B can be written as

B (t) = $\alpha_+ \exp(i\omega_+ t) + \alpha_- \exp(i\omega_- t)$. Calculate the values of $\omega_\pm$. Take as time origin the instant the atoms enter the laser wave and calculate a_+ and a_-. Assume that the intensity of the laser beam is uniformly distributed over its diameter and that the detector analyzes the light emitted by one point of the interaction zone. What can you say about the detected signal?

1. The states $|2s\rangle$ and $|3p\rangle$ do not have the same parity. Indeed, state $|2s\rangle$ has quantum number $l = 0$ while for state $|3p\rangle$ $l = 1$. Therefore $|2s\rangle$ is even with respect to spatial inversion, while $|3p\rangle$ is uneven. Now, operator $\boldsymbol{p}$ is uneven with respect to the same spatial inversion, so that $\langle 2s|\boldsymbol{p}|3p\rangle$ is even. This expression stands for the integral over all space with respect to r (i.e. from $-\infty$ to $+\infty$) of the product of the bra $\langle 2s|$ by the ket $\boldsymbol{p}|3p\rangle$. The fact that the expression to be integrated (the integrand) is even means that the integral cannot vanish. Therefore we can say $\langle 2s|\boldsymbol{p}|3p\rangle \neq 0$, meaning that the transition $|2s\rangle \rightarrow |3p\rangle$ is allowed within EDA. The interaction Hamiltonian is $V = -\boldsymbol{p}\,\mathbf{E}$. Its matrix representations are nondiagonal (the diagonal elements are uneven and therefore vanish).

2. The evolution equations are written as

$$\dot{A} = (-i2\pi V/h)\exp(-i\omega_0 t)B\,;\ \dot{B} = (-i2\pi V/h)\exp(i\omega_0 t)A$$

with $\omega_0 = \omega_{3p} - \omega_{2s}$. (where we define $V = V_{2s,\,3p}$). Applying RWA (*cf.* Problem VI.4.8.) yields

$$\dot{A} \approx (ip\pi E_0)\exp[-i(\omega_0 - \omega_L)t]B \quad ; \quad \dot{B} \approx (ip\pi E_0)\exp[-i(\omega_0 - \omega_L)t]A.$$

We look for a solution of the form B = exp $i\omega t$. Identifying the constants gives the second-degree equation $\omega^2 + (\omega_L - \omega_0)\omega - \Omega^2/4 = 0$ where Ω is the Rabi frequency ($\Omega = 2\pi p E_0/h$). The equation has roots $\omega_\pm$, equal to $\omega_\pm = [(\omega_0 - \omega_L)/2] \pm (1/2)\sqrt{(\omega_L - \omega_0)^2 + \Omega^2}$.

Therefore we can write $B = \alpha_+ \exp(i\omega_+ t) + \alpha_- \exp(i\omega_- t)$. At time t = 0, only state $|2s\rangle$ is populated, so A (0) = ± 1 and B (0) = 0. Writing out the initial condition B (0) = 0 yields $\alpha_+ + \alpha_- = 0$. Similarly, writing out A (0) = ± 1 yields

$$\alpha_+ = \Omega/2\sqrt{(\omega_L - \omega_0)^2 + \Omega^2}.$$

So that we finally obtain

$$B \approx \left[i\Omega/\sqrt{(\omega_L - \omega_0)^2 + \Omega^2}\right]\exp[i(\omega_0 - \omega_L)t/2]$$
$$\times \sin\left(\sqrt{(\omega_L - \omega_0)^2 + \Omega^2}\; t/2\right)$$

The detected signal (of spontaneous emission from state $|3p\rangle$) is proportional to the population of $|3p\rangle$, that is, to

$$|B|^2 \approx \Omega^2 \sin^2\left(\sqrt{(\omega_L - \omega_0)^2 + \Omega^2}\, t/2\right) / \left[(\omega_L - \omega_0)^2 + \Omega^2\right]$$

The characteristic time of the interaction is $t' = \Delta x / v_0$ (Δx is the distance between the detection point and the point corresponding to the origin of time, $t = 0$. v_0 is the velocity of the atoms). The detected signal, which is proportional to $|B(t')|^2$, depends on the point of interaction observed. The fact that there is a distribution of the velocities of the atoms implies that an average signal is observed whose amplitude is greatest when $\omega_L = \omega_0$. If the velocities are very dispersed, the observed average signal will practically not depend on time.

VII.5.5. Two-photon atomic spectroscopy

We wish to study the transition $|1s\rangle_{1^2S_{1/2}} \rightarrow |2s\rangle_{2^2S_{1/2}}$.

1. Show that this transition is forbidden within EDA.

2. In order to fill state 2s starting from the ground state 1s, a linearly polarized laser beam of amplitude E_0 is sent onto a vessel containing atomic hydrogen. The frequency ω_L of the wave is approximately equal to $(\omega_{2s} - \omega_{1s})/2$. The wave lifts the atoms, which are initially in state $|1s\rangle$, into a nonstationary state described by

$$|\psi(t)\rangle = a(t)|1s\rangle + b(t)|2s\rangle + \sum_n n(t)|n\rangle$$

where $|n\rangle$ stands for states of energy $E_n = h\nu_n$. States $|n\rangle$ all have the same parity, which is opposite to that of states $|1s\rangle$ and $|2s\rangle$ (i.e. dipole transitions are allowed between any state $|n\rangle$ and the states $|1s\rangle$ and $|2s\rangle$). Using EDA (interaction Hamiltonian $V = -\boldsymbol{p}\,\mathbf{E}$), show that b(t) vanishes for a first-order perturbation calculation. Calculate n(t) as a function of $\langle 1s|p_z|n\rangle$ and of $(\omega_n - \omega_{1s} - \omega_L)$, for this order of perturbation and within RWA. Then show that b(t) does not vanish for the second-order perturbation calculation, also within RWA. Calculate its value, assuming that the matrix elements $\langle n|p_z|2s\rangle$ are known. The result can be simplified by noticing that for all states under consideration, the laser frequency is such that $\omega_{2s} - \omega_{1s} - 2\omega_L \ll \omega_{2s} - \omega_n - 2\omega_L$. What does this result suggest? What practical comments can you make about the experiment?

1. For the same reasons as those mentioned in 1.a, since states $|1s\rangle$ and $|2s\rangle$ are of the same parity, the integrand of integral $\langle 1s|\,\boldsymbol{p}\,|2s\rangle$ is uneven so that the integral vanishes. That means that transition $|1s\rangle \rightarrow |2s\rangle$ is forbidden within EDA.

2. The evolution equation is

$(ih/2\pi)|\,\dot{\psi}\,\rangle = [(H_0 + V)]|\,\psi\,\rangle$ with $V = -\boldsymbol{p}\,\mathbf{E}$ within ADE.

Left-multiplying first by $\langle 1s|$, then by $\langle 2s|$, and last by $\langle n|$ yields

$$(ih/2\pi)\dot{a} = (h/2\pi)\omega_{1s}\, a + \sum_n n\, V_{1s,n} \quad ; \quad (ih/2\pi)\dot{b} = (h/2\pi)\omega_{2s}\, b + \sum_n n\, V_{2s,n}$$

$$(ih/2\pi)\dot{n} = (h/2\pi)\omega_n\, n + a\, V_{n,1s} + b\, V_{n,2s}$$

We took into account the fact that $V_{ab} = V_{nn} = 0$ (all transitions between states of the same parity are forbidden within EDA).

Let us change the variables as follows:

$$a = A\exp(-i\omega_{1s} t) \quad ; \quad b = B\exp(-i\omega_{2s} t) \quad ; \quad n = N\exp(-i\omega_n t)$$

Then we find

$$\begin{aligned}
\dot{A} &= (i2\pi E_0 \cos\omega_L t/h)\sum_n N\exp[i(\omega_{1s}-\omega_n)t]\langle 1s|p_z|n\rangle \\
\dot{B} &= (i2\pi E_0 \cos\omega_L t/h)\sum_n N\exp[i(\omega_{2s}-\omega_n)t]\langle 2s|p_z|n\rangle \\
\dot{N} &= (i2\pi A E_0 \cos\omega_L t/h)\exp[i(\omega_n-\omega_{1s})t]\langle n|p_z|1s\rangle \\
&\quad + (i2\pi B E_0 \cos\omega_L t/h)\exp[i(\omega_n-\omega_{2s})t]\langle n|p_z|2s\rangle
\end{aligned}$$

From this system of coupled equations we can conclude that since A (0) = 1; B (0) = 0 and N (0) = 0 at the zero order of field E_0, only N will be different from zero at first perturbation order. This in turn will induce nonzero terms for A and B in the second order. Let us calculate $N^{(1)}$, at the order 1 of the perturbation and within RWA:

$$\begin{aligned}
\dot{N}^{(1)} &\approx (i2\pi E_0 \cos\omega_L t/h)\exp[i(\omega_n-\omega_{1s})t]\langle n|p_z|1s\rangle \\
&\approx (i2\pi E_0/2h)\langle n|p_z|1s\rangle\exp[i(\omega_n-\omega_{1s}-\omega_L)t]
\end{aligned}$$

and integration gives

$$N^{(1)} \approx [\pi E_0 \langle n|p_z|1s\rangle / h(\omega_n-\omega_{1s}-\omega_L)]\{\exp[i(\omega_n-\omega_{1s}-\omega_L)t]-1\}$$

And now we can calculate $B^{(2)}$, taking account of the approximation mentioned and of the initial condition $B^{(2)}(0) = 0$:

$$\begin{aligned}
B^{(2)} &\approx [\pi^2 E_0^2/h^2(\omega_{2s}-\omega_{1s}-2\omega_L)]\sum_n (\langle 2s|p_z|n\rangle\langle n|p_z|1s\rangle)/(\omega_n-\omega_{1s}-\omega_L) \\
&\quad \times\{\exp[i(\omega_{2s}-\omega_{1s}-2\omega_L)t]-1\}
\end{aligned}$$

Again, spontaneous emission from the state is found to be proportional to

$$\begin{aligned}
|B^{(2)}|^2 &\approx (\pi E_0/h)^4\left|\sum_n (\langle 2s|p_z|n\rangle\langle n|p_z|1s\rangle)/(\omega_n-\omega_{1s}-\omega_L)\right|^2 \\
&\quad \times(1/2)[(\omega_{2s}-\omega_{1s})/2-\omega_L]^2 \sin^2[(\omega_{2s}-\omega_{1s})/2-\omega_L]t
\end{aligned}$$

Of course, this expression is only valid if the population of the state $|2s\rangle$ is not too large, i.e. at times t such that $t \ll 2\pi/[(\omega_{2s}-\omega_{1s})/2-\omega_L]$.

We can make two comments:

— The damping of the population of state $|2s\rangle$ should be taken into account by inserting the term $(-\gamma B/2)$ into the evolution equation of B. For interaction times which are very large compared to $2/\gamma$, we would then obtain $|B^{(2)}|^2$ = constant, meaning that the spontaneous emission no longer changes with time.

— Like saturated absorption spectroscopy, this spectroscopy is afflicted by Doppler effect. But, as already said, the use of two laser waves propagating in opposite directions is a good method to do away with this undesired broadening.

VII.6. Bibliography

7.6.1 General presentation of the laser:

BEESLEY M.J. — *Lasers and their Applications*, Taylor et Francis, London (1971).

DUCLOY M.— Les principes du rayonnement laser in *Des technologies pour demain*, JORLAND G., Seuil (1992).

HARTMANN F. — *Les lasers*, Presses Universitaires de France (1974).

LENGYEL B.A. — *Lasers*, J. Wiley, New York, 4th edition (1964).

ORSAZ A. — *Les lasers*, Masson (1968).

7.6.2. Specialized books concerning the laser:

BIRBAUM G. — *Optical Lasers*, Academic Press, New York (1966).

SIEGMAN A. — *Lasers*, University Science Books, Mill Valley (1986).

VERDEYEN J.T. — *Laser Electronics*, Prentice Hall, Inc., Englehood Cliffs (1981).

7.6.3. An introduction to laser-molecule interaction can be found in:

LALANNE J.R., DUCASSE A. et KIELICH S. — *Interaction laser molécule*, Polytechnica (1994), English translation: *Laser Molecule Interaction*, J.Wiley, New York (1996).

7.6.4. Nonlinear Optics can be found in:

BLOEMBERGEN N. — *Nonlinear Optics*, W.A. Benjamin Inc, New York (1965).

BOYD R. — *Nonlinear Optics*, Academic Press, New York (1992).

BUTCHER P. and COTTER D. — *The Elements of Nonlinear Optics*, Cambridge University Press, Cambridge (1990).

NEWELL A.C. and MOLONEY J.V. — *Nonlinear Optics*, Addison-Wesley, New York (1992).

SHEN Y.R. — *The Principles of Nonlinear Optics*, J.Wiley, New York (1984).

YARIV A. — *Introduction to Optical Electronics*, Holt, Rinehart and Winston, New York (1976).

7.6.5. Intra-cavity absorption spectroscopy is described in the article:

STOECKEL F. et KACHANOV A. — Spectroscopie d'absorption intracavité laser (ICLAS), in *Spectra Analyse*, 191, 31-34 (1996).

7.6.6. Saturated absorption can be studied in:

LETOKHOV M.D. and CHEBOTAYEV V.P. — *Nonlinear Laser Spectroscopy*, Springer Verlag, Berlin (1977).

LEVENSON M.D. — *Introduction to Nonlinear Laser Spectroscopy*, Academic Press, New York (1982).

7.6.7. Site selection spectroscopy is described in:

PERSONOV R.I.— Site Selection Spectroscopy of Complex Molecules in Solution and its Applications, in *Spectroscopy and Excitation Dynamics of Condensed Molecular Systems*, Chap.10, edited by Agranovich V.M. and Hochstrasser R.M., North-Holland, Amsterdam (1983).

7.6.8. Hole-burning is described in:

MOERNER W.E. — *Persistent Spectral Hole-Burning: Science and Applications*, Springer Verlag, Berlin (1988).

7.6.9. An introduction to selective spectroscopy is given in:

ORRIT M., BERNARD J. et BROWN R.— Faire de la spectroscopie molécule par molécule, *La Recherche*, 260, vol. 24, 1395-1397 (1993).

BASCHE T., MOERNER W.E., ORRIT M., and WILD U.P. — *Single-Molecule Optical Detection, Imaging and Spectroscopy*, VCH, Weinheim, (1997).

and in the two articles:

ORRIT M., BERNARD J., and PERSONOV R.I. — High-Resolution Spectroscopy of Organic Molecules in solids: From Fluorescence Line Narrowing and Hole Burning to Single Molecule Spectroscopy, *J. Phys. Chem.*, 97, 10256-10268 (1993).

ORRIT M., BERNARD J., BROWN R., and LOUNIS B. — Optical Spectroscopy of Single Molecules in Solids, *Progress in Optics*, edited by WOLF E., Elsevier Science B.V., 61-144 (1996).

Index